HYDRAULIC AND PNEUMATIC POWER FOR PRODUCTION

HYDRAULIC AND PNEUMATIC POWER FOR PRODUCTION

How Air and Oil Equipment Can be Applied to the Manual and Automatic Operation of Production Machinery of All Types with Numerous Existing Installations Explained in Step-by-Step Circuit Analyses

By
Harry L. Stewart

FOURTH EDITION — FIRST PRINTING

INDUSTRIAL PRESS INC.

200 MADISON AVENUE, NEW YORK, N.Y. 10016

Library of Congress Cataloging in Publication Data

Stewart, Harry L
   Hydraulic and pneumatic power for production.

   Includes index.
   1. Fluid power technology. I. Title.

TJ843.S74   1976   621.2   76-28238
ISBN 0-8311-1114-3

HYDRAULIC AND PNEUMATIC POWER FOR PRODUCTION

Preface to Fourth Edition

The fluid power industry annually produces products and systems valued at over $2.2 billion, an increase of 800 percent over the past three decades.

The principal reason for the great increase in the use of fluid power is the favorable "man–machine" relationship. By using fluid power systems, there is less demand upon the "man operator" who can direct and control great forces and many motions with very little effort and with a maximum amount of safety. Consider, for example, the use of fluid power in construction equipment. Here, mechanical power transmission systems which involve gears, chains, pulleys, belts, and cables have been made obsolete by hydromatic transmissions, servo controls, fluid motors and cylinders.

A great shortage of adequately trained fluid-power personnel still exists and educating people to use fluid power properly is vitally important. This text has been designed to assist in such a training program whether it be in the design of circuits, selection of system-matched components, installation of the components in the system, or maintenance of the system after it is installed. For those who use this text as a teaching aid, a set of questions for each chapter has been added at the end of the book.

The author wishes to extend his thanks to the many companies who graciously cooperated in supplying photographs, charts, and technical data for the new chapters: "Hydrostatic Transmissions and Fluid Motors," and "Hydraulic Servo Mechanisms." He also wishes to thank those who have furnished new information so that various other chapters could be updated.

Acknowledgment

The author wishes to express his appreciation to the following publications for permission to use material which has appeared in various articles written by him: "The Tool Engineer," "Machine and Tool Blue Book," "Plant Engineering," and "Product Engineering."

Acknowledgment is also made of the courtesy of "The American Society of Tool Engineers" in granting permission to reproduce the diagrams of air cylinder linkage movements which have been taken from a paper presented before the Society by Mr. J. C. Hanna of the Hanna Engineering Works.

Particular credit should also be given to the E. F. Houghton & Co. and the Mobil Oil Corp. for their fine cooperation in making available the considerable amount of technical information and illustrative material which forms the basis of Chapter 13, on "Packings and Seals."

For the assistance and help afforded by the many friends of the author and the companies who furnished innumerable technical bulletins, drawings, and illustrations, the author can only say that without their invaluable assistance, this book could not have been written.

Contents

CONTENTS

CHAPTER I

The Evolution of Fluid Power

It was near the end of the 17th Century that Torricelli, Mariotte and, later, Bernoulli carried on experiments to study the elements of pressure or force in the discharge of water through orifices in the sides of tanks and through short pipes. In that same period the French scientist, Blaise Pascal, evolved the fundamental law for the science of hydraulics.

Pascal's theorem was as follows: "If a vessel full of water, and closed on all sides, has two openings, the one a hundred times as large as the other, and if each is supplied with a piston that 'fits it exactly,' then a man pushing the small piston will exert a force that will equilibrate that of one hundred men pushing the large piston and will overcome that of ninety-nine." Given one piston of one square inch area on which is placed a pressure of one hundred pounds, sufficient pressure will be created to lift another piston, connected to the same hydraulic system, having an area of four square inches and carrying a weight of 400 pounds. Obviously, the conclusion is that a force of 100 pounds per square inch is transmitted in all directions raising four times as much in weight but only one-fourth of the distance.

However, in order for Pascal's law to be made effective for practical use, it was necessary to make a piston that would "fit exactly." Not until over one hundred years later was this accomplished. It was in that period that an Englishman, Joseph Brahmah, invented the cup packing which led to the development of a workable hydraulic press. Brahmah's hydraulic press consisted of a plunger pump piped to a large cylinder and ram. This new hydraulic press found wide use in England because it provided a more effective and more economical means of applying large forces to industrial applications.

About 1850 W. G. Armstrong, founder of the firm which grew into the large British Vickers-Armstrong Ltd., developed a hydraulic crane and invented the hydraulic accumulator. The accumulator was designed to store a large amount of fluid under an artificial head of pressure so that it would be available to supplement the fluid pressure during periods of abnormal demand for power.

Coming many years before the development of electric power and the electric motor, it was only natural that these developments in hydraulic machinery found ready acceptance in England's industrial activity. During the period from 1850 to 1860, London and other large English cities were piped for the transmission and sale of hydraulic power from central pumping stations to individual shops.

The first use of a large hydraulic press for forging work was made in 1860 by Whitworth. During the next 20 years many attempts were made to reduce the waste and excessive maintenance costs of the original type of accumulator. In 1872 Rigg patented a three-cylinder hydraulic engine, in which was embodied provision to change the stroke of the plungers to vary its displacement without a throttle valve. In 1873 the Brotherhood three-cylinder, constant-stroke hydraulic engine was patented and was widely used for cranes, winches, etc. Both of the above-mentioned engines were driven by fluid from an accumulator.

While the accumulator made it possible to overcome the slow speed of the earlier hydraulic power transmission system, it required throttle valves for controlling the fluid. These valves were inefficient and required costly maintenance. As a result, engineers were discouraged from attempting further applications to ordinary machinery.

At this point it seemed that the development and application of hydraulics to machines had reached the limits of its possibilities. Through a British patent by Hastie in 1873, it was indicated how it might be possible to design variable-stroke pumps and motor units that would deliver fluid efficiently and smoothly under controlled conditions without the use of an accumulator. Nevertheless, this invention was ignored and only in recent years have engineers devoted time and thought to the development of efficient hydraulic equipment.

Establishment of Industry Standards

During the last thirty years and particularly during the period of World War II the use of fluid power has grown by leaps and bounds and among many fields of application a most important one has been that of production machinery.

In fact, by 1951 applications in this field had increased so rapidly that a number of conferences were held in Detroit, Michigan, for the purpose of establishing a set of standards for industrial hydraulic and pneumatic equipment. These are now well known as the "Joint Industry Conference Standards for Industrial Equipment." Represented in these conferences were hydraulic equipment manufacturers, pneumatic equipment manufacturers, machine tool builders, packing and seal manufacturers, press manufacturers, resistance welding manufacturers, tubing and fitting manufacturers, and industrial users.

The J.I.C. symbols were adopted for the complete line of fluid power components. In 1966, a new set of graphic standards was released by the United States of America Standards Institute. The Institute changed its name to the American National Standards Institute, Inc., on October 6, 1969. The symbols, in the graphic standards, are known as ANSI symbols (see charts on end papers). They are similar in appearance to the J.I.C. symbols and have now replaced them. Through their use, the maintenance man can readily trace the flow of fluids through the controls without reference to the control manufacturer's catalogs. This is a great timesaver, especially on complicated layouts. Today, many machines containing fluid power systems have their circuit diagrams attached.

Advantages Offered by Air and Oil

Perhaps it would be in order that we clarify our thinking on one point. In using the term "fluid" we are referring specifically to air or oil, for it has been shown that water has certain drawbacks for the transmission of hydraulic power in machine operation and control. "Commercially pure" water contains various chemicals (some deliberately included) and also foreign matter, and unless special precautions are taken when it is used, it is nearly impossible to maintain valves and working surfaces in satisfactory condition. In cases where the hydraulic system is "closed" (i.e., one with a self-contained unit which serves one machine or one small group of machines) oil is commonly used, thus providing, in addition to power transmission, benefits of lubrication not afforded by water and increased life and efficiency of packings and valves. It should be mentioned that in some special cases soluble oil, diluted with water, is used with satisfactory results. Fire-resistant fluids are also coming more into use.

Usually hydraulic and pneumatic systems and equipment do not compete. They are so dissimilar that there are few problems of selection between them which cannot readily be resolved, all factors considered, with a clear preponderance in favor of one or the other. But what of the case where one of the two sources of power is not at hand, and compressed air is used to do something for which oil might be better adapted, and vice versa? Certainly, availability is one of the important factors of selection but this may be outweighed by other factors. In numerous instances, for example, air is preferred to meet certain unalterable conditions, i.e., in "hot spots" where there is an open furnace or other potential ignition hazard or in operations where motion is required at extremely high speeds. It is often found more efficient to use a combined circuit in which oil is used in one part and air in another on the same machine or process.

Perhaps the comparison between the use of compressed air and oil in fluid power circuits would be oversimplified by applying an efficiency factor of 80 per cent to the fluid power pump. However, the use of this more or less arbitrary efficiency would indicate that the employment of compressed air is 2.6 times as costly as the use of oil for fluid power. This latter comparison was made on the assumption that a large number of hydraulic systems today are saddled with the loss of fluid in use, which is inherent in the accumulator. If there is no accumulator in the system, air operation might well cost four times as much as oil.

The comparisons just drawn lead perhaps to wonderment at the tremendous amount of compressed air used throughout industry and particularly in the shop. The answer, of course, lies in the versatility of air, the simplicity of its application and the fact that air in many instances is not used in really large quantities.

There is poor economy in using an air cylinder for a long stroke wherein only a small portion of the stroke—for instance, at its end, or its beginning—is usefully employed. An example is the clamping cylinder with a long stroke to clear surrounding interferences. Another example is the case where high pressure is used at the beginning of the stroke for breakaway force, followed by a long clearance stroke. In this latter case, air can be conserved by not allowing air pressure to build up at the end of the clearance stroke.

Basic Elements of Fluid Power System

A hydraulic system can be broken down into four main divisions analogous to the four main divisions in an electrical system. First, the power device parallels the electrical generating station; second,

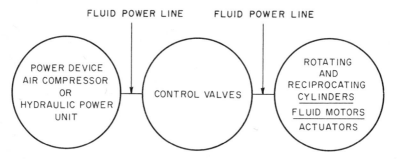

Fig. 1. Schematic diagram of a fluid power system.

the control valves parallel the switches, resistors, timers, pressure switches, relays, etc.; third, the lines in which the fluid power flows parallel the electrical lines; and fourth, the fluid power motor (whether it be a rotating or nonrotating cylinder, or a fluid power motor) parallels the solenoids and electric motors. These four main divisions in a fluid power system are shown schematically in Fig. 1.

Application of Fluid Power

The application of fluid power is limited only by the ingenuity of the designer, production engineer, or plant engineer. If the application pertains to lifting, pushing, pulling, clamping, tilting, forcing, pressing or any other straight line (and many rotary) motions, it is probable that fluid power will meet the requirements. The important part that fluid power plays in all phases of industry today is beyond calculation. To indicate its increasing importance, we need only to say that new uses are being found and adopted for air and hydraulic systems every day. This is true in large industries where economical mass production is a must and is equally true in small plants where highest efficiency is an absolute necessity for survival.

Whether a new piece of plant equipment is being designed or an old machine is being reconditioned for mass production, the addition of fluid power brings distinct advantages. Such a piece of equipment can be operated with very little effort and motion on the part of the operator. The fluid power controls can be placed in a central station so that the operator has, at all times, complete control of the entire production line whether it be a multiple operation machine or a group of machines. Such a setup is more or less standard in the steel mill industry.

Not only reduction in required manpower, but also reduction or elimination of operator fatigue, as a production factor, is an important element in the use of fluid power. Whether the operation is manual or automatic, fluid power steps in to shoulder a substantial part of the labor required—moves ahead with the job on a predeterminable time schedule of continuous motion. Exhaustive time studies have long since proved that taking away the job of lifting, pushing, pulling and turning from the operator and giving it to the machine has been a major factor in changing the color of a company's financial statement from red to black. With the substantial increase in the employment of women operators these operational advancements assume still further importance.

A fluid power system will also give flexibility to the equipment without requiring a complex mechanism. By merely adjusting a small flow control valve a large variation of speeds and feeds may be obtained. If several different pressures are required for the operation of a piece of equipment, they can easily be provided by the use of pressure-relief or pressure-reducing valves. If skip-feed is a desirable feature, a cam-operated flow-control valve may be added to the circuit and can be controlled by clamps or trips on the machine table. It is impossible to list and even difficult to visualize the widely diversified possibilities actually afforded through a combination of varied fluid power components. An important factor is that, with all this flexibility and simplicity of operation, the cost is relatively small.

Selection of the Fluid Medium

There is no hard and fast rule to follow in deciding which of the fluid mediums should be used in an application. However, here are some suggestions which, on a broad, general basis, are considered sound: If the application requires speed, a medium amount of pressure, and only a fairly accurate feed, use an air system; if the application requires only a medium amount of pressure and a feed of greater accuracy, use a combination of air and hydraulic systems; if the application requires a great amount of pressure or an extremely accurate feed, use an oil hydraulic system.

Obviously, there are other things to take into consideration when selecting the proper medium for the motivating force. First, if air is being considered it must be determined if there is enough air capacity available to handle the added air equipment. Many plants today have added so many air-operated devices that their compressor capacity is woefully inadequate. In numerous cases, at the beginning of a shift when all air devices are started at once, the pressure drops to such an extent that air clamping equipment will not hold the workpieces. The addition of a surge tank near the machine with multiple air devices often helps overcome some of the difficulty, but it is usually necessary to make provision for more compressor capacity.

The location of the equipment has much to do with the selection of the fluid medium. For example, if the equipment is subject to severe temperature variations (i.e., on a loading dock where temperatures may vary from 110°F to minus 20°F) an oil hydraulic system is usually considered best. If air is used and the weather becomes extremely cold, condensation in the lines would freeze and make the system inoperative. However, in hot locations it is sometimes better to use an air system.

Advantages of Unit Installations

Since the present-day trend is toward elimination of the central power station, it is advisable to consider the package unit. A pretty good example of this, although not in the category of fluid power, is the obsolescence of overhead line shafts. As is well known, failure of the central power unit meant temporary stoppage of the complete line of operations. By using a packaged hydraulic power unit to operate one machine or one group of machines, a failure affects only the one section of the total operation. Stand-by units are usually on hand in the maintenance department with which to get this one section back in operation with very little delay. Some machine tools are equipped with portable, packaged electro-hydraulic power devices. Quick-to-disconnect fittings on portable devices are valuable inasmuch as they eliminate the spilling of oil on the floor when making the power unit change.

Cost of Fluid Power Systems

Contrary to the belief of many individuals with little or no first-hand experience, fluid power systems are no more expensive in first cost or in maintenance than other types. In many cases they are actually less costly. If a simple program of preventive maintenance is followed, the fluid power system will give years of satisfactory service with very little attention. With even a minimum of mainte-

Courtesy of National Automatic Tool Co., Inc.

Fig. 2. Three-way boring and facing machine with hydraulic feed.

nance attention and good housekeeping the user can very easily eliminate unnecessary costs and increase the life of the equipment by many years. By the very nature of the oil hydraulic system, the moving component parts are working in a lubricating medium. As long as the oil is kept clean and free of condensation, the operating conditions will be ideal.

In air systems the air must be kept clean and properly lubricated for peak efficiency. There are many cases where fluid power systems have been in regular operation 10, 15 and 20 years. The records show that maintenance cost on valves, cylinders and power units has been amazingly low.

Typical Fluid Power Applications

Figure 2 shows a machine used in the manufacture of cast steel gate and globe valves. This machine consists of three units with 20-inch facing and boring heads, a loading device, a turntable, and two holding units. Note the large amount of hydraulic equipment employed on this machine.

The 17-spindle, automatic indexing machine shown in Fig. 3

Courtesy of Kingsbury Machine Tool Corp.
Fig. 3. Indexing-type automatic drilling and tapping machine.

Fig. 4. Die casting machine for casting electric motor rotors.

produces nine different part sizes in three different part configurations. For altering the machine operations to accommodate the different size parts, extra cams and speed and feed gears are employed. Production rates vary from 80 parts per hour gross for 1-inch size impulse steam trap bodies of No. 416 stainless steel, to 170 parts per hour for ⅜-inch size bodies of No. 420 stainless steel.

Pneumatic components also are used at various points on the stations to provide certain movements. Note, in particular, the two air cylinders in the upper center of the photo.

The machine illustrated in Fig. 4 is a die casting machine used for the casting of aluminum filled electric motor rotors. This machine has an extensive hydraulic system which makes use of several interlocks to provide the operator with maximum protection. Thus, be-

Courtesy of National Automatic Tool Co., Inc.
Fig. 5. Vertical hydraulic feed machines with special outside coolant systems.

fore the "shot" can be made, the large door in front of the machine must be closed. The machine illustrated is equipped with a large accumulator to provide an assist to the hydraulic pump at peak requirements. A heat exchanger is used to keep the oil temperature at a safe operating level.

As a rule, die casting machines operate at fairly high operating pressures in comparison to other types of hydraulic equipment. The operating system may be designed for use with petroleum fluids, fire resistant fluids, or water. The latter two are more popular due to their ability to resist fire but are more costly and often require more maintenance.

Figure 5 depicts two vertical hydraulic feed machines, each with an 18-inch stroke; fixed center head with 18 spindles; 12-inch column extension; and a 36-inch diameter, 4-position rotating table. There is a special central outside coolant system located between the two machines which services both of them.

These machines are designed to perform various drilling operations on plier halves. One machine drills, rough countersinks, finish countersinks, and reams the pivot hole. The other machine drills center, rough counterbores and finish counterbores. The production rate on each machine is between 815 and 1028 parts gross per hour, depending upon the parts being run. Note the clamping cylinders on each of the index tables and the hydraulic power units and controls which operate these hydraulic cylinders.

Courtesy of Sundstrand Machine Tool Div. of Sundstrand Corp.

Fig. 6. Automatic lathe using air-operated equipment.

Figure 6 shows an automatic lathe which is designed for high-speed, high-production applications.   The setup on a machine of this type is easy and convenient, making the machine adaptable for both short and long run lots.   As it is equipped with a twenty-five hp spindle-drive motor, it is also adaptable for a wide range of both shaft and chucking work.   Note the rotating air cylinder mounted on the back end of the machine spindle, at the left.   It is used for operating

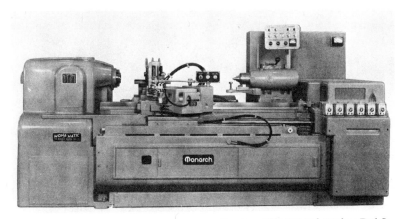

Courtesy of The Monarch Machine Tool Co.

Fig. 7. Modern production lathe with fluid power equipment.

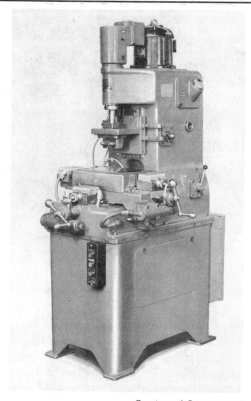

Fig. 8. Hydraulically operated shaving machine for shaving intricate parts for rifles.

a power chuck, a power operated collet chuck, or some other type of holding device.

An air-operated tailstock unit is available for shaft work where the workpiece is held between centers. The actuation of the tailstock cylinder is by dual push-button for forward, with a single push-button for return.

The lathe is designed with electrical interlocks so that the sliding sheet-metal guards must be in place before the machine will operate. Note the window in the front guard which provides visibility for the operator. All change gear covers are also interlocked and must be firmly in place before the machine will operate.

Figure 7 shows a modern production lathe with fluid power equipment. The air-operated spindle in the tailstock is mounted on large antifriction bearings to take heavy radial and thrust loads. This machine will accommodate a 12-inch-diameter air chuck; the machine will not start until the chuck jaws are closed and the tailstock spindle is in operating position.   The hydraulic unit incorporated in this machine supplies oil: to the carriage feed motor, to the facing and forming slide cylinder, and to the "Air-Gage Tracer" slide cylinder. Note the bank of hydraulic controls on the right side of the machine.

A special hydraulically operated shaving machine, illustrated in Fig. 8, is used in shaving intricate parts for rifles.   The hydraulic system on this machine consists of three cylinders, nine valves, and the pumping unit.   The shaving head which is controlled by one hydraulic cylinder operates with a maximum frequency of 130 strokes per minute.   The other two cylinders are used for operating the work table traversing feed mechanism.

Much fluid power equipment is employed on packaging machinery. Figure 9 shows a case packer that makes use of both air and hydraulic equipment to take care of many motions that are required in these versatile machines.   Note the hydraulic power unit and the control valve mounted on top of the case packer and also the air regulator, filter, and lubricator mounted near the base.   Cylinders are mounted on each end.   A machine of this type is especially efficient in the beverage industry where containers and case sizes tend to remain uniform.

Courtesy of Climax Products Div., The Lodge & Shipley Co.
Fig. 9. Case packer for handling a variety of containers.

CHAPTER 2

Hydraulic Fluids

The fluid which is selected for use in hydraulically actuated equipment will have a considerable effect on its performance, maintenance costs and service life. For the user of such equipment, the important question is: What basis should be used to evaluate the various hydraulic fluids which are available for the purpose intended?

There are two methods for selecting the fluid best suited for top performance in a specific hydraulic circuit. One is to choose the fluid by its adherence to set specifications. The other is to accept predetermined evaluation, based on performance rather than on physical specifications. Experience has proved the latter method to be safer and wiser, since some characteristics can only be determined under actual service conditions and over an extended period of use.

Three types of fluids are generally accepted as adaptable for use in hydraulic circuits: (1) petroleum oils; (2) synthetic fluids; and (3) water. While suitable for certain types of applications, water does not meet all the requirements for general hydraulic equipment use, as does petroleum oil or synthetic fluids. As a hydraulic fluid, water presents many problems, such as rust and corrosion, ineffective lubrication, temperature variations—both external and internal, and the hazard of foreign matter in the water itself which might cause an abrasive action on the smooth interior surfaces. All of these could act as factors detrimental to operating efficiency and long life of the equipment.

Due to the fact that petroleum oils and synthetic fluids (if properly refined and fortified to provide all of the basic service properties) more nearly meet all internal requirements for "perfect" hydraulic performance, the following discussion will deal specifically with those fluids.

It is well to remember that a hydraulic fluid must not only transmit power; it must also act as a sealant and a lubricant. To perform these three functions satisfactorily, the fluid must have certain basic properties which might be called *service properties*. Whether a plain hydraulic fluid or one containing additives is selected for use, these properties must be present in the proportions called for by the application, if service difficulties and possible equipment failure are to be avoided.

Six Important Service Properties

There are six service properties which are important to the specifier and user of hydraulic fluids: (1) viscosity; (2) viscosity index; (3) demulsibility; (4) oxidation stability; (5) lubricity or lubricating value; and (6) rust- and corrosion-preventive qualities.

Viscosity. The most important property of the fluid to be used in hydraulic equipment is its viscosity. Viscosity is the measure of the fluid's internal resistance to shear or flow at a definite temperature and pressure. It is defined as the shearing force required to move two plane surfaces relative to one another with a film of fluid between them. Expressed in quantitative terms (metric system) the absolute viscosity of a fluid is that force in dynes required to move a plane surface one square centimeter in area over another plane surface at a velocity of one centimeter per second, when the two surfaces are separated by a layer of fluid one centimeter in thickness. The unit is the *poise,* which is equal to one dyne-second per square centimeter.

The viscosity of an oil is measured in an instrument called a viscosimeter. There are several types, but for commercial purposes the viscosimeter is essentially a container in which the oil is held and allowed to run out through a hole or tube of established size at the bottom. The reading is taken for a standard quantity of oil at a standard temperature. In North America, the Saybolt Universal Viscosimeter is the recognized standard for oils of normal viscosity. The time of flow is taken in seconds and the viscosity reading is expressed as Saybolt Universal Seconds (SUS or SSU).

Viscosity has a direct bearing on hydrodynamic losses, frictional and pressure losses, leakage, and pump and valve noises within the circuit. For any given hydraulic circuit there is a specific oil viscosity range which will assure maximum results.

The majority of hydraulic systems operate most efficiently with fluids falling within the ranges: 135 to 165 SSU; 185 to 230 SSU; and 275 to 315 SSU. Some hydraulic systems require oils having lighter or heavier viscosity ranges than those shown but they are in the minority.

Viscosity Index. It might be said that one of the properties of an ideal hydraulic fluid would be that of retaining the same viscosity under all temperature and pressure conditions to which it is subjected. Unfortunately, no petroleum-base oils have this characteristic. As temperatures increase, these oils thin out, i.e., their viscosity decreases; as temperatures decrease, these oils thicken and their viscosity increases. The rate of change of viscosity with temperature is indicated on an arbitrary scale called the viscosity index (V.I.). The higher the numerical value of this index, the less the viscosity changes for a given change in temperature, and vice versa. Thus, two oils may have the same viscosities at 100°F, but may show entirely different viscosities at some higher temperature. The one that has the highest viscosity index will show the least change in viscosity and, from that standpoint, is preferable for hydraulic machine use.

An example showing the variation in viscosity indexes of several oils would be:

Sample No.	Saybolt Viscosity at 100°F	Saybolt Viscosity at 210°F	Viscosity Index
Oil A	300	52.75	100
Oil B	300	48.50	50
Oil C	300	46.25	0

Oil A has the least change in viscosity between 100°F and 210°F and, therefore, has the highest viscosity index. Oil C has the greatest change in viscosity between 100°F and 210°F and, therefore, has the lowest viscosity index.

The numerical scale of viscosity index is arbitrarily based on a value of 100 assigned to a specific paraffinic oil and a value of 0 assigned to a specific asphaltic oil. Since there are some oils whose rate of viscosity change is even less than that of this paraffinic oil and there are others whose rate of viscosity change is more than that of this asphaltic oil, it follows that the viscosity index extends beyond 100, and below 0 into the minus values.

Figure 1 shows the effect of temperature on the viscosity characteristics of four hydraulic fluids having viscosities of 150 SSU, 300 SSU, 525 SSU and 950 SSU respectively at 100°F and a viscosity index of 100. This chart may be used where a change in hydraulic oil is required to maintain the recommended viscosity at an elevated temperature.

It is possible to improve, i.e., raise the V.I. of an oil, up to values as high as 170 by the addition to it of certain synthetic agents called additives. When the use of a viscosity index improver is contemplated, however, the effect of these additives on other characteristics must be studied, as in some cases they may be adversely affected from the standpoint of intended use.

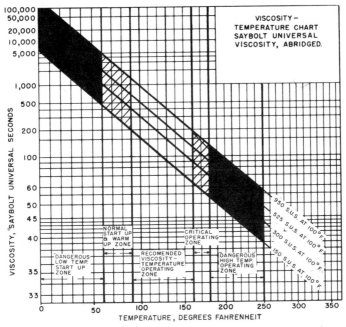

Courtesy of E. F. Houghton & Co.

Fig. 1. Effect of temperature on the viscosity characteristics of four hydraulic oils, each having a Viscosity Index of 100.

Demulsibility. The service property of a hydraulic fluid which enables it to separate rapidly and completely from moisture and to successfully resist emulsification is known as demulsibility. This property is particularly important since the general design, construction, and operation of a hydraulic system is conducive to the forming of moisture and of stable water-in-oil emulsions within the system. Highly refined hydraulic oils are basically water resistant in nature and have excellent demulsibility characteristics when new. They are inclined to maintain this important property for long periods of service.

Use of a hydraulic fluid without effective demulsification qualities is, as a rule, reflected in the destruction of lubricating value and sealant properties. Not only does such a condition greatly reduce the life-expectancy of the fluid, it also reduces the service life of the working surfaces. Such a fluid is also more susceptible to water absorption.

In considering causes of ineffective demulsibility, attention should be given to the organic chemicals which have been added to fortify some other physical property, as well as to compounds which may have found their way into the system during operation.

Oxidation Stability. Here is another service property which is exceedingly important in judging a hydraulic oil. It is defined as the fluid's ability to resist oxidation and deterioration for long periods. It is based on the inherent chemical structure of the oil itself, which can be further fortified by additives developed to help resist breakdown.

Many underestimate the effect of heat on lubricating oil. It has been said that the life of an oil is reduced by about 50 per cent for every 15-degree rise in temperature above 140°F under average operating conditions; however, that specific claim may be subject to question. Heavy oils do have greater resistance to heat than light oils which have been derived from the same source, and the average hydraulic oil is a light oil. Fortunately, there is a wide choice of oils available for use in the viscosity range required by hydraulic fluids.

It should be noted that the temperature of the oil in the reservoir of a circulating oil system does not always represent a true state of operating conditions. Localized hot spots occur on gear teeth, bearings, or at the point where oil under pressure is forced through a small orifice. Continuous passage of an oil through these points may produce local temperatures high enough to carbonize or sludge the oil, yet the oil in the reservoir or circulating system may never register temperatures above 100°F to 130°F.

Some metals also have an undesirable catalytic effect on oils, particularly at elevated temperatures. They usually tend to increase the rate of acid formation in the oil. Zinc, lead, brass and copper are among the common metals which have the greatest effect. Metals are most active as catalysts when present in a finely divided form. In this latter state, iron sometimes becomes an active catalyst.

When oxidation of an oil takes place, sludge is formed. As soon as a small amount of sludge is formed, the rate of formation generally increases more rapidly thereafter. As sludges are formed, certain changes in the physical and chemical properties of the oil take place. The oil becomes progressively darker in color, heavier in viscosity, and organic acids are formed as evidenced by an increase in the neutralization number.

The extent to which changes take place in different oils depends upon the type of oil, type of refining, and whether it has been treated to provide further resistance to oxidation. A great difference in stability will be found in varying types of oils. One oil may break down and sludge excessively after a 24-hour run in a hydraulic machine, while another will perform satisfactorily for 1000 hours.

Oxidation Tests. Laboratory men have worked for years in an effort to design laboratory tests which will predict the stability of an oil. One of these tests is the Indiana Oxidation Test. It is carried out by heating 300 cubic centimeters of an oil to 341 °F in a glass tube and bubbling air through the oil at the rate of liters per hour.

To find the amount of sludge present at a given time, a 10-gram sample of the oil is diluted with 100 cubic centimeters of petroleum naphtha and, since only the oil goes into solution, the sludge or oxidized oil precipitates and may be filtered and measured. The amount of sludge is reported in milligrams per 10 grams of oil. The test is normally run for 150 hours. Sludge determinations are run on samples taken after 50, 100 and 150 hours. Viscosity determinations are also run on samples taken after 100 and 150 hours. After determining its viscosity, the sample taken for the 100-hour test is returned to the oxidation tube within one hour. Some laboratories also determine the neutralization number at the beginning and conclusion of the test.

Upon conclusion of the test, the sludge figures are plotted against time in hours. Table 1 is an example of the way in which results are reported.

TABLE I

INDIANA OXIDATION TESTS

	Sample No.					
	A	B	C	D	E	F
Initial S.U. Viscosity at 100°F, sec.	190	200	315	306	295	307
Viscosity after 50 hours, sec.	202	236	363	460	320	324
Viscosity after 100 hours, sec.	211	*	442	650	356	348
Viscosity after 150 hours, sec.	225	*	552	†	392	386
Orig. Neut. No.	0.03	0.10	0.05	0.05	0.03	0.05
Final Neut. No.	0.18	1.56	0.64	4.68	1.05	0.74
Hrs. to form 10 mg. of sludge	68	12	20	11	45	54
Hrs. to form 100 mg. of sludge	166	48	102	68	206	over 300

* Test discontinued when 100 mgs. of sludge were obtained.
† Material semi-solid at 100°F.

Samples A and B represent typical figures obtained on hydraulic oils of the 200 SSU viscosity at 100°F type, Sample A being considerably more stable than Sample B. Samples C and D are slightly heavier oils. These two samples both began to oxidize at a fairly rapid rate but, as the test continued, Sample D oxidized at an increasingly rapid rate until, at the end of 150 hours, it was solid at room temperature and semi-solid at 100°F. Sample E is a highly refined oil. The addition of an inhibitor to this oil improves the stability further as represented by Sample F.

The same apparatus is used in the Continental Oxidation Test Method. In this method, a 9-foot length of No. 16 soft iron wire is polished with emery paper, wound into a coil and placed in the oxidation tube so that is entirely covered by the oil. The same test temperature and rate of air flow used in the Indiana Method are used in this method. The use of the iron wire accelerates the oxidation of the oil with the result that sludge is formed at a faster rate. Iron is used because it is the metal present in largest amount in a hydraulic system.

This test is normally run for 48 hours. Samples are removed for sludge determmation after 8, 24 and 48 hours and, for viscosity determinations, after 48 hours. Results are reported as shown in Table 2. Curves illustrating the different rates at which some hydraulic oils oxidize are given in Fig. 2.

The extent to which an oil exhibits stability is partially due to the presence of minute quantities of certain chemical compounds which occur naturally in the oil. These compounds are called oxidation inhibitors. They occur in different amounts in different oils. As the

TABLE 2

CONTINENTAL OXIDATION TESTS

	A	B	C	D	E
Original S.U. Vis. at 210°F	51	53.5	51	50.5	53
Final S.U. Vis. at 210°F	54	60.5	57	55.2	55.5
% Increase in Viscosity	5.9	13.1	11.7	9.3	4.7
Original Neut. No.	0.084	0.028	0.056	0.056	0.028
Final Neut. No.	0.74	2.0	1.42	0.96	0.56
Original Carbon Residue	0.15	0.16	0.24	0.10	0.15
Final Carbon Residue	0.64	1.15	0.87	0.64	0.31
Naphtha Insolubles					
(a) 10 mg. point (hrs.)	13	18	3	26	48+
(b) 100 mg. point (hrs.)	48+	48+	25	48+	48+
(c) Final Insoluble					
mg./10-gram sample	92.6	54.0	183	27.2	6.9

oils go through the different refining processes, some of these inhibitors are occasionally removed from the oils.

The stability of an oil can be measurably improved by the addition of synthetic oxidation inhibitors. The amount of inhibitor used depends on many factors but it is usually added in quantities of from

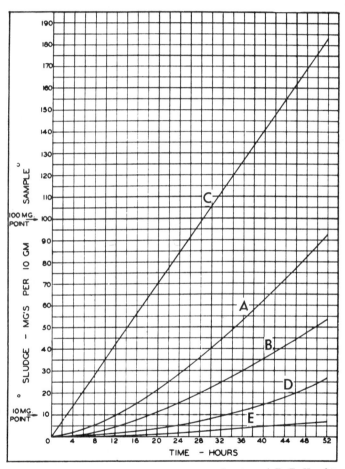

Courtesy of E. F. Houghton & Co.
Fig. 2. Comparative oxidation rates of various hydraulic oils.

0.005 to 0.5 per cent. The helpful effect that these inhibitors have on the stability of an oil is illustrated by reference to Table 2.

Sample B is an oil obtained from a mixed base or Mid-Continent stock and has been subjected to a heavy acid treatment and clay filtering. Sample E is the same oil as Sample B to which has been added 0.15 per cent of an oxidation inhibitor. The effect of the inhibitor on increasing the stability of this oil as measured by this oxidation test is shown by the lower percentage of increase in viscosity, final neutralization number, and naphtha insolubles obtained with Sample E.

All five oils could be used successfully in many hydraulic systems, but in those systems where the oil must function under severe operation conditions, Sample D and E would perform satisfactorily. Sample B would have a chance of performing satisfactorily and Samples A and C would probably deteriorate rapidly in service.

The five samples given in Table 2 would, on the strength of these tests, be rated with respect to potential oxidation stability as follows:

1. Sample E
2. Sample D
3. Sample B
4. Sample A
5. Sample C

Some laboratories have recently begun to use the ASTM "Proposed Method of Test for Oxidation Characteristics of Steam Turbine Oils" as a means of measuring the stability of hydraulic oils. This test is found under Appendix 2 of the 1943 ASTM Committee D-2 Report. Essentially, the test consists of subjecting a sample of oil to a temperature of $95°C$ in the presence of water, oxygen, and an iron-copper catalyst, and the time required to build up a neutralization number of 2.0 in the sample is determined. One large user considers an oil to have excellent oxidation characteristics if the neutralization number does not exceed 0.25 at the conclusion of 1000 hours.

There are certain objections to some of these test methods which should be stated. Originally, some were designed to determine the oxidation characteristics of engine oils. Therefore, the test temperatures were set for $341°F$ to $400°F$. Obviously, few hydraulic oils ever attain these temperatures.

Furthermore, engine oils are usually considerably heavier oils than hydraulic oils and, therefore, are inherently more stable. To illustrate, it is not uncommon for a hydraulic oil with a viscosity of 150 Saybolt Universal seconds at $100°F$ to have a flash point of $350°F$. This means that some of the constituents in the oil will vaporize below this temperature. Blowing oxygen or air through this oil at that temperature will assist in driving off more of the light "ends." This results in obtaining a higher per cent increase in viscosity readings than would be obtained in service where the oil is operating at a lower temperature in an enclosed system.

Lubricating Value. Lubricity, or oiliness, is another factor not determinable by specification. Some machine tools use the same oil both as an hydraulic medium and way lubricant, thus requiring a high degree of oiliness. Additives to increase this facility are often added, overcoming jerky motion when sliding over the ways. This is particularly important when the slide moves at slow speed and when cuts and feeds are severe.

The ability of an oil to perform satisfactorily, i.e., to minimize friction and wear, is due to that property termed "lubricating value." Unlike viscosity, which is a purely physical property, the lubricating value is directly related to its chemical nature and its chemical affinity for metals. Lubricating value depends on oiliness, extreme pressure or anti-weld properties, and on the composition of rubbing surfaces. Oiliness agents are sometimes used to improve that property. Extreme pressure additives build up resistance to welding, forming an adherent strong film to resist high unit pressure. Superfinished metal surfaces reduce the friction and rate of wear.

Film Strength. Seldom included in the hydraulic oil specifications because the lubrication function of such an oil has been generally considered as secondary. Nevertheless, film strength can be important, for lack of it will result in excessive wear of moving parts and eventually in increased clearance. Adequate film strength, for example, will minimize wear on vanes and on annular rings of vane pumps.

Additives used to impart lubricity or film strength to an oil are essentially organic chemicals which may or may not contain sulphur, chlorine or phosphorus. They function in two ways: (1) due to molecular structure, some of these materials exhibit a tendency to "plate out" on a metallic surface and require a greater force to remove them from the surface of the metal than molecules of mineral oils, or (2) other materials chemically react with the metal under certain conditions of temperature and/or pressure to form a chemical compound on the metal surface which prevents seizure or welding.

Various test machines have been invented to evaluate these additives. The Almen Machine, the Cornell or Falex Machine, the Timken Machine, and the Shell Four-Ball Machine are some of the more prominent ones. As a general rule, the use of these machines in evaluating mild film strength additives of the type required in hydraulic oils has not been too successful. Most machine tool manufacturers have had to rely on actual testing of the lubricant under operating conditions in order to obtain a satisfactory evaluation.

Rust- and Corrosion-preventive Qualities. Freedom from corrosion is another factor to be considered. The neutralization number of the oil (indicating its degree of acidity or alkalinity), when new, may be satisfactory, but after use the oil may tend to develop corrosive tendencies as it begins to deteriorate. The base stock should have been carefully processed with the specific aim of inhibiting harmful acid formation which would attack bearings or finished metal surfaces in the system.

Another form of corrosion is *rusting,* which must be combated in a different manner. Many systems are idle for lengthy periods after a run at elevated temperatures. This permits moisture to be condensed in the system, resulting in rust. Also, a machine tool using a water-soluble coolant may not be entirely tight, allowing water to leak into the hydraulic system. Rust-preventive additives, usually synthetic chemicals, are often used in hydraulic oils.

One of the most popular tests for determining the rust-preventive properties of a hydraulic oil is the ASTM test for rust-preventing characteristics of steam turbine oils in the presence of water, ASTM No. D-665-42T. This test involves stirring a mixture of 300 cubic centimeters of the oil under test with 30 cubic centimeters of distilled water at a temperature of 140°F with a cylindrical steel specimen immersed therein. After 48 hours' contact with the oil-water mixture, the specimen is examined for signs of rust. The temperature and time limit given herein can be changed to increase or decrease the severity of the test. For example, if the test temperature is raised to 180°F, many oils which satisfactorily pass the test at 140°F will fail at the higher temperature. In either event, this test is considered to be a very mild one by concerns engaged in selling rust-preventive additives.

Another method of testing the rust-preventive qualities of an oil is the humidity cabinet test. This test is usually run at either 100° or 120°F for periods of time ranging from 48 to 500 hours in a cabinet which is maintained at 95 to 100 per cent relative humidity. The apparatus required for this test is quite expensive and it is often quite difficult to obtain duplication of results between different laboratories. The test consists of dipping polished or sandblasted steel panels, usually 2 × 4 inches in size, into the oil to be tested for a period of 5 seconds to one minute. The coated panels are then allowed to drain for a period ranging from 2 hours to 24 hours and suspended from glass cross-rods in the humidity cabinet by means of a glass hook or a specially prepared string for the desired period of time. Great care must be used in preparing the panels in order to insure duplicate results.

The humidity cabinet test is much more severe than the previously mentioned ASTM test. Generally speaking, most oils which will pass the 200-hour humidity cabinet test will easily pass the ASTM steam turbine oil rust-preventive test but the reverse is not always true.

Standard Specifications

It should be pointed out that there are other specifications which are also used as a basis for selecting the proper fluid, such as: color, gravity, pour point, flash and fire points, carbon residue and neutralization number. While these specifications have been pretty generally accepted as standard in evaluating the various qualities of a hydraulic oil, yet their importance as compared to the importance of the six service properties places them in a minor role as the following definitions will show.

Color. The color of an oil usually is an indication of the type of crude from which it has been obtained. However, some additives tend to give the oil a darker color. Before modern refinery methods and the use of additives came into being, the color of an oil was quite often used as a measure of its degree of refinement. This is no longer a reliable criterion.

Gravity. Gravity is of no particular importance to the user of hydraulic oils. In any one viscosity range, oils obtained from different base stocks will vary in API gravity readings. Oils of the naphthenic type have low API gravity readings; oils of the Pennsylvania type have high API readings; while oils of the Mid-Continent type have readings between the two. However, since certain modern refining methods have an effect on gravity as well as other properties, this determination cannot be used as a means of identification.

Pour Point. The pour point of an oil indicates the temperature below which the oil will not flow freely. This is usually taken to be 5°F above the "cold test" or "setting point," which is the temperature at which the oil congeals. Except for oils used in low temperature equipment, the pour point is not an important characteristic for hydraulic systems. For cold weather use, hydraulic oil should have a cold test at least 25°F below the lowest expected operating temperature. Also, in order that the oil may respond to the pump suction, it must be sufficiently fluid. Oil having a viscosity of 4000 seconds (Saybolt) at the lowest operating temperature is considered to be the heaviest which can be pumped efficiently.

Flash and Fire Points. Flash and fire points indicate the temperatures at which the oil begins to volatilize. The former is the oil temperature at which the oil will give off sufficient vapors to ignite momentarily; the latter is the oil temperature at which sufficient vapors are given off to ignite and continue to burn when exposed to spark or flame. In some special applications where oils of very light viscosity must be used, these temperature points are of importance.

Carbon Residue. The carbon residue test is a measure of the percentage of carbon that remains in a sample of the oil after the volatile content has been driven off by means of heat. It is one of the tests which has been made on mineral oils for many years and it has been carried over to physical property or specification sheets for hydraulic oils more as a habit than as a necessity.

Neutralization Number. The neutralization number of an oil denotes its degree of acidity or alkalinity. After an oil has gone through the various refining processes, it usually contains very small amounts of organic acids. Hydraulic oils rarely contain any free alkali. The neutralization number of hydraulic oils usually runs between neutral and 0.1. Some oils which contain additives show a higher figure. An untreated hydraulic oil having a neutralization number higher than 0.1 will often show a tendency to deteriorate more rapidly in use under severe operating conditions than one which has a lower neutralization number.

The preceding tests are of little value when given individually. It is conceivable that a consumer could order a "300-at-100°F" viscosity oil, place it in a piece of hydraulic equipment and have it perform satisfactorily. However, that consumer could not use the color, viscosity index, gravity, or any of the other tests individually as a means of obtaining a suitable oil for that particular hydraulic system. Individually, these tests are meaningless. They must be used in connection with one or more of the others to have any real value. This point is further clarified in Table 3. The four oils shown in this table all have the same viscosity at 100°F. However, there is considerable variation between the other physical properties of these oils.

Basic Types of Mineral Oils

The terms that are commonly used to describe the basic types of mineral oils are rather loose and arbitrary since there is no exact line of demarcation between them. The three basic types are:

1. Pennsylvania or paraffin base oils
2. Naphthenic, asphaltic or Gulf Coast base oils
3. Mixed base or Mid-Continent type oils which contain both paraffin and naphthenic compounds.

TABLE 3

VARIATIONS IN PHYSICAL PROPERTIES OF SAMPLES OF HYDRAULIC
OILS WITH SAME VISCOSITY

	A	B	C	D
S.U. Vis. at 100°F	295	296	295	295
S.U. Vis. at 210°F	54.5	47	51	52
Viscosity Index	112	23	85	96
Gravity, A.P.I.	29.8	23.4	26.2	29.5
Flash Point, °F	420	370	405	420
Fire Point, °F	470	420	465	475
Color, N.P.A.	5	2	2½	2
Pour Point, °F	25	− 15	10	− 5
Carbon Residue	0.52	0.015	0.15	0.07
Steam Emulsion Number	175	275	240	125
Neutralization Number	.02	.05	.05	.02

In Table 3, A is a Pennsylvania oil; B is a naphthenic oil; and C is a Mid-Continent oil. It can readily be seen that Oil A has the highest gravity, flash point, fire point, pour point, carbon residue, and viscosity index; Oil B has the lowest values for the same tests and Oil C lies between Oils A and B.

Oil D is included in this table to illustrate the change in physical properties which can be obtained by subjecting an oil to special refining processes. Oil D is a Mid-Continent oil similar to Oil C which has been subjected to a solvent-treatment process. This process has made it more similar to Oil A with respect to gravity, flash point, fire point, and viscosity index and more similar to Oil B with respect to pour point and carbon residue.

Other Factors Affecting Oil Performance

Foaming. Certain hydraulic systems, due either to the construction of the machine or to the type of work performed, cause excessive foaming to take place as the oil is returned to the reservoir. In some cases, the presence of additives in the oil also causes foaming.

When necessary, the tendency of a hydraulic oil to foam excessively can be greatly reduced by the use of anti-foam agents. These agents are usually synthetic organic chemicals which are added in proportions varying from 0.001 to 0.5 per cent. The Cooperative Lubricants Research Committee of the Cooperative Research Council has designed a test for determining the foaming characteristics of lubricating oils. The test is conducted by placing the oils to be tested in a large graduate. Low-pressure air flows through a glass tube to a gas-diffuser stone at the bottom of the graduate for a specified time.

The volume of foam formed and the rate of the collapse of the foam are determined after the air has been turned off.

Gums and Sludges.   Gum solvency is another factor for which no definite specification can be written.   When an oil begins to break down, as even the best will eventually do under prolonged service, it is important that the oil has the ability to hold gums and sludges in solution.   If the oil has been processed with active solvents, it will tend to hold incipient formations in solution, preventing their being deposited on control valves, in pipe lines, on filtering units, or in reservoirs.

Additive Treatment of Hydraulic Oils.   From the foregoing, it will be clear that science has improved on oil as it comes from the refinery by means of a variety of treatments given that oil.   To sum up, by the use of additives it is possible to strengthen many proper-

TABLE 4

EXAMPLE OF PROCESSING EFFECTS ON A HYDRAULIC OIL

Type of Test	Original Oil	Same Oil with Special Processing
Viscosity Change Due to Oxidation		
Original viscosity (before oxidation)		
at 210°F, SSU	47.6	45
Final viscosity after oxidation, SSU	51.5	45.9
Increase in viscosity, per cent	8.2	2.0
Conradson Carbon Change Due to Oxidation		
Final Conradson carbon value	.59	.20
Acid Number Change Due to Oxidation		
Final Acid number	1.15	.55
Precipitation—Naphtha Insolubles Due to Oxidation		
Total milligrams after 48 hours	63	2.4

ties: viscosity index, anti-foaming, gum solvency, oiliness, film strength, anti-corrosion, and most important of all, oxidation stability.

As an example of what has been done along this line, the comparison in Table 4 shows the improvement in stability of a very high-grade straight petroleum oil as a result of special processing. In the left-hand column are values for the original oil.   In the right-hand column are corresponding values for the same oil after processing.

Analyzing the comparisons given, attention is directed to the following:

1. Viscosity change is several times less for the processed oil. Consequently, its performance in a hydraulic system will be more uniform throughout its service life.

2. Lower final Conradson carbon (carbon residue) value and acid number illustrate again the stabilizing effect of special processing and treatment by the greater resistance to change which it imparts to the petroleum oil.

3. The lower value shown for precipitation of naphtha insolubles (sludge) plainly indicates the better stability characteristics of the processed oil. Less sludge formation may be expected in proportion to the figures given.

As the use of additives in hydraulic oils increases, new means of evaluating them in the laboratory are being sought. The gap which exists between what is expected of a hydraulic oil on the basis of laboratory tests and the actual performance of that oil under severe operating conditions is rather large. Efforts to reduce this gap must continue in order that the manufacturer of hydraulic oils may successfully keep pace with the engineers as they, in turn, design more complex hydraulic equipment in which oil is used as the hydraulic medium.

It should be made clear that, while additives will improve a good base stock, they will not necessarily make a good hydraulic oil out of a cheap or ill-fitted stock. They will make good oil better, but they cannot make a poor oil equal to a good one.

Fire-Resistant Hydraulic Fluids

In recent years fire-resistant synthetic fluids have been developed for use as hydraulic media where fire hazard exists. While hydraulic oils made from petroleum stocks are almost ideal for the purposes intended, they are inflammable under normal conditions and can become explosively dangerous when subjected to high pressures and a source of flame or high temperature. This hazard has occupied the attention of industry and of the suppliers of fluids who have spared no expense to overcome the danger to personnel and property.

Safety engineers and personnel responsible for uninterrupted production have, therefore, turned to the use of fire-resistant fluids in such danger spots. Two general types of fire-resistant fluids are available: (1) water-base; and (2) synthetic mixtures.

Acceptable fluids of the water-base type consist of water, a glycol-type thickener and antifreeze compounds, rust and corrosion inhibitors, and additives to impart lubricity and anti-wear qualities. Because of their high water content, they are fire resistant. Work is still being directed towards the perfecting of water emulsion fluids. These do not fall into the exact classification of thickened water-base fluids of the glycol type.

Non-water-base fluids now commercially marketed are chemically described as phosphate esters. These fluids depend for their fire resistance on a high percentage of phosphorus-containing compounds in their molecular structure.

Both types of fire-resistant fluids possess advantages and disadvantages. The present types of products offer definite advantages in their oxidation resistance and low rate-of-change of viscosity with temperature. The fire-resistant fluids do not break down to form sludges and petroleum gums which would interfere with the proper functioning of valves, pump parts, controls, etc. Also, because of their high viscosity index, they can be used over a range of temperatures without thinning out or thickening up excessively. They do not carbonize under the temperatures normally encountered. On the other hand, the water-base fluids do not have the high lubricity of a petroleum oil, although additives have increased the oiliness factor and reduced wear. The synthetic-base fluids have been found to attack packings normally used in hydraulic systems, so that special packings are required when they are employed.

Work is still progressing in the development of fire-resistant fluids. For example, because of their greater mass and higher specific gravity, their high-velocity flow characteristics are now being particularly studied with relation to pump design, screens, etc.

It is a matter of record that the water-base fluids were approved by the United States Navy for use on catapults on aircraft carriers following explosions attributed to the petroleum-base media.

When installing a fire-resistant fluid, care should be taken to clean the system and remove all trace of petroleum oils which may formerly have been used. The manufacturer of the fire-resistant fluid should be consulted as to both installation and maintenance.

Trouble Shooting

No discussion of hydraulic circuits and the fluids used would be complete without mention of typical operating difficulties.

Not all of these troubles should be attributed to the oil by any means, although improper selection of a hydraulic oil, either with reference to type, fortification, or fluidity, may lead to other troubles listed herein. A slight deficiency in operation of the hydraulic system may easily cause a fall-off in production. This deficiency may be due to design or to mechanical troubles, which in turn may be traced back to the fluid medium used.

For the benefit of the man charged with proper maintenance of hydraulic mechanisms, some of the more common difficulties and remedies are listed in Table 5.

Too much stress cannot be placed upon the factor of contamination of the hydraulic oil. Where soluble oils and other cutting fluids are used on the machines, and design is such as to permit leakage, there is always a possibility that some of this material will find its way into the hydraulic system and, under the conditions imposed by the hydraulic service cycle, this fluid may oxidize and form gums, sludge, or even corrosive materials. Despite the fact that some oils

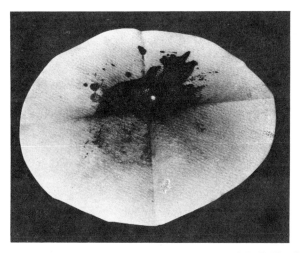

Courtesy of E. F. Houghton & Co.

Fig. 3. Contamination surrounding a ball check valve in a hydraulic system on a milling machine.

are additive-treated and have the ability to keep certain gums and sludges from harming the system, there is a limit to the amount of such gums which can be carried in solution by the additive and any excess amount will, of course, settle out and cause trouble. Figure 3 shows contamination surrounding a ball check valve taken from a milling machine which had become sluggish and "chattery," so that an overhaul job was required.

Soluble oil contamination is usually accompanied by a change in color of the oil, this color changing from the original petroleum

TABLE 5

HYDRAULIC MECHANISM—COMMON DIFFICULTIES AND THEIR REMEDIES

Trouble	Remedy
Pump Chatter	
1. Sticking Vanes	Check contaminants in oil. Install or inspect filter. Analyze oil for tendency to oxidize.
2. Worn Parts	Inspect valves, pistons, rings, vanes, gears, etc. Renew where worn.
3. Cavitation	Check inlet. Area should be approximately twice that of discharge pipe. Use lower viscosity oil. Use oil of lower cold test (based upon ambient temperature). Check inlet screen for plugging.
4. Air Leaks	Inspect suction-point connections. Common practice is to flood joints with oil during operation, noting point where pump runs quietly. Make certain intake is below oil surface at all times.
Overheating	
1. Discharge Pressure too High	Reduce relief valve setting.
2. Improper Oil—Sluggishness	Use lower viscosity oil.
3. Excessive Internal Leakage	Use heavier and higher viscosity index oil. Inspect packings for wear. Check for worn valves, etc.
4. Undue Friction in Pump	Inspect setup—readjust.
5. Oil Cooler	Check for clogged passages or inadequate capacity.
6. Oil Capacity	Increase tank size so that the rise in oil temperature will be lessened.
Loss of Pressure	
1. Relief Valve	Inspect relief valve for spring breakage or gummed parts.
2. Excessive Internal Leakage	Inspect valves, pistons, rings, gears, vanes, etc., for wear; replace.
3. Pump Underrunning	Check motor drive for drop in rated speed.
4. Pump Failure	Inspect for presence of foreign materials, oxidation deposits, etc., clogging internal orifices and valves (see Fig. 4).
5. Sluggishness	Use lower viscosity oil.

Trouble	Remedy
Erratic Ram or Table Motion	
1. Valves Binding or Sticking	Examine for metal particles or foreign matter (see Fig. 3). If gummed, check for contamination of oil or oil oxidation. Inspect for air in system.
2. Sluggishness during Warm-up Period	Analyze oil for cold test and viscosity at the surrounding room temperature. If too high, change to oil of low cold test and high V.I.
3. Binding of Ram, Rods, etc.	Check for misalignment of piston rods. Packings should be examined for "curling" or binding.

General Causes of Inefficient Operation

1. Lack of proper drain periods. Should be determined by periodic sampling and analysis of oil.
2. Contamination, usually through leakage, by cutting or grinding fluids.
3. Inadequate knowledge of functioning of hydraulic circuits being used.
4. Air leaks.
5. Improper hydraulic oil.
6. Poor or decomposed packings.

bloom to a milky, streaked appearance. Where straight cutting oils are entering the hydraulic system, the extent of this condition may be noted by a change in odor and an uneven layer-like appearance which is caused by imperfect blending of the hydraulic oil with the cutting oil.

Water in System

In order to prevent the occurrence of problems attributed to the entrainment of water in a hydraulic fluid, or minimize the effects of such water, several suggestions are offered:

1. Hydraulic oil reservoirs should be designed to be of sufficient capacity to allow for proper settling of water and/or water-in-oil emulsions. Baffle plates of proper size welded to the bottom of such tanks are extremely helpful. Facilities for periodically draining settlings from the bottom of reservoirs should be provided.
2. The intake line to the hydraulic pump should be at least 2 to 3 inches above the floor of the reservoir. It should also be located at the farthest point from the discharge line.
3. Efficient filters of proper size should be installed. The filtering element should in no way cause the removal of additives from

the hydraulic oil. Filters should be periodically inspected, drained and cleaned.

4. The physical condition of the oil should be checked periodically, either visually or by laboratory methods.

5. Hydraulic systems should be thoroughly flushed with a very light type of hydraulic oil that has been processed with gum solvents after every drain period. This is particularly important after noticeable quantities of water or other types of impurities have been detected in the used oil.

In most of the trouble encountered in hydraulic circuits, experience has proven that, as a general rule, one or more of three basic rules have been broken. The first of these rules is proper selection of the hydraulic oil; the second, cleanliness of hydraulic systems; and the third, reasonably careful maintenance. When all three items have been given attention, the user may be almost certain that his equipment will function to his entire satisfaction. It is only an occasional occurrence where the design of the machine is at fault or where the oil supplier can rightfully be blamed.

Qualities of a Good Hydraulic Oil

Having considered the data which original specifications provide, as against the other more important factors which affect service and which are not revealed in "specs," it should be clear that hydraulic oils need more thorough evaluation than merely buying by routine specification. That is why plant men are more interested in what the oil will do in service.

Here are the "musts" and "must nots" which a good hydraulic oil should meet:

1. It must be of the correct viscosity to provide the right seal so that working pressures can be developed and maintained.
2. It must have a high viscosity index to resist changes in fluidity as temperatures change.
3. It must not permit wear on working parts.
4. It must not foam.
5. It must be stable, resisting oxidation.
6. It must not precipitate gums or sludges on working parts.
7. It must retain all of its original properties in use.
8. It must have a long service life.
9. It must protect parts against rust.

Without some of these properties, the hydraulic oil used in an expensive mechanism can shorten the life of equipment and spoil fine work.

Flushing of Hydraulic Systems

There are several procedures used in the flushing of hydraulic systems which have become contaminated and in which gums have formed to such an extent that purging is necessary to maintain normal operation. Experience has proved that some of these have certain weaknesses which merit discussion of the whole problem. The result of improperly cleaning out the hydraulic system of an injection molding machine, for example, was the filter deposit of resinous material, oxidized oil, and asbestos shown in Fig. 4.

Courtesy of E. F. Houghton & Co.

Fig. 4. Filter deposit of resinous material, oxidized oil, and asbestos taken from an improperly cleaned out injection molding machine hydraulic system.

One procedure used in the past has been to drain the hydraulic system, replace the oil with a light flushing oil (100 SSU at 100°F viscosity) and operate the mechanism without work for periods of from 20 minutes to 3 hours, depending upon the degree of contamination. The flushing oil is then drained and a new charge of hydraulic oil installed.

However, some manufacturers of hydraulic pumps warn the user never to use any oil which is lighter in viscosity than that recommended for usual operation. Therefore, the pump manufacturer or the oil supplier should be consulted to determine what materials may be safely used without danger of damage to the mechanism.

Because it is virtually impossible to remove all liquids from a hydraulic system in normal draining, no material should be used that would affect the operation of the mechanism or tend to contaminate the fresh charge of oil. Consequently, material such as kerosene, naphtha, alcohol, steam or water should not be used in the flushing operation.

Still other procedures are based on the practice of following the flushing medium with a charge of the "same oil used in service" then refilling with new oil. This is to insure that the lines, "pockets" and other portions of the system will not be contaminated by flushing oil or that the final viscosity is not reduced below a safe limit.

In recent years special flushing oils have been developed which contain gum solvent chemicals. These in combination with light mineral oils tend to clean the gum formation from the working parts of the system. Such solvents will take the oxidized oil into solution and, following a purging charge, do a good job of cleaning hydraulic systems without dismantling.

CHAPTER 3

The Hydraulic Power Unit

The oil hydraulic power unit is the key to the whole hydraulic system. It is to the hydraulic system what the electrical generating plant is to the electrical system—it is the source of fluid motivation. Basically all industrial hydraulic power units are pretty much the same, although there are vast differences in the components which make up these units. The basic unit contains a hydraulic pump, an oil reservoir with a cover, a suction strainer or filter, a motor coupling, an electric motor, pressure gage, relief valve (see Chapter 6), hydraulic oil (see Chapter 2), and the necessary internal piping. Each of these components is extremely important in the design of an efficient functioning unit.

Typical hydraulic power units are shown in Figs. 1, 2 and 3. The American National Standards Institute (ANSI) symbols have been used in the diagrams (see charts on endpapers). Figure 1 (upper) represents a unit equipped with a constant volume pump. Figure 1 (lower) illustrates such a unit. Figure 2 (upper) represents a unit with a variable volume pump. Figure 2 (lower) illustrates such a unit. Figure 3 (upper) represents a unit with a constant-volume pump and variable-volume pump operated by a double shaft extension electric motor. Figure 3 (lower) illustrates a hydraulic power unit with L-shaped reservoir.

The Hydraulic Pump

The heart of the power unit is the hydraulic pump. Numerous standard pumps are available. These may be classified mainly into three types: the gear, the vane, and the piston type.

The principle upon which vane, gear and piston pumps operate is that a partial vacuum is formed as the internal parts go through their cycle; oil is forced up into the pump due to the atmospheric pressure on the oil; this oil is then forced out of the pump under pressure as the cycle progresses. At sea level, atmospheric pressure

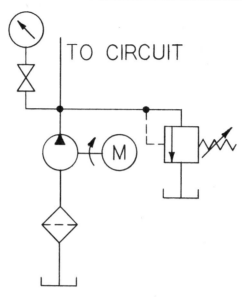

TO CIRCUIT

Fig. 1A. Schematic diagram showing basic elements of a hydraulic power unit
utilizing a constant-volume pump.

Fig. 1B. Hydraulic power unit with a constant-volume pump.

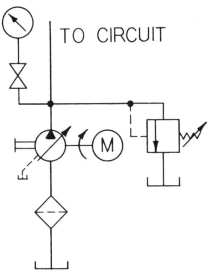

Fig. 2A. Schematic diagram showing basic elements of a hydraulic power unit utilizing a variable-volume pump.

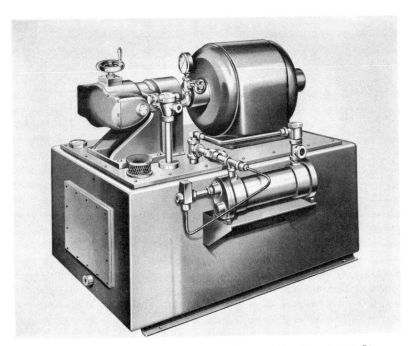

Courtesy of Abex Corp., Denison Div.

Fig. 2B. Hydraulic power unit with a variable-volume pump.

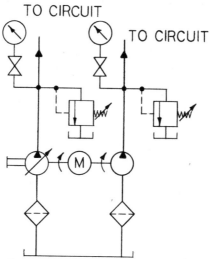

Fig. 3A. Schematic diagram showing basic elements of a hydraulic power unit utilizing both variable-volume and constant-volume pumps driven by one electric motor.

Courtesy of Double A Products Co.

Fig. 3B. Hydraulic power unit using a double-shaft extension electric motor to drive two sets of pumping equipment. Note the control valves and filters mounted on the side of the reservoir.

on the oil is in the order of 14.7 lb per sq in. The idea that pumps cause a "suction" is not correct; all they do is to form a partial vacuum and the atmospheric pressure does the rest in getting the liquid from the supply tank into the pump.

When considering the selection of a hydraulic pump, the service for which it is intended should be carefully considered. On some applications an inexpensive pump may meet all the requirements necessary to do a satisfactory job while on others an expensive, complicated pump may be required.

Operating pressure is an important factor to be considered when selecting a pump. There is no reason to invest in an expensive pump if the operating pressure for the system is only going to be 100 to 200 lb per sq in. A simple gear-type pump would probably amply meet the requirements. On the other hand, if an application requires operating pressures up to 5000 lb per sq in., then an expensive pump would be required to meet the rugged service.

Gear pumps of certain designs are capable of producing pressures up to 3000 lb per sq in. The maximum operating pressure of vane-type pumps usually does not exceed 2000 lb per sq in. Some piston-type pumps are capable of producing 10,000 lb per sq in. but the majority are designed for pressures between 2500 and 3000 lb per sq in. The trend is definitely toward higher pressures.

Frequency of operation is another factor to be considered. An application where the pump is only in operation a few minutes out of every hour, or one in which the pump is only running very infrequently at full peak load, will not require the same type of pump as one that is to be constantly under full load.

Operating conditions such as temperatures, oil condition, shock loads, etc. are also important factors to be considered in the selection of a pump.

Hydraulic pumps are built in a very wide range of sizes extending from a delivery capacity of less than one gpm to that of many hundreds of gpm. Figure 4 illustrates a large pump capable of delivering 1520 gpm at 300 rpm. However, the majority of applications in the industrial field require pumps with a delivery of less than fifty gpm.

Gear Type Pumps

The gear-type pump may be either of the external gear design or of the internal gear design. The principle upon which this type of pump operates is that of a pair of gears being driven by an external means (electric motor, gasoline engine, etc.), the oil moving into

Courtesy of The Aldrich Pump Co.

Fig. 4. Nine-plunger reciprocating hydraulic pump rated at 900 brake hp having a pumping capacity of 1520 gpm at a speed of 300 rpm.

Fig. 5. Hydraulic power unit of the "high-low" type with a gear pump to the left of the electric motor and a vane pump to the right.

the pump as a partial vacuum is formed and then being discharged by the meshing of the gear teeth on the discharge side of the pump. Oil is actually squeezed or forced out as the gears mesh.

Of all the different hydraulic pumps manufactured, the gear type is produced in the greatest quantities. Various gear designs, such as helical, spur, spiral, or herringbone may be employed for the rotating element. Depending upon the type of service expected, the gears may be made of bronze, cast iron, or heat-treated steel. Figure 5 shows a hydraulic power unit of the "high-low power" type using a gear pump at the left for delivering a large volume at low pressure and a vane pump at the right for delivering a small volume at high pressure.

There are several variations in the gear-type design, one being based upon the "Gerotor" principle. This consists of a pair of gear-shaped elements, one shaped as an external gear that moves within the other which is shaped as an internal gear as shown in Fig. 6 (lower). The tooth form of the inner "Gerotor" is generated from the tooth form of the outer "Gerotor" so that, as shown in Fig. 6 (upper), each tooth of the inner "Gerotor" is always in sliding contact with the outer "Gerotor" and provides a fluid-tight engagement.

As the teeth disengage, the space between them increases in size, thus creating a partial vacuum into which oil flows from the suction port. When the space or chamber reaches its maximum volume, it is then exposed to the discharge port and as the space diminishes in size with the meshing of the teeth, it forces the oil from the pump. Each "Gerotor" tooth space gradually opens during 0 to 180 degrees of shaft rotation and gradually closes during 180 to 360 degrees of shaft rotation, which allows for good filling and provides a smooth discharge.

The outer rotor of the "Gerotor pump" is designed with one tooth more than the inner rotor. As a result, low tooth sliding velocity is achieved. As an example, when the outer rotor has nine teeth and the inner rotor has eight teeth and the shaft which drives the inner rotor runs at 1200 rpm, then the outer rotor will rotate at eight-ninths this speed or $1066\frac{2}{3}$ rpm. The difference in speed between the two rotors will be $133\frac{1}{3}$ rpm. Since the pump has close fitting tooth engagements and close running clearances, low speeds of below 100 rpm at pressures of 1000 lb per sq in. are possible.

Advantages of gear-type pumps are: (1) they are inexpensive; (2) have few moving parts, and (3) feature simplicity in design and construction.

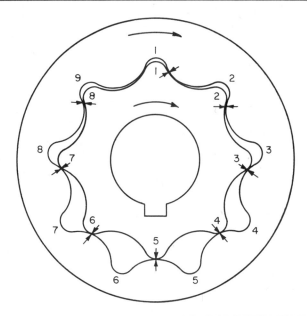

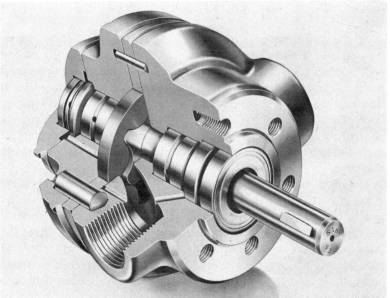

Courtesy of Double A Products Co.

Fig. 6. (*Upper*) Cross-sectional view of the "Gerotors" that move the fluid and create the pressure in the hydraulic pump shown below. (*Lower*) Cutaway view of a gear-type hydraulic pump where the gearing consists of one internal and one external gear, both called "Gerotors."

Vane-type Pumps

This is a rotary type and also operates on the principle of increasing and diminishing volume. It consists of a shaft, rotor, vanes, cam ring, pump housing, bearings, and seals.

Vanes are held in a series of slots around the rotor which either rotates in an oval-shaped "cam ring" or is itself located eccentrically. In either case, as the rotor turns, these blades move in and out of their respective slots as they ride against the enclosing ring. This ring is hardened and ground to close tolerances to provide a suitable

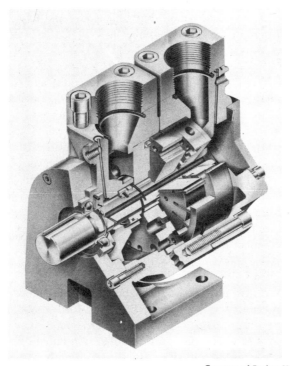

Courtesy of Parker-Hannifin Corp.

Fig. 7. Cutaway view of dual-vane pump.

seal. In the design shown in Fig. 7, each slot has two vanes and this dual vane construction provides multiple sealing for reducing slippage to a minimum. As the pump rotates, oil enters where the volume between the blades is increasing and is forced out where the volume between the blades is decreasing. This cycle occurs twice in each revolution, the two inlet sections being in opposed positions and the two outlet sections being similarly located with respect to each other.

The opposing inlet sections and opposing outlet sections are cross-connected through the cam ring. Inlet sections are interconnected to provide equal flow of oil and pressure to the pumping chambers. The radial load on the rotor is always balanced due to the fact that the pressure forces exerted by the pressurized fluid in one direction are equal to the forces acting in the opposite direction. This hydraulic balance eliminates "cocking" of the rotor and helps to prevent wear in general.

There are several advantages to the vane-type design:

1. Simplicity in construction, high efficiency, and low cost makes this type of pump exceedingly popular in the 1000 to 2000 lb per sq in. range.

2. Several modifications of this design are available for different applications. It is manufactured as a single unit; as a double unit which provides a high gallonage rating yet keeps the outside diameter of the pump at a minimum; as a two-stage pump for producing higher pressure; and as a combination unit where one vane-type pump of large-volume delivery operates at low pressure until resistance is met, then a second vane-type pump of small-volume delivery set at high pressure cuts in for producing the required pressure to finish the operation. This latter type is used on many press applications where it is necessary for a low-pressure fast approach, high pressure hold, and a low-pressure fast return of the ram.

3. It is compact in design as can be seen from Fig. 8.

Rotary Piston Pumps

In rotary piston pumps the mechanism which actuates the piston has a rotary motion rather than a back-and-forth motion as in the reciprocating piston pumps. Rotary piston pumps may be of the *radial* type or the *axial* type. In the former, a number of pistons and cylinders are arranged radially around the rotor hub, while in the axial type they are located in a parallel position with respect to the rotor shaft.

In the radial type, each piston rides in its cylinder with the base of the piston pressing out against an eccentric ring or the eccentric surface of the casing. As the rotor revolves, the eccentricity of the ring or casing causes an in-and-out or pumping motion of the pistons.

In the axial type, the base of each piston is connected by a piston rod to the driving plate. This driving plate is free to tilt along any

Fig. 8. Vane-type pump mounted on a hydraulic power unit.   Note compactness of design.

diameter through its center.   Across the surface of this driving plate rotates a so-called wobble- or cam-plate which is tilted at an angle to the shaft.   The rotation of the wobble-plate produces an in-and-out motion of the pistons in their cylinders.

Constant-displacement rotary piston pumps contain no means of changing the volume of oil discharged at any given speed.

Variable-displacement rotary piston pumps, on the other hand, contain means of varying the length of piston stroke and hence the volume of discharge for a given speed by changing the eccentricity or angularity of the devices which actuate the piston plungers or connecting rods.   Zero discharge is obtained when the centerline of the cylinder and the centerline of the rotor coincide in the case of the radial type and when the plane of the wobble- or cam-plate is at right angles to the rotor shaft in the case of the axial type.

Figures 9 and 10 show cross sections of pumps of the radial piston type.   Figure 9 illustrates the variable volume style, while Fig. 10 illustrates the constant displacement style.   Note the likeness in design between the two styles.   One difference is that the variable volume style has the slide block arrangement which is mounted between four horizontal ways in the case and is used to vary the stroke of the pistons which in turn changes the volume.   The volume control shown is of the screw type but many other types are available such as

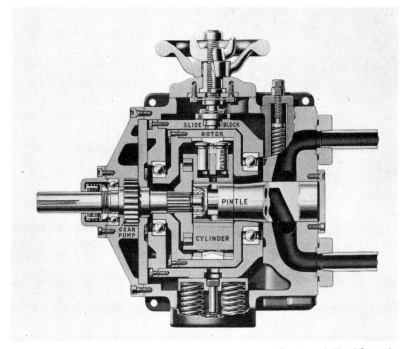

Courtesy of The Oilgear Co.

Fig. 9. Variable-volume style of radial piston pump.

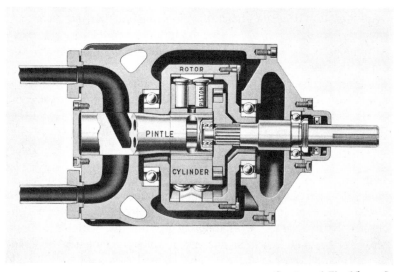

Courtesy of The Oilgear Co.

Fig. 10. Constant-displacement style of radial piston pump.

remote control, servo-motor control, micro-screw control, servo-motor pressure unloading control, two-speed control, and several others —making it easy for the user to select the control that best fits his application.   Another difference is that the variable volume pump is equipped with a gear pump for partially supercharging the main system and for operating hydraulic controls.

The pistons in these two pumps rotate as well as reciprocate.   Note the conical surface of the reaction ring and how the convex surface of the heads of the pistons fit against it.   This allows for a compact design and reduces the number of moving parts in the assembly.

Figure 11 illustrates a variable-displacement, rotary-piston pump of the axial type.   The maximum desired volume to be pumped can be controlled by the volume limit adjustment.   The pressure compensator control is shown in the upper-left-hand corner of the photo. When the control has been set at the desired pressure, the pump will automatically deliver maximum flow until the preselected pressure has been reached.

In general the advantages of the rotary piston pump, due to its design possibilities, are as follows:

1. It is capable of delivering high operating pressures.   There are constant displacement piston-type pumps available for operating pressures up to 10,000 lb per sq in.

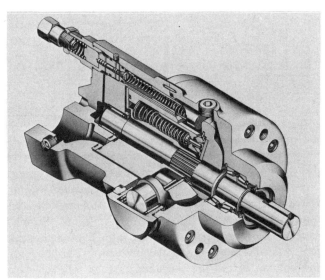

Courtesy of Delavan Manufacturing Co.

Fig. 11. Cutaway view of an axial-type variable-displacement rotary piston pump.

2. It will handle oils in a wide viscosity range. The range usually recommended is from 70 to 300 SSU at 100°F.

3. It will provide a variable delivery of oil. By the use of stroke change mechanisms, the volume may be varied from 0 to maximum, which allows for great flexibility in application.

4. It is quiet in operation.

The Oil Reservoir

The oil reservoir must be designed to have sufficient capacity for the system which it serves. The sizes of the reservoirs on standard hydraulic power units are based upon pump delivery. The data in Table 1 show the reservoir sizes as listed by one manufacturer when using constant-volume pumps capable of producing 1000 lbs per sq in. pressure. Note that several pump sizes are grouped with each reservoir size.

TABLE I

RESERVOIR SIZES FOR CONSTANT VOLUME PUMPS

Reservoir Size (Gallons)	Pump Size (Gallons per min.)	Reservoir Size (Gallons)	Pump Size (Gallons per min.)
8	0.32 0.66 1.42	70	17.1 20.8 27.7 33.5
20	1.9 2.5 4.7 7.5	100	36 44 55
40	10.3 13.2		

A good rule to follow is that the capacity of the reservoir should never be less than twice the delivery capacity of the pump in gpm. This, however, has some exceptions. For example, in a simple press circuit using a 16-inch bore by 60-inch stroke cylinder with a 2-to-1 piston rod differential, the speed of the outstroke of the ram may only require a pump capable of delivering 7.5 gpm. Normally this size of pump would require a reservoir with a 20-gallon capacity. Now, let us look more closely at the problem. The data for computing the volume of the press cylinder are as follows:

Area of 16-inch cylinder = 201.1 sq in.; Stroke = 60 in.; Volume of cylinder—rod extended 201.1 × 60 = 12,066 cu in.; Volume of cylinder—rod retracted 12,066 cu in. ÷ 2 = 6,033 cu in.; Differential volume of cylinder = 12,066 — 6,033 = 6,033 cu in. or 26 gallons.

It is readily seen that the differential volume of the cylinder is greater than the capacity of the oil reservoir. This does not take into consideration the additional oil required by the piping. If the reservoir was filled at the beginning of the forward stroke of the piston, the adding of oil to the reservoir as the piston progressed on its forward stroke would permit filling the blind end of the cylinder. But on the return stroke with the oil directed to the rod end of the cylinder, the reservoir would soon overflow due to the fact that only half as much oil is required on the return stroke of the cylinder. This is a situation that can easily arise if a thorough study is not given to the circuit layout and its components. Of course the remedy is the use of a larger reservoir.

A similar circumstance often arises when a power unit is placed in the basement of a plant and the motivating equipment is placed on the second or third floor. Extremely long pipe lines consume a large volume of oil and the standard size reservoir may be much too small for the requirements. A check valve placed near the pump will eliminate flow back through the pump but still a large reservoir is usually required. Case histories reveal that piping layouts of the above type are often made without considering the difficulties which may be encountered.

Another factor which must be taken into consideration when figuring capacity is heat. The reservoir must be of sufficient capacity so that the hot oil returning from the system or the oil spilling over the relief valve will not cause the oil in the reservoir to exceed 130° to 140°F depending upon the type oil used. Lower temperatures are even more desirable. At high oil temperatures packings may deteriorate and start to leak and the pumps will have more slippage as the viscosity of the oil decreases. A larger capacity reservoir is not always the answer to the heat problem. Often it is necessary to use coolers. (See Chapter 8 for details on coolers.)

There are instances where the oil becomes so hot in the unit that it will actually boil over the top. Of course, this is an exception rather than the rule. However, immediate steps must be taken to overcome such a situation, usually by changing the circuit.

The reservoir must be designed so that it can easily be cleaned. Since the oil should be changed at regular intervals, clean-out holes of ample size should be provided in the sides of the reservoir. Also a drain plug should be provided in the lowest section of the reservoir. The oil is drained from the drain plug connection, then the sludge and other foreign matter which accumulates is removed through the clean-out holes. These clean-outs are usually a rectangular hole and are covered with bolted-on gasketed steel plates. The plates are easily removed. Both drain plugs and clean-out cover are shown in Fig. 12.

Extreme care should be used when cleaning a power unit reservoir for any foreign matter left in the tank will contaminate the new supply of oil. Often maintenance men will use rags or waste to wipe out the interior thus leaving a deposit of lint which will soon

Courtesy of The Oilgear Co.

Fig. 12. The oil reservoir located in the base of this unit has numerous drain plugs and clean-out covers for draining the oil and removing the sludge.

clog up the filter. The bottom of the reservoir should slope toward the drain plug so that as much of the oil as possible can be drained out before the clean-out plates are removed.

A third item to consider is the use of baffles in the reservoir. Baffles are used to temporarily separate the incoming oil from the outgoing oil. Often when the oil returns from the system it comes into the reservoir with such force that it causes a foaming action. If the oil in this condition were allowed to move directly to the suction strainer a considerable amount of air would be forced into the system causing cavitation in the pump and jerky movement of the cylinder pistons. Baffles are of various designs. Some designers separate the reservoir into two sections by only using one baffle plate, others divide the reservoir into three sections by using two baffle plates. Some baffle plates are cut out in the two bottom corners so as to allow any foreign matter to drain to the lowest section in the unit, other baffle plates have a number of cutouts in order to break up the oil surge. Almost any type of barrier will be successful as long as it eliminates the possibilities of the exhaust oil immediately contacting the suction strainer. The baffle also helps deflect any foreign matter that is returned from the system.

The return line on the relief valve should also be placed as far away from the intake strainer as possible. This would place the end of it close to the return connection on the reservoir.

Oil level gages are a "must" on the oil reservoir. There are several types available that are being used successfully. Some users prefer the glass disc type while others have preference for the tubular type. The paramount point to be kept in mind is that these gages are functional rather than ornamental. Many power unit complaints can be attributed to the fact that the user pays no attention to the amount of oil that he places in his unit, not realizing that the gage level is of important concern.

The oil reservoir top or cover must be structurally sound in order to carry the weight of the pump and electric motor and also resist any torsional forces when the electric motor is driving the pump. Mounting pads for the motor and the pump are usually machined to close tolerances for assuring better alignment of the shafts.

The reservoir cover usually contains the filler arrangement which may have a built-in air filter so that clean air at atmospheric pressure will contact the oil. The main requirement of a good intake filter is to collect all foreign particles attempting to pass through it over a reasonable period of time and still allow the air to pass through freely. Thus it is important that this filter be of sufficient size so that it will not become easily clogged. A number of designs of air filters are available including those made with fine mesh wire, those using a braided wire cloth and a type using a synthetic filtering material. When units are placed in highly contaminated atmospheres, it may be necessary to use special filters.

Oil reservoirs should have a protective coating inside and out to prevent rusting. When the user reconditions such a unit after it has been in service, caution should be used in selecting the type of coating to be applied, particularly if synthetic oil is to be used in the reservoir. Some synthetic oils will remove certain paints.

Furthermore, the user should also be certain that the paint which he uses will not be affected by hot oil as this may cause the paint to flake off and clog up the oil filter. A reputable paint dealer can recommend the type of paint to use for this service.

The Oil Filter

The statement, "Seventy-five per cent of all hydraulic maintenance trouble is caused by dirty oil" is not an exaggeration. Dirt can cause untold damages to the entire hydraulic system. The use of filters is an important step in keeping dirt out of the system and in reducing hydraulic maintenance. Filters are listed under two types: the sump type and the line type. The sump type or immersion type, as it is often called, is placed in the oil reservoir and is connected to the intake line of the pump. The line-type filter is mounted outside of the tank either in the intake line or in the return line from the system.

Sump-type filters range in design from a fine mesh wire box to a synthetic element type which will remove particles as small as one micron (0.00004 inch). The ideal filter would be one which would remove all foreign matter in the oil and yet permit the full volume of oil demanded by the pump to flow without any restriction over a reasonable service period. This would certainly remove any possibilities of cavitation in the pump which occurs when the filter becomes so clogged that oil cannot be forced into the pump by atmospheric pressure, thus causing the pump to be "starved."

The sump-type filter should be located near, yet well off, the bottom of the reservoir. In no case should it ever touch the bottom. The reason for this is that any sludge or other foreign matter collects at the bottom of the reservoir and if the filter is too near the bottom, the sludge or foreign matter would soon be forced into the filter. On the other hand, the filter should not be placed too high in the reservoir so that it is not amply covered with oil. There should always be several inches of oil above it.

When using a large volume pump it is often advisable to use several filters placed on a manifold connected to the intake line. By doing this, more filter area is made available since, when several smaller diameter filters are used, they can be arranged so that there is plenty of oil over the top of them. If only one large filter is used, the top of the filter may come so close to the surface of the oil that air may be drawn into the filter. Several filters can also be placed advantageously in regard to their position in the reservoir.

A question which often arises is "How can one determine when the sump-type filter becomes dirty or clogged?" Visual inspection is difficult unless the strainer is removed from the reservoir. However, if the pump becomes noisy, then one should immediately check the filter, as this is an indication that it is dirty. Oil filters should be thoroughly cleaned and inspected each time the oil is changed providing that the oil is changed at regular intervals. Some plants let the filters go without being cleaned or the oil being changed until trouble really develops but these plants are by far in the minority. If a pump becomes noisy or is not producing its rated capacity, one should immediately check the filter. Since the submerged-type fil-

Courtesy of Marvel Engineering Co.

Fig. 13. An easily-cleanable filter composed of a heavy screen cloth (not shown) around which is placed a fluted fine Monel wire cloth insert which in turn is protected by an expanded metal housing.

ters are not visible and often not too easily accessible they are sometimes forgotten.

Figure 13 shows a filter designed with a fine Monel* wire cloth insert. Note how the wire cloth is formed to give maximum filtering area in a given space. This wire cloth is supported on the inside by a substantial heavy screen cloth and is protected on the outside by an expanded metal housing. Its design facilitates easy installation and cleaning.

Figure 14 (lower) shows various sizes and shapes of impregnated fiber filter elements. Figure 14 (upper) shows a complete line-type filter with the impregnated fiber filter element in place. Note the thickness of the filter cartridge. The fibers are firmly bonded by impregnation with a resinous material and rigid structural strength is

* Registered trade-mark.

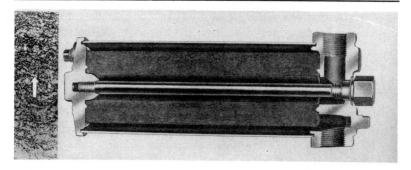

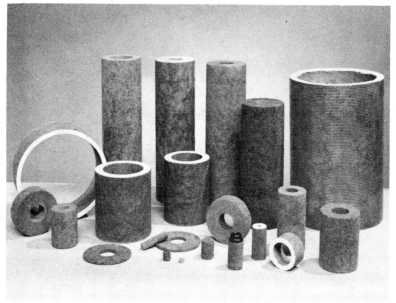

Courtesy of Cuno Engineering Corp.

Fig. 14. (*Upper*) Line-type filter containing an impregnated fiber filter element. (Inset on left shows greatly magnified cross section of the filter material. Note the increasing compactness of the filter material in the direction of oil flow.) (*Lower*) Various sizes and shapes of impregnated fiber filter elements.

obtained by polymerization of the resin through the application of heat. This type filter receives the oil from the outside and allows it to pass to the inside. As can be seen in the cross-section insert of Fig. 14 (upper), the density of the fibers increases toward the center and the screening area becomes finer and finer. This fiber type of filter has a high percentage of porosity, however. By controlling the density of the filter cartridge, various degrees of filtration may be obtained.

Courtesy of Marvel Engineering Co.

Fig. 15. Line-type filter (shown circled) is installed at the suction end of the pump on this 10,000-pound automatic pipe tester.

Figure 15 shows a line-type filter used on a hydraulic power unit. Note that it is in the intake line between the reservoir and the pump.

Figures 16 and 17 show two types of oil purifiers that are used in large hydraulic systems for reconditioning hydraulic oil. Figure 16 illustrates the cartridge-refill type while Fig. 17 illustrates the bulk-refill type. These filters or purifiers make use of fuller's earth which is a clay-like material having a high absorption power. It is highly porous and the total particle surface areas range up to 590,000 to 690,000 sq ft per lb. Although fuller's earth will remove some of the additives that are placed in hydraulic oils, it will maintain these oils free from all contaminants, both soluble and insoluble. All oil deterioration products will be removed from the oil as quickly as they are formed and no asphaltenes, gums, resins, lacquers, etc., will settle in the hydraulic system to cause trouble. Filters using fuller's earth are suitable for purifying both mineral and synthetic oil.

Connecting Motor and Pump

There are several methods of connecting the hydraulic pump to the electric motor. One method is the direct adaptation of the pump to the motor by using a special end bell on the motor for flange mounting. The pilots on the pump and motor must be held to very close tolerances in order to keep the two shafts in line. The ends of the shafts are designed for direct connection.

Another method is the use of a pulley drive. When using a pulley drive it is usually necessary to use an outboard support in order to relieve the strain or side thrust on the pump bearing.

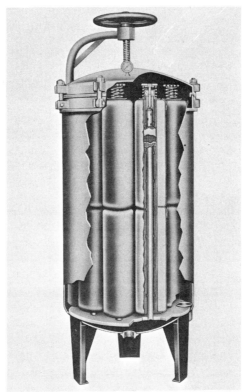

Courtesy of Commercial Filter Corp.

Fig. 16. Multi-cartridge type oil purifier.

The third method—which is by far the most widely used—is to make the connection by means of a flexible coupling. The purpose of such a coupling is to take care of parallel or angular displacement of shaft centerlines, endwise displacement of the shafts, vibration laterally or endwise and the absorption of shock loads. Thus, where a flexible coupling is used, a certain degree of misalignment which may develop during service can be tolerated and the impact blow of a sudden load change, starting or stopping, is cushioned and dissipated partly as heat, and partly as elastic deformation for transmission after a short time lag.

The large number of designs of flexible couplings may be roughly grouped into two broad classes: Those that are non-resilient and

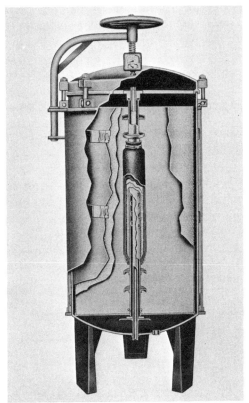

Courtesy of Commercial Filter Corp.

Fig. 17. Bulk-refill type oil purifier.

are designed primarily to take care of shaft misalignment and those that are resilient and provide considerable cushioning for torsional stress as well as providing automatic adjustment for shaft displacement.

One example of the former type is the roller chain type of flexible coupling shown in Figs. 18 (left) and 18 (right). This coupling consists of two hardened steel sprockets and a length of double strand roller chain. Clearances between the chain and the sprocket teeth provide freedom of movement for accommodating misalignment of shafts and end float of the electric motor shafts. A limited degree of torsional flexibility is also furnished. Holes in the sprockets can be bored to accommodate the required shaft sizes. It is not necessary that both the motor shaft and the pump shaft be of the same diameter. In installing or disconnecting the roller chain type coupling it is only necessary to remove a single pin in the chain itself. Removal of the roller chain completely disengages the two sprockets. Figure 1 shows this type of coupling installed on a hydraulic power unit.

A coupling casing such as that shown in Fig. 18 (left) is recommended for the chain-type coupling. It not only protects the teeth of the sprocket and the chain from abrasives which may be present

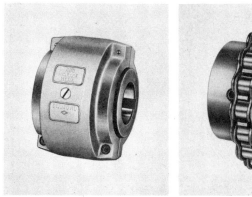 

Courtesy of Diamond Chain Co., Inc.

Fig. 18. Roller-chain type of flexible coupling. (*Left*) With casing. (*Right*) Without casing.

in the air but also provides a housing so that lubricants can be used. These lubricants materially reduce the friction and wear as the teeth move across the chain rollers when under a heavy load. In highly abrasive areas, without the use of a casing, records show that the teeth on the sprockets have been worn away after a few weeks' service.

A resilient type of coupling is shown in Fig. 19. This is a rubber-bronze bushed flexible coupling. The flexible rubber bushings with their resiliency provide a very quiet drive which is cushioned in both directions. The rubber bushings also provide the necessary amount of freedom of movement between each flange. The use of oil-less bronze bearings for the connecting studs makes lubrication unnecessary.

Free-end float is another feature of this design so that electric motor shaft end-play is not a detrimental factor.   End thrust and other stresses are absorbed by the coupling.   With this type of coupling it is very easy to determine with a feeler gage and straight edge whether or not the pump and motor shaft are in alignment. Figure 20 shows one of these rubber-bronze bushed flexible couplings mounted on the hydraulic power unit of a broaching machine.

Whatever type of flexible coupling is used, the pump and motor shaft should be placed in as close alignment as possible.   Excessive

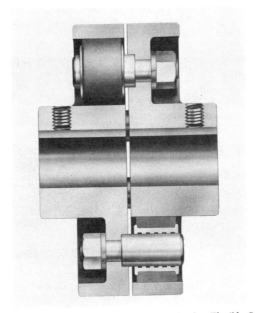

Courtesy of Ajax Flexible Coupling Co., Inc.

Fig. 19. Cutaway view of a rubber-bronze bushed flexible coupling.

misalignment causes undue heat, power loss, vibration, and bearing wear, all of which can be kept at a minimum if caution is used when mounting the pump and electric motor.

Pressure Gages

Power unit manufacturers use one of two types of pressure gages on their equipment.   One has a round dial face similar to the gage shown in Fig. 21 (left).   The dial is graduated into lb per sq in.

markings and figures are placed at certain intervals. The extent of the gage scale depends upon the pump pressures. The gage may be built for pressures of from, say, 0 to 100 lb per sq in. or it may be built for pressures from 0 to 20,000 lb per sq in., all depending upon the requirements. A bourdon tube and a mechanism for converting its movement with pressure changes to rotary movement of the pointer, as shown in Fig. 21 (right), is used in most round-faced pressure gages. The second type is the plunger type as shown in Fig. 22. This type of gage is also built in various pressure ranges. One big advantage of this type of gage is that it can be rebuilt very easily, since by merely replacing the plunger assembly the gage is made as good as new.

Between the pressure gage and the power source a shut-off valve is recommended. After the operating pressure is set on the unit

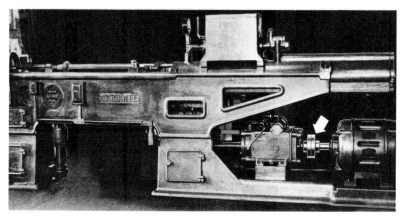

Courtesy of Ajax Flexible Coupling Co., Inc.

Fig. 20. Rubber-bronze bushed flexible coupling (arrow) connects electric motor to pump on this hydraulic broaching machine.

the gage should be shut off. This not only protects the gage from any shock resulting from sudden pressure changes but will greatly increase the life of the gage as well. On gages using a bourdon tube, shock alleviators or snubbers are recommended.

Electric Motors

The type of electric motor required for a power unit largely depends upon the requirements of the application. In hazardous locations an explosion-proof motor is normally required. Other applications may call for open drip-proof, totally enclosed, or totally

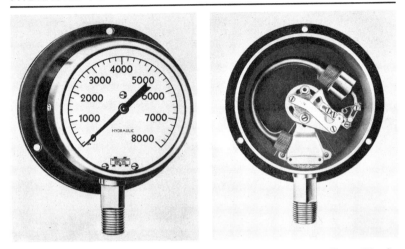

Courtesy of Marshalltown Mfg. Co.

Fig. 21. (*Left*) Round-faced hydraulic pressure gage. (*Right*) Gage with dial face and pointer removed reveals bourdon tube and a mechanism for converting the tube's movement with pressure changes to the rotary movement of a pointer.

enclosed fan-cooled types. Totally enclosed fan-cooled motors are recommended for sizes ranging from 5 horsepower to 50 horsepower.

The electric motor must be securely mounted to the power unit cover and its shaft must be in alignment with the shaft of the pump. For long efficient service the motor must be given proper maintenance. Always select a motor that is large enough to do the job. The following table is a guide to the selection of the proper size of motor for various rates of pump delivery at different pressures. Thus if a pump were required to deliver 50 gallons per minute against a pressure of 500 lbs per sq in., then the approximate motor size required would be 20 horsepower.

TABLE 2

MOTOR SIZES FOR GIVEN PUMP DELIVERY

Pump Delivery Gallons per Min. (approx.)	Maximum Fluid Pressure		
	250 psi	500 psi	1000 psi
	Motor Size, Horsepower		
1½	½	¾	1½
3	¾	1½	3
5	1	2	3
10	2	3	7½
21	5	7½	15
55	10	20	40

Courtesy of Scovill Fluid Power Division

Fig. 22. Plunger-type hydraulic pressure gage.

Internal Piping

Last but not least is the piping of the unit itself. As in all hydraulic piping (see Chapter 5) the lines should be large enough to carry the oil with the least amount of friction. The line leading into the pump is usually larger than the one leaving the pump so as to allow more than ample entry of fluid. Many times the outlet is made one pipe size smaller than the inlet. The piping should be as straight as possible with the smallest possible number of bends. The piping joints should, by all means, be free of any leaks. Seamless steel tubing or steel pipe is normally used in piping up the power unit.

Disadvantages of Multiple Duty Power Unit

While a single unit may be used to supply fluid power for a group of machines, the trend is toward a power unit for each machine or each primary circuit. The disadvantages of using one unit for two or more machines are:

1. *Loss of speed and pressure.* Perhaps the best way to illustrate this is to consider two hydraulic molding presses. Usually when figuring a power unit for two such presses only a single pump large

enough to take care of one press is provided. This is based on the supposition that the two rams would not often be called upon to move at the same time. However, if they do move at the same time, their speed is reduced to one-half that of a single ram.

Unless additional valving were placed in the basic circuit, whenever the master valve is shifted on one press with both rams operating, the pressure on the ram of the other press will drop until the ram of the first press reaches the end of the stroke. On the other hand, with extra valving added, such as counterbalance valves, to hold up the pressure, one may run into the difficulty of losing fluid volume through operation of the relief valve, thus cutting down the ram speed.

2. *Additional piping.* When using one power unit for two machines it is usually wise to place the power unit in a central location and then pipe to each machine. Even when this is done, long runs of expensive piping are often required and it is sometimes necessary to tear up the floor in order to properly place the exhaust lines.

When using only one machine per power unit, the power unit should be placed as close as possible to the machine and in a position where the exhaust lines can drain freely into the power unit reservoir. This eliminates back pressure in the exhaust lines, which often causes hydraulic control valves to malfunction. Exhaust lines from the circuit to the reservoir should be large enough to carry the oil at practically zero pressure.

3. *Limitation on pump selection.* With one power unit for two circuits, such as for two molding presses where long holding periods are required at high pressure, the most suitable pump cannot be used, i.e., the high-low pump with a changeover valve. This is because if one press is opened or closed while the other press is holding, the holding pressure drops as soon as a master valve is shifted and the low-pressure pump cuts in. (See advantages of high-low pump as listed under the pump section of this chapter.)

4. *Loss of production.* If for any reason the power unit should fail when it is used for two or more machines, all of the machines will be inoperative until the unit can be put back in service. When using a unit for each machine, only one machine will be out of service. With heavy production schedules, a plant can rarely afford to have a group of machines out of service.

Power Unit Difficulties

Here are some of the ways a power unit may fail to operate properly:

1. *Pump shaft rotating in the wrong direction.* Make certain that the rotation of the shaft of the electric motor is in the direction indicated by the arrow on the pump.

2. *Relief valve improperly set.* The relief valve may be set so that all of the oil is spilling through the relief valve and no oil is going into the circuit at any pressure. A broken relief valve spring will cause the same condition.

3. *Worn pump.* If the parts of the pump become severely worn there will be so much slippage that the pump will not build up pressure. This is usually easily detected.

4. *Clogged filters.* This causes reduction in volume and considerable pump noise.

5. *Insufficient oil in reservoir.* A low oil supply can certainly be a source of trouble. This also is true of dirty oil. It is often difficult to make the user realize that oil must be changed regularly and that only the right amount should be used. There are case histories where the user disregarded this basic fact with the result that the unit was completely ruined. (See also Chapter 2 on hydraulic fluids.)

6. *Loose intake or suction line.* A small air leak in the suction line will cause the pump to become extremely noisy due to cavitation, i.e., air-filled spaces in the oil. An easy way to check this is to pour a small amount of oil over each joint between the pump and the strainer. If the pump noise stops, it is certain that an air leak is present.

Intensifier or Pressure Booster

An intensifier or pressure booster may be used in conjunction with the power unit. Its operating principle is very simple. The unit has a piston which automatically reciprocates. The large piston is exposed to the force delivered by the low-pressure pump and this force moves the large piston back and forth as the valve stem shifts. The small area pistons on each end of the large piston forces oil into the system under high pressure. The increase in pressure is directly proportional to the difference in areas of the two sizes of pistons. Low-pressure oil can be delivered to the system by the low-pressure pump until resistance is met, then by the use of a sequence valve the low-pressure pump delivers to the booster and the booster delivers high-pressure oil to the cylinder for a high-pressure squeeze on the workpiece.

Figure 23 shows a pressure booster designed to deliver both low- and high-pressure oil from a single low-pressure pump. It converts

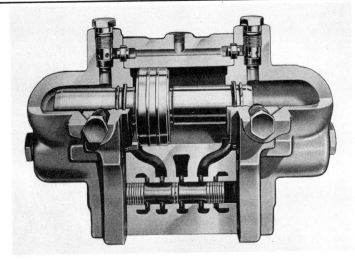

Courtesy of Rexnord, Inc.

Fig. 23. Pressure booster designed to deliver both low- and high-pressure oil from a single low-pressure pump.

a large-volume, low-pressure oil supply to a proportionately small-volume, high-pressure output. It may be used with either a constant-volume or a variable-volume pump. A device of this type has a number of advantages:

1. It automatically cuts in the circuit when high pressure is called for and cuts out when the demand is satisfied.

2. The production of heat is held to a minimum, since the high pressure is only produced by a small volume of oil.

3. It eliminates the need of high-pressure pumps which are expensive.

4. Its horsepower requirements remain constant. As the pressure increases the volume decreases.

5. It is much more compact than adding another power unit for high-pressure cycling.

For air-oil boosters see Chapter 16.

CHAPTER 4

Hydraulic Accumulators

An accumulator is essentially a pressure storage reservoir in which a noncompressible hydraulic fluid is retained under pressure from an external source. The fluid, under pressure, is readily available as a quick secondary source of fluid power and even though it may have been pumped into the accumulator in pulsations or an uneven flow, it will be discharged from the accumulator in a smooth even flow.

Among the types of applications for which accumulators are suitable are: (1) hydraulic shock suppression; (2) fluid make-up in a closed hydraulic system; (3) leakage compensation; (4) source of emergency power in case of power failure; and (5) holding high pressures for long periods of time without keeping the pump unit in operation. Accumulators are used in conjunction with hydraulic systems on large hydraulic presses, farm machinery, diesel engine starters, hospital beds, power brakes and landing gear mechanisms on airplanes, hatch covers on ships, lift trucks, and other devices and machinery too numerous to mention.

Accumulators are usually divided into three classes: the dead-weight, the spring-operated, and the air-operated types. Each type is entirely different in design, but their over-all performance and end results are quite similar.

The Dead-Weight Accumulator

The basic elements of a dead-weight accumulator are shown in Fig. 1. It consists, essentially, of a piston loaded with a dead weight and moving within a cylinder to exert pressure on the hydraulic oil. The mass of dead weight may be concrete, pig or scrap iron, steel, or any other heavy material. The piston of this type of accumulator must be a precision fit in the accumulator tube so that the leakage past the piston will be at a minimum. The tube usually has a honed or ground finish to insure long life to the piston packing and to reduce friction.

4-1

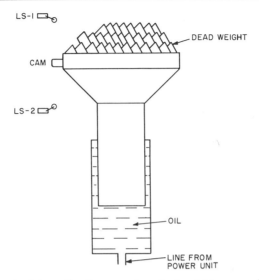

Fig. 1. Schematic diagram of a dead-weight accumulator.

One method of starting and stopping the pumping unit is shown in Fig. 1. When the accumulator is emptied, the cam on the trip arm contacts limit switch LS-2, turning on the pumping unit. As the pump fills the accumulator, the cam trips limit switch LS-1, stopping the pumping unit.

The big advantage of this type of accumulator is that the pressure remains constant for the full stroke or until all the fluid is spent. This is not true of other types. For example, an accumulator of 12-inch bore and 60-inch stroke with a dead weight of 50 tons will deliver a continuous fluid pressure of 885 lb per sq in. (disregarding friction) and a volume of approximately 29 gallons.

Another advantage of this type is that it can supply large volumes of fluid under high pressure. The volume is only limited by the manufacturer's ability to produce huge precision-made parts; i.e., tubes and rams. The large volume of fluid makes it possible to supply pressure to several hydraulic circuits. In other words, the accumulator can act as the central station and distribute pressure to the circuits on a number of machines.

The disadvantages of the dead-weight type accumulators are: (1) They are extremely bulky, especially the ones that deliver high pressures and volumes. The section holding the dead weight must be amply supported. (2) The larger sizes are expensive. They are usually fabricated on the job due to their size.

The Spring-Loaded Type Accumulator

Several types of spring-loaded accumulators are available. They are usually smaller and less expensive than the dead-weight type. Whether they are of a single-spring type or a multiple-spring type, the spring or springs act against a hydraulic piston forcing the fluid into the hydraulic system. Figure 2 shows a diagrammatic sketch of a spring-type accumulator. This type of accumulator is often built directly into the power unit.

This type of accumulator delivers only a small volume of oil at relatively low pressure. Furthermore, the pressure exerted on the

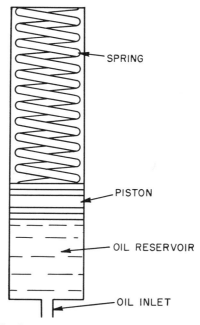

Fig. 2. Diagrammatic sketch of a spring-type accumulator.

oil is not constant as in the dead-weight type. As the springs are compressed, the accumulator pressure reaches its peak and as the springs approach their free length, the accumulator pressure drops to a minimum. The volume depends, again, on the bore and stroke of the accumulator. A pressure switch, with a mechanism for starting the pumping unit when the pressure drops to a certain level and stopping it when the peak pressure is reached, is often used in a circuit employing this type of accumulator.

Air or Gas Operated Accumulators

The pneumatic bladder-type accumulator consists of a bag or bladder of synthetic material which is precharged with air or nitrogen to a determined pressure. This bladder is placed within the accumulator shell and the balance of the space filled with oil. When additional oil is forced into the shell from the pumping unit, the gas in the bag is compressed, thus forming a reservoir of pressure. When oil is called for from the accumulator, the bladder expands and forces

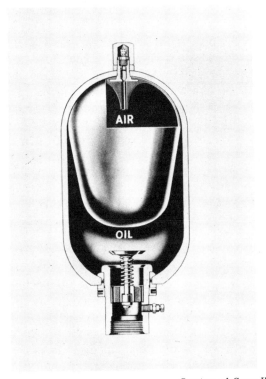

Fig. 3. Bladder-type accumulator.

the oil out into the circuit. Neglecting the effect of any temperature change, since the mass of gas remains the same as the bladder expands —increasing the volume, the pressure which it exerts on the oil decreases. Figure 3 shows a cross section of a bladder-type accumulator. Note the connection on top for charging the bladder with air or nitrogen.

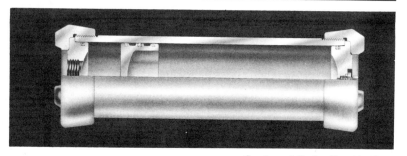

Courtesy of Parker-Hannifin Corp.

Fig. 4. Pneumatic piston-type accumulator.

The pneumatic piston-type accumulator shown in Fig. 4 consists of a precision machined tube, heavy-duty end caps with high-pressure seals, and a piston with a packing that will keep the oil and air separated. A gas precharge valve is used in one end cap. This accumulator may be mounted in any position although the preferred position is with the axis vertical and the gas connection down. The piston moves toward the air end as the hydraulic fluid forces its way into the accumulator.

The advantages of the air-operated accumulators are that they are compact, have only a few functional parts and are not expensive. The one drawback of this type, however, is that, like the bladder type, the pressure does not remain constant while the whole oil charge is being displaced.

Use of Accumulator as Leakage Compensator

Circuit No. 1 (Fig. 5) shows a bag-type accumulator being used as a leakage compensator in a basic hydraulic circuit consisting of a power unit, a check valve, master valve, pressure switch, power cylinder, and an accumulator. With this circuit, the only time the power unit operates is when the pressure drops to an unsafe operating level. This saves electric power and reduces heat in the system which would otherwise be created by constant operation of the pump. This arrangement is especially helpful in circuits, such as are used for curing presses, which require high pressure for long periods.

The circuit functions as follows: Operator places workpieces on large press platen and shifts handle of four-way, two-position master valve (2). Oil flows to blind end of press ram (1) and ram descends contacting work. Pressure builds up and oil fills accumulator. When maximum pressure is reached, contacts on pressure switch open stopping electric motor.

The press ram may set for a long period of time before enough leakage replacement oil is drained from the accumulator to lower the operating pressure to the extent that the contacts on the pressure switch will close, starting the motor on the power unit. The maximum length of time is determined by the volume of the accumulator, and the leakage of oil past the piston of the valve and the press ram. Often the addition of an "O" ring on the ram is of benefit to reduce the ram leakage.

The check valve (3) is placed between the pump and the accumulator so that the pump will not reverse when the electric motor is stopped and permit all of the accumulator charge to drain back into the power unit.

When pressing cycle has been completed, operator shifts handle of master valve (2) to original position and oil flows to rod end of ram and ram is returned to starting position completing the cycle.

Accumulator Used as Secondary Source of Energy

Circuit No. 2 (Fig. 6) shows one of the most common uses of accumulators, which is to make available a quick source of energy. It is actually the secondary source, while the power unit is the primary source. For example, large bore cylinders operating at even a moderate speed require a large volume of oil. It is not economical to

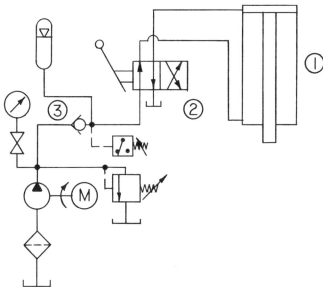

Fig. 5. Circuit employing a bag-type accumulator as a leakage compensator.

supply a pump of such capacity for only intermittent usage due to the size of the pumping unit required and to the heat that may be involved. The accumulator has been the answer in many such applications, where there is a sufficient interval or intervals during the working cycle for the accumulator to store up fluid to have it ready when needed.

The circuit shown makes use of a spring-loaded type accumulator. Operator places workpiece on slide-table (1) and shifts handle of four-way, two-position valve (2) and oil flows from accumulator A to blind end of slide cylinder (1). Piston advances at a speed set on flow control valve (3) until slide-table (1) reaches end of stroke

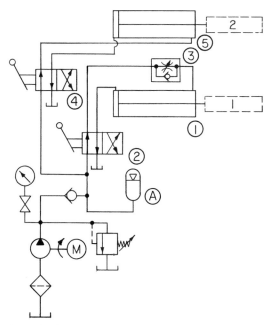

Fig. 6. Circuit employing a spring-loaded type accumulator for making available a quick source of energy.

and curing operation takes place. While operator is loading slide-table (2), accumulator A is being charged. Operator then shifts handle of master valve (4) and oil flows from accumulator A to blind end of slide cylinder (5) and loaded slide-table moves into curing position. Operator then shifts handle of valve (2) and oil from accumulator A and also from pumping unit rapidly flows to rod end of cylinder (1) and slide-table retracts quickly to loading station.   Operator unloads work and reloads and advances slide-table (1) into curing position.   When curing time has elapsed on slide-table (2), operator shifts handle of valve (4) and oil flows from accumulator and pumping unit to rod end of cylinder (5) returning slide-table to loading position at a rapid rate.

An exceptionally small pump operating in conjunction with a large accumulator can produce the required results.

Accumulator Used as Fluid Make-up Device

Circuit No. 3 (Fig. 7) shows an accumulator being used in a closed hydraulic circuit as a fluid make-up device.   This device makes up for the difference in volume between the rod end and the blind end of a hydraulic cylinder.   When a high external force contacts end of piston rod the oil is forced from the back side of the piston (1) through the orifice in the cover to the rod end of the cylinder.   When

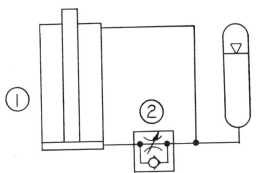

Fig. 7. A closed hydraulic circuit with accumulator acting as a fluid make-up device.

the cushion nose on the end of the piston rod enters the cushion recess in the blind cylinder cover, oil is partially trapped slowing down the movement of the rod.   Since there is more oil in the blind end of the cylinder than can be diverted around to the front end, due to the differences in volume resulting from space taken up by the piston rod, this excess oil flows to the accumulator and pressure is built up.   When the external force is removed from the rod the stored-up energy in the accumulator forces the piston forward, first by unseating the ball in the ball check assembly (2) so the oil can flow through the recess section of the blind cover, then, when the cushion nose leaves the cushion recess, the oil flows through the large orifice.

In order to obtain satisfactory performance from this device, the piston packing and piston-rod packing must be as nearly leakproof as possible. "O" rings make a good piston packing in this instance.

Synchronizing Ram Movements of Two Cylinders

Circuit No. 4 (Fig. 8) makes use of two accumulators for synchronizing the ram movements of two hydraulic cylinders. If the accumulators are not to be out of proportion, from the standpoint of size, with the rest of the equipment, the cylinder strokes must be short. This circuit functions as follows: Operator places workpiece

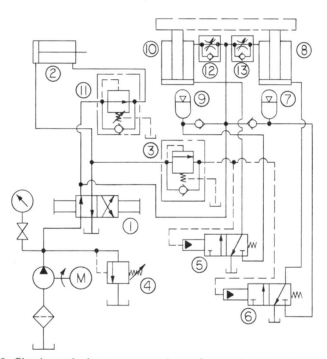

Fig. 8. Circuit employing two accumulators for synchronizing the ram movements of two hydraulic cylinders.

on machine bed and shifts handle of four-way, two-position valve (1). Oil flows to blind end of clamp cylinder (2) and workpiece is clamped onto machine bed. After workpiece is clamped, pressure builds up and opens sequence valve (3) and oil flows to pilot connections of valves (5) and (6), shifting the valves and allowing oil to flow from accumulator (7) through valve (6) on to blind end of cylinder (8), and oil also flows from accumulator (9) through valve (5) on to blind end of cylinder (10). Pistons of both cylinders (8) and (10) move out uniformly because the accumulators (7 and 9) act as a reservoir in collecting the same amounts of oil in each one when they are charged and then dispel the same amounts at the same time as they discharge. In other words they act as a flow equalizer and are helped by flow controls (12 and 13). When operation is completed, operator shifts handle of valve (1) to original position, and oil flows to rod end of cylinders (8) and (10) returning the rams, then pressure builds up opening sequence valve (11) and oil flows to rod end of clamp cylinders (2) and workpiece is released. The accumulators are charged during the return stroke of the rams of cylinders (8) and (10). All components should have very little internal leakage if satisfactory operation is to be expected.

Accumulators Provide Emergency Source of Power

Circuit No. 5 (Fig. 9) shows accumulators being used as an emergency source of power. Even though there may be an electrical

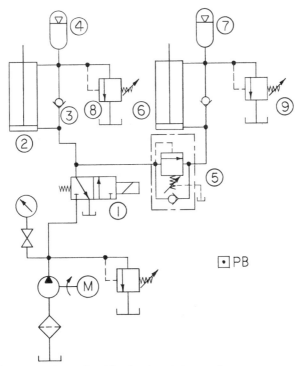

Fig. 9. The accumulators in this circuit act to supply an emergency source of power.

power failure caused by storms, fire, or flood, it is necessary in some hydraulic applications to be able to return the pistons of certain cylinders to their starting position. For example, where a hydraulic system is operating a flood gate and the power fails, if some provisions were not made to operate the gate, considerable damage may result. In this application, the accumulator must have sufficient volume to fill the cylinder at a working pressure that will return the ram. The two accumulator systems function like this: Operator depresses push button energizing solenoid of master three-way control valve (1) and oil flows to blind end of cylinder (2). This cylinder has a 2:1 differential piston rod, i.e., the cross-sectional area of the piston rod is one-half that of the cylinder tube. At the same time the oil pressure flows to blind end of cylinder it also unseats check valve (3) and oil under pressure also flows to rod end of cylinder and into accumulator (4), charging accumulator as the piston of cylinder advances. When piston reaches end of stroke, pressure builds up and sequence valve (5) opens and oil flows to blind end of cylinder (6) and the same procedure is followed in cylinder (6) and accumulator (7) as in cylinder (2) and accumulator (4). Relief valves (8) and (9) prevent any excess pressure build-up which might cause stalling of the cylinders on the outstroke. If a pressure switch is provided in the system (not shown), then after the piston of cylinder (6) has reached its outstroke, the pressure switch will open stopping the electric motor of the power unit. This will keep the power unit from operating during long stand-by periods.

If there should be a power failure, the piston in valve (1) will shift to its original position when current is no longer supplied to the solenoid and the oil in the blind ends of cylinders (2) and (6) then exhausts through the open port. This shift is caused by the spring pressure on the opposite end of the valve. Oil then is forced from the bag-type accumulators to the rod end of cylinders (2) and (6) and their pistons retract to the starting position.

Accumulator Used as Holding Device

Circuit No. 6 (Fig. 10) depicts an accumulator used as a holding device. The accumulator maintains high working pressure on the workpiece during a long stand-by period while other operations are being performed on the workpiece. In this circuit, the operator shifts handle of four-way, three-position master valve (1) directing the oil to blind end of slide cylinder (2). Slide is moved forward at a speed set by flow control valve (4). When piston of slide cylinder reaches end of stroke, operator releases handle of valve (1) and its piston returns to neutral. Since valve (1) has all ports blocked when in neutral, the pressure then flows to valve (3) which is a two-position, four-way valve. Valve (3) delivers two operating pressures since relief valve (6) is set at 500 lb per sq in. while the relief valve (5) on the power unit is set at 1000 lb. per sq in. When the valve handle is at one position it passes oil at 500 lb per sq in. and when in the other, at 1,000 lb per sq in. Oil then passes to three-position, four-way valve (7). Operator shifts handle of valve (7) allowing oil to flow to blind end of press cylinder (8) and press ram descends.

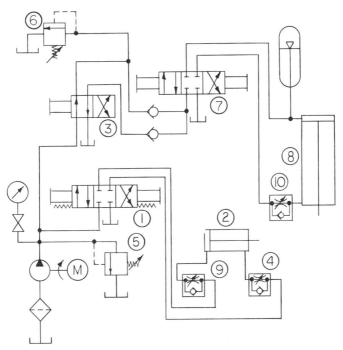

Fig. 10. The accumulator in this circuit acts as a holding device.

Oil pressure charges accumulator and operator returns handle of valve (7) to neutral position blocking cylinder ports and allows pump to dump the oil to exhaust. In the meantime pressure in accumulator maintains a holding pressure on the blind end of the ram.

After completion of operations on workpiece, operator shifts handle of valve (7) to opposite position and pressure flows to rod end of ram cylinder (8) and piston returns. Operator returns handle of valve (7) to neutral position. He then shifts handle of valve (1) to opposite position and oil flows to rod end of slide cylinder (2), retracting slide and completing cycle.

By using the accumulator as shown, there is no heat problem at the power unit during the stand-by periods.

Accumulator Used as Shock Suppressor

Circuit No. 7 (Fig. 11) shows an accumulator being used as a shock suppressor. In a high-pressure hydraulic system the sudden stoppage of high velocity oil can cause considerable damage to the

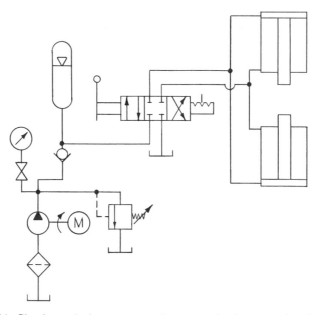

Fig. 11. Circuit employing an accumulator as a shock suppressing device.

piping. The shock may snap heavy pipes, loosen fittings and cause leaks. By the addition of a small accumulator, nearly all of the shock can be eliminated.

A dead-weight type accumulator is recommended for the most severe applications since a sudden surge into the accumulator only causes the dead weight to ride up and down freely. When using the bag-type accumulator in circuits with extremely heavy surges, there have been instances when the bag failed to stand up in service. However, on less severe applications the bag type has proved satisfactory. Of course, the big advantage of the bag type is that it is very compact and can easily be installed.

As shown in Circuit No. 7 the accumulator is placed near the shutoff point in order to be more effective in quickly absorbing the shock wave.

Accumulator Used in Dual Pressure Circuit

Circuit No. 8 (Fig. 12) makes use of the accumulator in a dual-pressure circuit. This circuit can be satisfactorily used on a hydraulic press. The accumulator will give the increased capacity when needed.

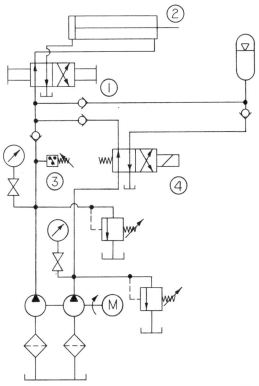

Fig. 12. Use of the accumulator in this dual-pressure circuit will give increased capacity when needed.

In this circuit the pumping unit is equipped with a large-volume, low-pressure pump and a small-volume, high-pressure pump. The operator moves the handle of the four-way valve (1), shifting its piston so that the delivery of both pumps plus the oil stored in the accumulator discharges into the blind end of press cylinder (2) and the ram moves down rapidly at low pressure. When the workpiece is

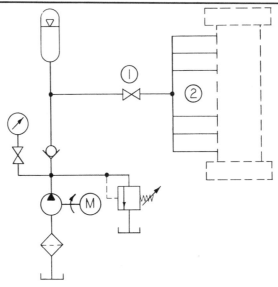

Fig. 13. The accumulator in this circuit acts to dispense lubricants to the various bearing areas of the machine or mechanism.

met, pressure immediately builds up closing pressure switch (3) which in turn energizes the solenoid of valve (4) shifting its piston and directing the large-volume pump delivery to the accumulator to recharge it. The small-volume, high-pressure pump continues to pump but now at a high pressure and the press ram finishes its downward travel, in this case pressing a bushing into the workpiece.

Operator shifts handle of valve (1) back to original position and both pumps and the accumulator delivers oil to the rod end of the press cylinder and the ram retracts. At the retracted position, pressure switch (3) again is closed energizing solenoid of valve (4) and low-volume pump maintains oil pressure to hold ram up while large-volume pump is recharging accumulator. Check valves are employed to keep the high and low pressures from mixing when they are being applied separately during the cycle. Two relief valves are used in the circuit; one on each pump line.

By employing an accumulator in this manner the size of the low-pressure pump can be greatly reduced with considerable savings due to lower horsepower required.

Accumulator Used as Lubricant Dispenser

Circuit No. 9 (Fig. 13) would scarcely be classed as a fluid power circuit, yet it has a definite place in modern industry. The application is that of a fluid dispenser, dispensing lubricant to a number of bearings of a complicated mechanism. Any type of accumulator can be used for the application. The lubricant is pumped up into the bottom section of the accumulator. A shutoff valve (1), as shown, is advisable so that it can be closed during the recharging period. Also it can act as a throttling device in metering the lubricant to the system (2).

If a bag-type accumulator is used, the air chamber need not be precharged. After the bag has been filled with lubricant, an air line can be connected to the valve on top of the accumulator and, through a timer, short blasts of air can be shot into the air chamber giving intermediate shots of lubricant into the system, thus supplying sufficient lubricant to take care of the mechanism.

CHAPTER 5

Fluid Power Lines

Although fluid power lines play an important part in the fluid power system, too often very little consideration is given to the piping layout, with the result that the efficiency of the entire system is greatly reduced. This is especially true in hydraulic systems for machine operation and control.

Layout Requirements

When laying out or piping up such a system, four basic requirements should be taken into consideration:

1. The lines must have a sufficiently large cross-sectional area to allow the fluid to flow with the least amount of friction. Friction causes heat and also loss of efficiency. In a hydraulic system, a flow rate of 15 feet per second should be maximum, while lower flow rates are more desirable.

Lines which carry the exhaust oil must be as large as possible so as not to build up too much back pressure. When several valves are exhausting into one manifold, oftentimes enough back pressure will build up to cause certain valves to malfunction. Restrictions from pipe fittings must also be kept to a minimum.

2. Fluid lines must be as short as possible and with the least number of bends. In order to reduce the time lag in a system, short lines between the master control valve and the power cylinder are advisable. The ideal condition would be to have the control valve mounted right beside the cylinder but this is not always possible, especially if it is necessary to make use of a manually-operated valve. Long lengths of high-pressure hydraulic piping are expensive, especially in the larger diameters. Long lengths of piping also create more friction.

In laying out or "piping up" a hydraulic system the radius of any bend, whether in tube or pipe, should be in the neighborhood of three times the pipe diameter in order to assure the highest efficiency. If the bend or radius is less than three times the pipe diameter, a pressure drop is caused by the sudden change in direction of the oil. In other words, it is like approaching a sharp corner with a car, then having to make a sudden turn. If the bend is greater than three times the pipe diameter, the pressure drop is caused by turbulences set up in the pipe. A longer distance must be traversed before the direction is changed. Elbows, street-ells and sharp bends should be eliminated wherever possible.

Tube benders are a necessary piece of equipment in properly making tube bends for a fluid power installation. Although very small diameter tubing can be bent by hand, one should always use a tube bender, not only to obtain better appearing bends, but also to eliminate the risk of collapsing the tubing. The open-side bender as shown in Fig. 1 is designed so that it can be slipped over the tubing

Courtesy of Parker-Hannifin Corp.

Fig. 1. Open-side tube bender calibrated in degrees to facilitate making bends of various angles.

at the exact spot where the bend is to be made. The bender is graduated in degrees so that a bend of any angle may be formed. Some benders only bend one size of tubing while others bend several sizes of tubing by changing certain parts.

For bending tubing on a production basis, various types of tube-bending machines have been designed. These machines are operated by air or hydraulic means and eliminate any effort on the part of the operator.

3. Hydraulic pressure lines, especially those carrying high-pressure oil, should be securely fastened. When oil flowing at high velocity

through a pipe is suddenly stopped due to the shifting of a valve or a cylinder reaching the end of its stroke, any bends in this pipe have a tendency to be straightened out. There is a tremendous whipping action created and if the pipe is not properly anchored, it can easily snap off a fastening bracket. Wherever vibration is present to any great degree, supporting brackets should be placed close together.

Courtesy of Wagner Electric Corp.

Fig. 2. Both large and small piping in this hydraulic installation are well supported by brackets to impart additional rigidity.

Figure 2 shows a fine example of a clean piping layout. Note provision of supports for both large and small pipe.

4. Fluid power lines must be free of foreign matter and leaks. It is not uncommon for new fluid power components to be hooked up with some old pipe that has been lying around a maintenance department for years. The result is that dirt, rust and scale quickly get into the system causing considerable damage to the precision-made

components. Even if new piping is used, it should be thoroughly
inspected for pipe scale and other foreign matter. Hydraulic piping
should be pickled and if made of steel that has been spellerized will
be much more resistant to pitting due to corrosion.

Common Difficulties

When using pipe for connecting up a system, extreme care should
be used to make certain that no shavings from pipe threads are left
in the pipe. Even small shavings can quickly ruin a valve or a
cylinder. It is sometimes difficult to impress upon installation men
the importance of the use of clean piping. Any small piece of foreign
matter can quickly cause trouble. This also applies to pipe compound which is often used to excess and which, again, causes trouble.

Leaks in pipe or tubing connections not only reduce the efficiency
of the system but are costly. Since leaks in air systems are not
visible, they often go unheeded for a long period of time and thus
result in considerable expense due to wasted air. In hydraulic systems,
leaks are expensive due to the amount of oil lost and in money
spent for maintenance in cleaning up the oil. When oil leaks onto
the floors of plants, it is dangerous not only from a fire hazard
standpoint, but also because it results in a slippery, hazardous foot-
ing for the workmen.

The sight of leaky systems makes many skeptical about using fluid
power for their production problems, but there is really no excuse for
leaks in piping connections if proper installation is made and maintained. Leakage is often caused because the installer does not put
enough tension on his wrench when he screws the fitting onto the
pipe or tightens the tube connection onto the tubing. On high pressure systems, it is particularly necessary to make all joints perfectly
tight. Often, on applications where extreme vibration takes place,
the vibration will loosen the piping connections. One way of eliminating this difficulty is by welding the joints.

Selection of Pipe or Tubing

Fluid power lines are listed under three classifications: rigid, semi-
rigid, and flexible.

A variety of pipe and tubing may be used for fluid power lines
but the important thing is to use the right type for the intended
service. Too often, hydraulic systems operating at 1,000 pounds per
square inch are piped with galvanized water pipe.

Recommended for air service are standard weight steel pipe, steel
tubing or copper tubing. For hydraulic service use extra heavy steel

pipe, seamless steel tubing, or aluminum tubing. The latter is used very little in industrial hydraulic applications. While copper has been used successfully on low-pressure hydraulic systems, its use seems to be waning due to work hardening and breakage caused by vibration.

Hose is used where there is a necessity for flexibility. Annealed seamless steel tubing is ideal for connecting up a hydraulic system. It affords smooth clean lines for these systems and is readily flared for making the connection with the tube fitting. Figure 3 shows one method used to flare seamless steel tubing. The installer carefully cuts tubing at right angles, then burrs end of tube. He then slips nut and sleeve over end of tube and shoves them down a few inches. Next he grips the tube in one hand and places the flaring tool in end of tube with the other. He then hammers on the end of the flaring tool until a uniform flare is made on the end of the tube. The sleeve and nut are pulled up into place and the connection to the system is ready to be made.

Figure 4 illustrates a different type of flaring tool. After properly preparing the end of the tube (i.e., making certain that the end is cut off squarely) so that the end of the tube is not out of round and that all burrs are removed, the end of the tube is placed in the flaring bar with just a small amount of the tube protruding above the face of the bar. It is then securely locked by the screw on the right end. A small amount of oil placed on the end of the tube will eliminate the possibility of the cone tearing the tubing during the flaring operation. The cone is advanced by turning the large wing nut on the top of the yoke assembly. This one tool will flare several diameters of tubing. Tube flares usually form an angle of either 45° or 37° with the axis of the tube. The 45° flare conforms to S.A.E. Standards while 37° flare conforms to ANSI Standards.

The data in Table 1 show safe working pressure for cold drawn seamless steel tubing.

On heavy-duty hydraulic systems where pipe of large sizes is used, the connections are often made with welded fittings. This type of fitting is welded to the pipe and has a flange which is fastened to the valve or cylinder with screws.

Tube Fittings

Tube fittings provide safe, strong, dependable connections without the necessity of threading, welding, or soldering the tubing. They provide connections that will hold up beyond the strength of the tubing itself. Tube fittings are made of steel, aluminum, or bronze bar stock or forgings. They are made in straights, unions, 45°- and

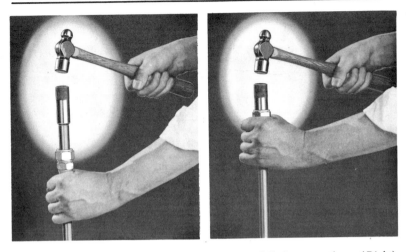

Fig. 3. (*Left*) Flaring tool in end of tube at start of flaring operation. (*Right*) Same tool at the completion of the operation.

Courtesy of Imperial Eastman Corp.

Fig. 4. Flaring tool for flaring various sizes of tubing.

TABLE I

SAFE INTERNAL WORKING PRESSURE FOR COLD DRAWN SEAMLESS STEEL TUBES*

Wall Thick.	Size, Inch								
	¼	5⁄16	⅜	7⁄16	½	⅝	¾	⅞	1
	Pounds per Square Inch								
0.028	2240	1795	1493	1281	1120	896	747	640	560
0.035	2800	2244	1867	1602	1400	1120	933	800	700
0.042	3360	2692	2240	1922	1680	1344	1120	960	840
0.049	3920	3141	2613	2243	1960	1568	1307	1120	980
0.058	4640	3718	3093	2654	2320	1856	1547	1326	1160
0.065	5200	4167	3467	2975	2600	2080	1733	1486	1300
0.072	5760	4615	3840	3295	2880	2304	1920	1646	1440
0.083	6640	5321	4427	3799	3320	2656	2213	1897	1660
0.095	7600	6090	5067	4348	3800	3040	2533	2171	1900
0.109	. . .	6987	5813	4989	4360	3488	2907	2491	2180
0.120	. . .	7692	6400	5492	4800	3840	3200	2743	2400
0.134	. . .	. . .	7147	6133	5360	4288	3573	3063	2680

* Carbon content, 0.10 to 0.20; soft annealed; Rockwell B–50; yield point 30,000 pounds per square inch. Pressures calculated by Barlow's Formula.

90°-elbows, and tees. The straights have the tube connection on one end and may have either a male or female pipe thread on the opposite end. Figure 5 shows the cross section of a fitting that uses the "Collet Grip" principle. Note the slit in the collet sleeve similar to that employed in a lathe collet.

Unions have tube connections on each end; elbows have tube connections on one end and a male pipe thread, female pipe thread, or a tube connection on the opposite end; tees have several different combinations—tube connections at all three outlets, tube connections

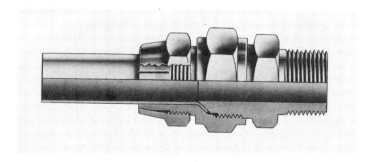

Fig. 5. Cross section of a fitting using the "Collet Grip" principle.

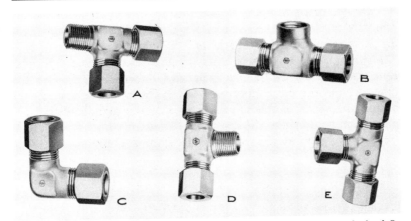

Courtesy of The Weatherhead Co.

Fig. 6. Some standard-type tube fittings. (A) Male run tee. (B) Female branch tee. (C) Union elbow. (D) Male branch tee. (E) Union tee.

at two ends and male or female pipe thread on the side, or pipe thread on one end, a female or male pipe thread on the opposite end and a tube connection on the side. Figures 6 and 7 illustrate a number of the standard-type tube fittings. There are a number of special fittings available which are used on installations where a standard fitting just won't meet the space requirements or where the component connection which the fitting is to join is not of a standard nature.

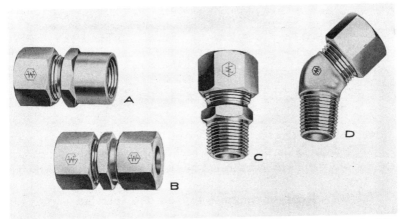

Courtesy of The Weatherhead Co.

Fig. 7. Some standard-type tube fittings. (A) Reducer. (B) Union. (C) Male connector. (D) 45° male elbow.

Tube fittings are of two types—the flare type and the flareless type. The flare type usually uses one of two angles of flare: 45°-angle for S.A.E. type fittings and 37°-angle for ANSI type fitting.

Figure 8 (upper) illustrates a 90°-elbow with a flare-type tube connection on one end and a male pipe thread connection on the other. Note the heavy-duty tube nut. The sleeve which extends beyond the nut in the photograph is free floating and aligns the flare portion of the tubing to the seat in the fitting. It also gives an excellent bearing support to the tube and dampens vibration. Figure 8

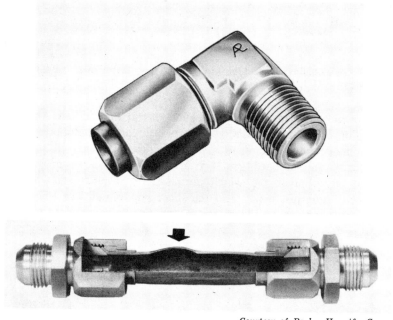

Courtesy of Parker-Hannifin Corp.

Fig. 8. (*Upper*) Widely used flare-type hydraulic tube fitting. (*Lower*) Cutaway view of a tube and fitting assembly after being subjected to a high enough pressure to cause tubing to rupture (at arrow).

(lower) shows a cross-sectional view of this flare tube connection on a straight fitting after the tube has been subjected to enough pressure to produce a rupture as indicated by the arrow. The connection is still tight.

Figure 9 (upper) illustrates a flareless-type tube fitting. As the upper view shows, this fitting is made up of a body, a sleeve and a nut. As the nut is tightened, it forces the sleeve forward into the body taper. The pilot on the end of the sleeve is contracted, thus forcing the cutting edge of the sleeve into the outer surface of the tube making a groove which forms a tight joint between the fitting and the tube. The nut presses on the bevel of the sleeve forcing it to clamp very tightly to the tube. The resistance to vibration is concentrated at

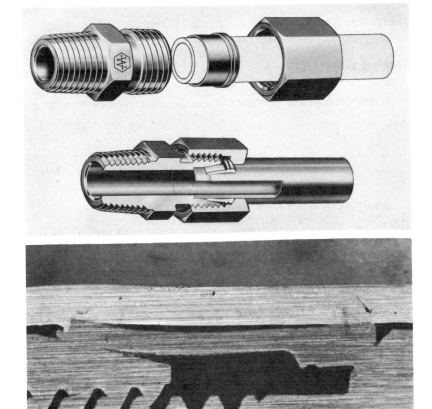

Courtesy of The Weatherhead Co.

Fig. 9. (*Upper*) Cutaway and exploded views of a flareless-type tube fitting. (*Lower*) An unretouched magnified cross section of a portion of this fitting.

this point rather than at the point where the cutting edge of the sleeve is contacting the tube. When the nut is fully tightened, the sleeve is bowed a bit at the midsection and acts as a spring. The spring action of the sleeve maintains constant tension between the body and the nut, thus preventing the nut from loosening. After the first assembly, the sleeve becomes permanently attached to the tube.

Figure 9 (lower) shows an unretouched magnified cross section of the tube fitting in Fig. 9 (upper). Note how the sleeve makes connection with the tube. These fittings conform to the ANSI Standards.

Flareless fittings eliminate the need of flaring tools of any kind. It is only necessary to cut off the tubing so that it is reasonably square.

Flexible Lines

Flexible lines, made up of flexible hose, play an important part in many fluid power systems. Hose for this type of service are

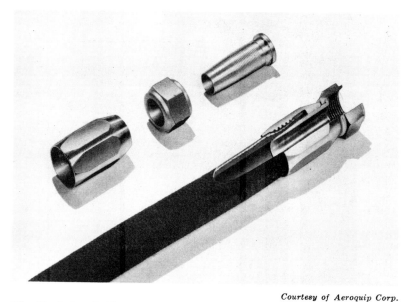

Courtesy of Aeroquip Corp.

Fig. 10. A detachable, reusable hose fitting for use with both low- and high-pressure hose lines is shown disassembled and cutaway.

manufactured in four pressure ranges: *low-pressure hose* for use in air systems, low-pressure hydraulic systems and for exhaust lines of high-pressure systems; *medium-pressure hose* for medium-pressure hydraulic systems up to 1200 lb per sq in.; *high-pressure hose* for high-pressure systems up to 3000 lb per sq in.; and *extra high-pressure hose* for hydraulic systems up to 5000 lb per sq in. Fluid power hose for low pressure may be made of cloth braid and rubber while the extra high-pressure hose has one or more plies of steel braid, cloth braid, and rubber. Note the hose as shown in Fig. 10 which has a rubber inner tube, a braid and a cover. The inner tube is designed to withstand the attack of the materials which pass through it. The braid which may consist of several layers is the determining factor in the strength of the hose. The cover is designed to withstand external abuse. Also note the coupling which is designed so as to be reusable. Fluid power hose is now available for use in applications where there are extreme temperature variations.

While the majority of fluid power hose is used on pressure applications of various degrees, there are some applications which require hose to be used on suction or vacuum lines. In selecting hose for this service, it should be remembered that the construction must be such that it will not collapse or go out of round under any working condition.

Computing Length of Flexible Hose Assemblies

In applying hose to an installation the data and chart of Table 2 should be of great benefit. Note that this information covers both stationary installations and relative motion installations.

Stationary Installation. In most hydraulic hose installations the hose has at least one bend along its path. It is advisable to keep the start of this bend slightly away from the coupling. This distance plus the length of the coupling (and adapter if used) will be designated by A (as shown in Table 2, left) and remains constant for each size hose. R is the bend radius and should never be less than the minimum value shown in the table. Then the length of the hose assembly in Table 2 (left) will be

$$L = 2A + 3.1416R$$

The following is the general formula:

$$L = 2A + X$$

where X is the length determined by design experience or actual measurement.

The foregoing two formulas are to be used where there is little or no movement. Short hose assemblies are used to dampen vibration from a power unit to other parts of a machine. Wire-braided hose is highly recommended for this use as it will absorb vibration to a much greater extent than pipe or tubing.

Relative Motion Installation. When there is relative motion between the two end connections of a hydraulic hose assembly, this motion or travel must be considered. As shown in Table 2 (right), let T equal amount of travel. This amount of travel is added to the length of the hose as figured for a stationary installation $(L = 2A + X)$ so that the formula becomes

$$L = 2A + X + T.$$

This will permit the proper flexing in all positions of travel.

TABLE 2

DATA FOR CURVED AND STRAIGHT FLEXIBLE HOSE ASSEMBLIES

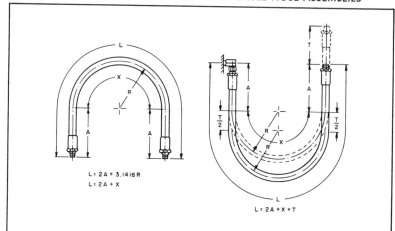

$$L = 2A + 3.1416R$$
$$L = 2A + X$$

$$L = 2A + X + T$$

Courtesy of Anchor Coupling Co., Inc.

Int. Diam. of Hose	Dimen. A	2A	Min. Radius, R	Min. Over-all Length*
$\frac{3}{16}$	4	8	4	$4\frac{5}{16}$
$\frac{1}{4}$	5	10	4	$5\frac{1}{4}$
$\frac{5}{16}$	5	10	5	$5\frac{1}{2}$
$\frac{3}{8}$	6	12	5	$5\frac{7}{16}$
$\frac{1}{2}$	7	14	7	$5\frac{13}{16}$
$\frac{3}{4}$	8	16	$9\frac{1}{2}$	6
1	10	20	11	$7\frac{1}{16}$
$1\frac{1}{4}$	10	20	16	8
$1\frac{1}{2}$	10	20	20	$8\frac{1}{4}$
2	12	24	22	10

All dimensions in inches.

* Minimum over-all length for straight assembly. This includes two couplings and $\frac{1}{2}$ inch of exposed hose between.

It is good practice to have the total length of the hose assembly lie in one plane and bend in that plane. This will prevent any torsional stress from being set up when the movement occurs. This also eliminates any tendency to loosen the fittings.

The question often arises as to where flexible lines can be used.

The applications are too numerous to enumerate but there are a few which should be called to the reader's attention.   One air-cylinder application which is noteworthy is the use of flexible hose in making the connection between the rigid piping and the distributor on a rotating air cylinder.

The flexible hose allows the distributor to float rather than to have undue tension thrown upon it as when rigid piping is used.   Even if a perfect connection were made by using rigid piping, when the distributor is repacked it would be nearly impossible to duplicate the original piping lengths.   In other words, if any one pipe connection was made tighter or looser after the distributor was repacked it will likely cause a binding action.   When using flexible hose there is no problem.

Flexible hose is used with trunnion and pivoted mounted air and hydraulic cylinders.   These cylinders are found on machine tools, lifting devices, steel mill roll-over devices, cranes and many other devices.

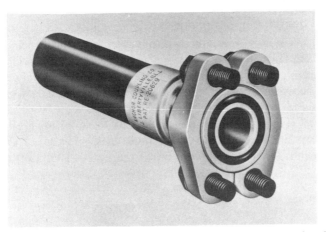

Courtesy of Anchor Coupling Co., Inc.

Fig. 11. Rubber-covered hose with pressed-on fitting.

Flexible hose is also used as a surge dampener.   A length of flexible hose in·a hydraulic system can often reduce shock in the system and spares rigid lines from injury.

Flexible hose assemblies (i.e., the hose equipped with fittings) are of two types: the ones with "renewable" couplings and the ones with "pressed on" couplings.   The advantage of the "renewable" coupling is that "on the job" cutting of the hose can be made without special tools.   Figure 10 illustrates the renewable type.   Figure 11 illustrates

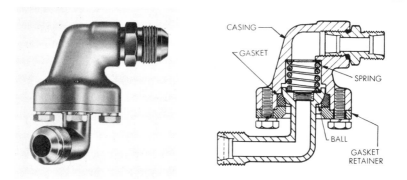

Courtesy of Barco Mfg. Co.

Fig. 12. Joint suitable for use in a temperature range of −100°F to +500°F and a maximum pressure rating of 3000 psi.

a hose with a 4-bolt split-flange pressed-on fitting. This is not a reusable type connector. Note the "O" ring seal.

Special Joints and Couplings

While one does not usually think of swivel and rotating joints, self-sealing couplings, quick disconnect couplings and other miscellaneous couplings in connection with fluid power lines, they certainly are important accessories in this field. The swivel joint is surely growing in popularity as improvements have been made in these devices. There are many industrial applications for these where flexible all-metal lines are required instead of hose lines. Figure 12 shows a joint which is suitable for a temperature range of −100° to +500°F.

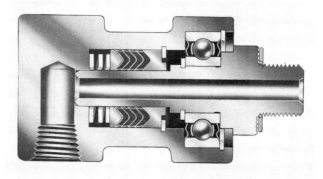

Courtesy of Barco Mfg. Co.

Fig. 13. Single-acting rotating joint capable of being operated at 2500 rpm under normal conditions.

The pressure rating is 3000 lb per sq in. It can easily be seen that the use of flexible joints in "hot spots" around a furnace or rolling mill or in other places of intense heat would be desirable. The joint not only allows a swivel action of 360°, but also allows 15° side flexibility for movement in any direction due to the ball and socket design. This eliminates wear when side thrust is present.

Rotating joints are used on many present-day machines. Primarily a rotating joint is used where there is need for a stationary distributor housing and a rotating distributor shaft. The joint may be of either of the single-acting or double-acting type. The latter provides an axial connection and a radial connection for two fluid lines. About the only application for a double-acting joint is in connection with double-acting air or hydraulic cylinders. Single-acting rotating joints are used for passing coolants through hollow center rotating cylinders, for operating clutches, for operating air or hydraulic distributors and for numerous other fluid power applications. Shown in Fig. 13 is a cross section of a single-acting rotating joint capable of rotating at 2500 rpm under normal conditions.

Self-sealing couplings as used with hydraulic lines make it possible to separate the lines without losing oil. This eliminates the need of draining the line before disconnecting. It also eliminates spilling oil and causing a "messy" situation which is so common in breaking the average connection. Such a coupling is especially beneficial on testing fixtures where the lines must be continually connected and disconnected. Figure 14 shows a self-sealing coupling.

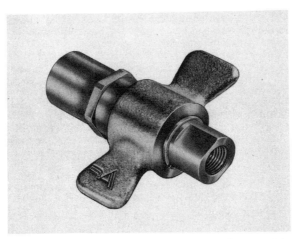

Courtesy of Aeroquip Corp.

Fig. 14. Self-sealing coupling.

ONE PUSH COUPLES   ONE PULL UNCOUPLES

Courtesy of Lincoln Engineering Co.

Fig. 15. Illustrates ease of coupling and uncoupling.

Quick connection couplings of several types have been designed for both air and hydraulic service. By very little effort on the part of the operator these connections can be immediately separated or connected as shown in Fig. 15. These, when used with portable air tools, are valuable inasmuch as the tool can be moved around the shop and an immediate connection made at any station where a mating connection is located. This eliminates the need for a pipefitter every time an air tool is to be moved. Figure 16 (left) illustrates the two sections of a quick disconnect coupling. The male sections would be connected to the lines of portable air tools and the female sections which contain the shutoff valve would be installed at various stations. Figure 16 (right) illustrates an all purpose coupling designed to handle positive pressure in hydraulic and pneumatic systems. The seal is of the U-Packer design and sealing effect increases as the pressure increases. The fluted valve stem design provides for an ample flow of the fluid. When the coupling is disconnected an immediate seal is provided to eliminate leakage from either section.

In conclusion it should be emphasized: whether tubing, pipe, or hose are used, make certain that it is of the correct size and strength, and perfectly clean on the inside.

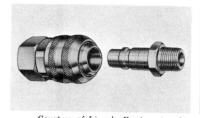

Courtesy of Lincoln Engineering Co.   *Courtesy of Snap-Tite, Inc.*

Fig. 16. (*Left*) Quick disconnect coupling for air lines. (*Right*) Cutaway view of coupling showing valved nipple and valved coupler.

CHAPTER 6

Hydraulic Valves
and Their Functions

Hydraulic valves control and direct the oil from the time it leaves the pump until it starts on its return to the reservoir in a hydraulic circuit. The relief valve is almost always the first one encountered by the oil after it leaves the pump. From that point on throughout the circuit, the type of valve employed depends upon the design of the circuit. In a very simple circuit the second and only other valve used will probably be the master control valve. In a complex circuit the oil may flow through a sequence valve, master control valve, reducing valve, flow control valve, check valve and several others before it finally returns to the reservoir.

Following is a brief outline of the functions and applications of various standard valves and also some of the things to check if the valves fail to perform satisfactorily. Often some very minor adjustment may be the difference between a smoothly functioning circuit and one which is a continual source of trouble. Hydraulic valves as used in industrial applications are built in several pressure ranges. Some have a range from 0 to 1500 lb per sq in., some range from 1500 to 2500 lb per sq in. and still others from 1000 to 5000 lb per sq in. The majority of present applications are in the range of 0 to 1500 lb per sq in., but the trend is toward higher pressures.

The Relief Valve

The relief valve is the protector of the hydraulic circuit. While in the majority of applications it is considered a part of the power unit, it may also be used in other parts of the circuit. The relief valve not only protects the pump but also the electric motor and other components of the circuit. If it were not for the relief valve, pressure would build up until some component would burst or the electric motor would overload to a point where it would stall.

Relief valves may be of the pilot-operated type or the direct-acting type. The pilot-operated type employs a small control spring, while the direct-acting type uses a large one. Either type usually uses different springs for different pressure ranges. The ranges, in pounds per square inch, run something like this: 0 to 750, 700 to 1500, and 1500 to 2500; or 0 to 250, 200 to 1000, and 900 to 1500; or 1 to 1500, 1000 to 3000, and 2500 to 5000.

The relief valve is normally closed, i.e., the valve piston keeps the exhaust core or connection closed until such a time as the pressure exceeds the spring setting, when the valve opens and oil is returned

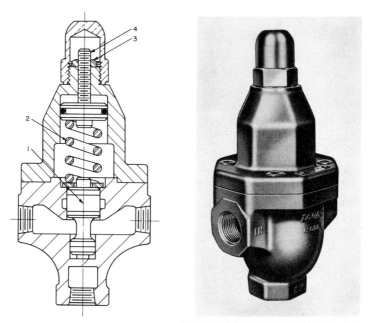

Courtesy of Logansport Machine Co., Inc.

Fig. 1. Direct-acting type of relief valve.

to the oil reservoir. After the excess pressure is relieved, the exhaust passage is quickly and positively closed.

Figure 1 shows a direct-acting type of relief valve. When the pressure on the valve piston (1) becomes great enough to overcome the pressure exerted by the hold-down spring (2), the discharge port is opened and the pressure is relieved.

Figure 2 shows a relief valve of the pilot-operated type. As in the direct-acting type the movable main poppet allows a large volume of oil to escape to the oil reservoir when the system pressure exceeds the setting of the valve. However, in the pilot-operated type, the action of the large, main poppet is controlled by a much smaller poppet called the control poppet. System pressure acts on both sides of the main poppet because of the small orifice shown in Fig. 2. Since there is a greater area exposed to system pressure on the left side, the main poppet is held firmly on its seat, thus reducing leakage. System pressure also acts on the control poppet by way of the orifice just mentioned. When pressure is great enough to overcome the adjustable spring pressure bearing on the control poppet, oil will flow to the tank. Now the forces acting on the main poppet are changed, because flow past the control poppet causes a pressure difference

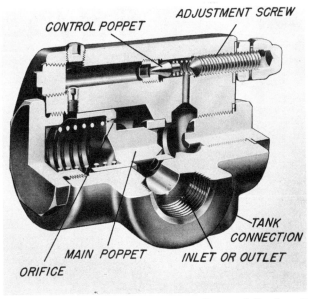

Courtesy of Abex Corp., Denison Div.

Fig. 2. Relief valve of the pilot-operated type.

across the orifice. Pressure on the right side now exceeds that on the left and the main poppet moves to the left. A large volume of oil escapes to the oil reservoir at atmospheric pressure, thus reducing pressure in the system. When reduced pressure allows the control poppet to reseat, the main poppet again closes. Pressure adjustment is easily made by means of a socket head set screw.

Figures 3 and 4 show high-pressure relief valves of the pilot-operated type suitable for operating at pressures up to 5000 pounds per square inch.

Courtesy of Abex Corp., Denison Div.

Fig. 3. High-pressure relief valve of the pilot-operated type. This valve can be used for pressures up to 5000 pounds per square inch.

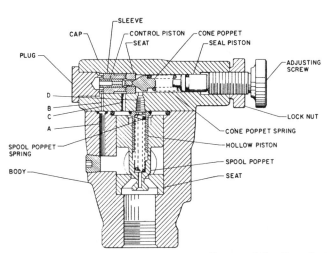

Courtesy of Abex Corp., Denison Div.

Fig. 4. Cross-sectional view of a high-pressure relief valve of the pilot-operated type suitable for operating at pressures up to 5000 pounds per square inch.

Relief valves, as well as other valves, are rated by the size of their pipe ports. For example, a Model XXXX-¾ would be a valve having pipe ports with ¾-inch American (National) Dryseal pipe threads or American Standard pipe threads. Large, high-pressure valves are usually built with flange connections. For the majority of installations, valve sizes are ¼-, ⅜-, ¾-, 1-, 1¼-, 1½-, and 2-inch. For exceedingly large volumes, larger size ports are used.

In selecting a relief valve, make certain that the capacity is large enough for the application and that it has the correct operating range. One manufacturer of hydraulic valves has set up the following table for fluid flow through its line of valves:

TABLE I

RATE OF FLOW THROUGH RELIEF VALVES

Port Size, inches	Volume, gals. per min.
¼	3
⅜	6
½	10
¾	16
1	28
1¼	50
1½	70
2	100

Figure 5 shows a simple hydraulic circuit in which a relief valve is being used in connection with a hydraulic power unit.

Figure 6 illustrates the use of two relief valves in a circuit. This circuit was designed so that the oil in the system would not overheat when a long stand-by period was required during the loading and unloading period. High-pressure oil spilling through the exhaust of a relief valve causes considerable heat especially if a fairly large volume pump is being used. In this circuit, however, the high pressure is only used on the quick downstroke where very little time is consumed, then when the ram reaches the top of the stroke, the oil spills through the low-pressure relief valve (2).

Valve (2) is set so that there is just enough pressure in the system to return the ram. Oil spilling through the low-pressure relief valve causes very little heat. This circuit is being used extensively on applications where a single high-pressure relief valve would cause excessive heat.

Relief Valve Difficulties. If the relief valve fails to function properly it may be caused by one of the following:

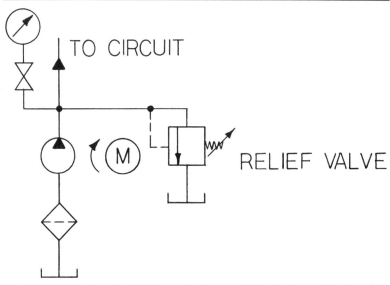

Fig. 5. Simple hydraulic circuit employing a relief valve in conjunction with a hydraulic unit.

1. Dirt or air in the oil will cause a pressure fluctuation. Air in the oil may be due to: a leak in the inlet line caused by improper piping; an insufficient amount of oil in the oil reservoir; or a foaming action of the oil. This latter will cause quite a pressure fluctuation. Lint, pipe compound, scale or other foreign matter in the oil can foul the relief valve and not allow the piston to properly seat itself, thus causing leakage and a fluctuation of pressure. While there is usually a strainer on the intake line to the pump for removing impurities, there still may be some within the piping or within the valve itself. Core sand in the valve may also be a source of trouble.

2. An extremely noisy relief valve may be the result of air passing through the valve. It is sometimes difficult to determine whether the noise is caused by air in the pump or air in the relief valve.

3. If pressure fails to register on the pressure gage after tension is placed on the valve spring, the valve spring may be broken or the piston may be stuck due to some foreign substance lodging between the body and piston. It is also possible that the valve and piston may be worn to such a point that the amount of leakage will make the valve inactive.

Should the relief valve be disassembled for any reason, caution should be exercised when reassembling to make sure that all of the parts are replaced correctly. A mistake which is often made is to place the cover gasket so that a drain hole will be cut off. This quickly stops the valve from functioning.

When piping from the exhaust port of the relief valve to the reservoir, make certain that no restrictions are present. Back pressure can be troublesome in any hydraulic system.

The Directional Control Valve

The directional control valve may be a three-way valve or a four-way valve. A three-way valve is one with three ports: an inlet port, a cylinder port, and an exhaust port. The four-way valve has four ports: an inlet port, two cylinder ports, and an exhaust port. Three-way valves are used primarily for actuating single-acting hydraulic cylinders and four-way valves for actuating double-acting hydraulic cylinders and fluid motors.

Directional control valves of the three- or four-way type may be actuated by several methods; manually (hand or foot), hydraulic pilot pressure, air-pilot pressure, electrically with solenoids, or by mechanical means.

Three-way Valves. The three-way valves may be normally open or normally closed. In a normally open valve (also called "open-

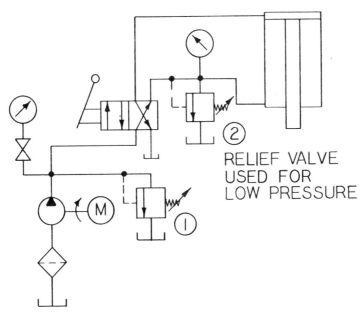

Fig. 6. Hydraulic circuit employing two relief valves to prevent overheating of the oil in the system.

center" type) the inlet and the cylinder port are connected in the
normal position while in the normally closed valve ("closed-center"
type) the inlet is blocked and the cylinder port is connected to the
exhaust port when the piston is in the normal position. Figure 7
shows a hydraulic circuit with a three-way, solenoid-operated, spring-
return type valve actuating a single-acting press ram. The ram is
returned to the starting position by its own weight as oil is released
from the bottom of the cylinder.

Four-way Valves. Four-way valves are of the two-position or
three-position styles. In the two-position valve the spring is usually

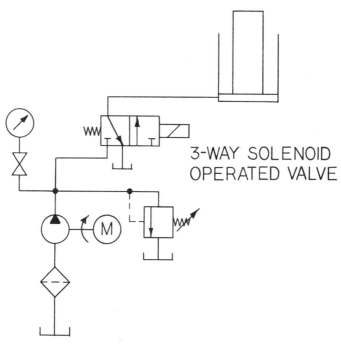

Fig. 7. Hydraulic circuit with a three-way solenoid-operated spring-return
type of valve actuating a single-acting press ram.

offset so that it returns the piston to the starting position when the
operating means is released. In the three-position valve the spring
is usually centered so that when the operating means is released the
piston is returned to the "neutral" position. Two- and three-position
valves are also built with ball detents for accurately locating the positions. Even though there are numerous standard-type valves available, there are still numerous applications for special master control

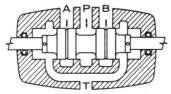

Closed center valve with all ports blocked when piston is in intermediate position.

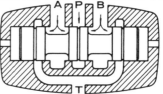

Closed center valve with all ports blocked when piston is in intermediate position.

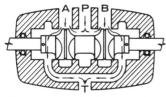

Open center valve with all ports open to exhaust when piston is in intermediate position.

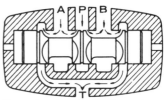

Open center valve with all ports open to exhaust when piston is in intermediate position.

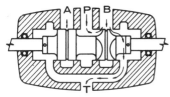

Open center valve with inlet and cylinder port **B** open to exhaust and cylinder port **A** blocked when piston is in intermediate position.

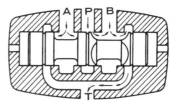

Open center valve with inlet and cylinder port **B** open to exhaust and cylinder port **A** blocked when piston is in intermediate position.

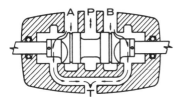

Closed center valve with both cylinder ports open to exhaust when piston is in intermediate position.

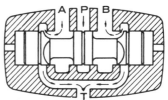

Closed center valve with both cylinder ports open to exhaust when piston is in intermediate position.

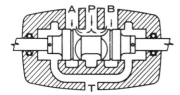

Open center valve with both cylinder ports open to inlet but closed to exhaust when piston is in intermediate position.

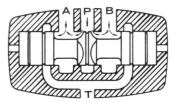

Open center valve with both cylinder ports open to inlet but closed to exhaust when piston is in intermediate position.

Courtesy of Rivett Division of Applied Power Inc.

Fig. 8. Standard types of three-position four-way valves. Left-hand column illustrates the manually operated, solenoid-operated, and mechanically operated valves. Those in the right-hand column illustrate the pilot-operated valves.

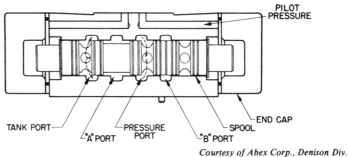

Courtesy of Abex Corp., Denison Div.

Fig. 9. Open-center type hydraulic valve.

valves. Some requirements call for special metering grooves, some need special internal drilled passages, still others need two four-way valves hooked together to act as an eight-way type for operating two cylinders for special positioning.

Figure 8 shows some of the standard available types of three-position, four-way valves. The diagrams in the left-hand column illustrate the manually-operated, solenoid-operated, and mechanically-operated valves, while those in the right-hand column illustrate the pilot-operated valves.

A very important open-center type valve is shown in Fig. 9. With the piston in the neutral or intermediate position, the cylinder ports are blocked and the inlet is connected to the exhaust. This valve may

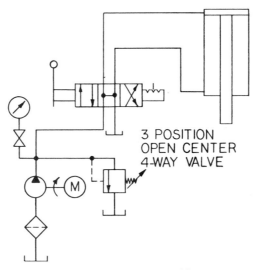

Fig. 10. Hydraulic circuit using a three-position open-center valve.

Courtesy of Logansport Machine Co., Inc.

Fig. 11. Hydraulic press with manually operated three-position open-center valve shown at right.

be used on press applications or other applications where there is a long stand-by period for loading or unloading. By the use of this valve, the oil is exhausted back to the reservoir at no pressure, thus eliminating the heat factor. Figure 10 shows this valve as it is used in a hydraulic circuit and Fig. 11 shows such a valve arranged for manual operation on a hydraulic press.

The closed-center valve, shown at the top left of Fig. 8, is used successfully in "inching" applications, especially if the piston has metering grooves. It is often desirable to "inch" heavy objects or when raising a platform it may be advantageous to raise it in short steps. Figure 12 shows the interior of a closed-center valve. This type of valve has the disadvantage, however, of causing excessive heat in the system due to the pressure being blocked when the piston is in the neutral position. The oil must then spill through the relief valve at high pressure. This is especially true where a large-volume, high-pressure pump is being used. However, Fig. 13 shows a hydraulic circuit in which it does have an advantage over the open-center valve. Several closed-center valves are here used in a multiple circuit for the operation of a bank of presses. If open-center

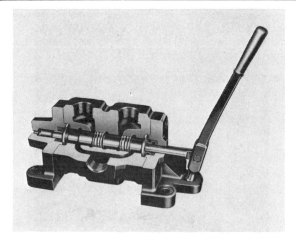

Courtesy of Logansport Machine Co., Inc.

Fig. 12. Cutaway view of a closed-center valve.

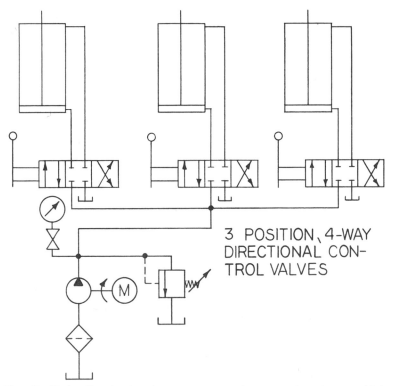

3 POSITION, 4-WAY
DIRECTIONAL CON-
TROL VALVES

Fig. 13. Hydraulic circuit using several closed-center valves in a multiple
circuit for the operation of a bank of presses.

valves were used, all the oil would return to the reservoir and there would be no available pressure for the circuit when the valve pistons are in the neutral position.

The closed-center type valve having both cylinder ports connected to exhaust in the neutral position finds use on various types of feeding mechanisms where the piston rod of the cylinder must be free to "float" as the valve is shifted to the neutral position.

Directional control valves are of rugged construction to meet the needs of their many applications. In Fig. 14, a manually operated, subplate mounted, four-way, directional-control valve is shown. The valve is designed to be mounted on a subplate of the NFPA standard mounting pattern. A subplate mounted valve has the advantage, from a maintenance standpoint, that there are no breakable pipe connections involved when the valve is removed for servicing.

The majority of hydraulic directional-control valves are of the sliding metal spool design. The spools are hardened and ground, then closely fit into a high-tensile-strength valve body. Depending upon the operating pressure and the type of hydraulic service, spools may

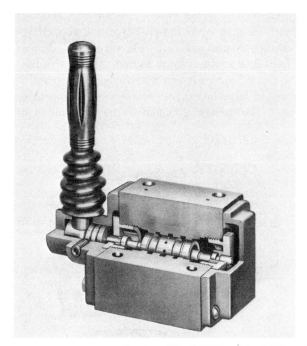

Courtesy of Rivett Division of Applied Power Inc.
Fig. 14. Manually operated subplate mounted control valve.

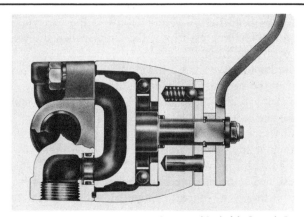

Courtesy of Barksdale Controls Division

Fig. 15. Cutaway view of a shear-type master control valve.

be made of steel; stainless steel; or bronze; while valve bodies are made of high tensile cast iron; steel; or bronze. In order to assure minimum internal leakage, spool-to-body fit is held to within a few ten thousandths of an inch.

Other directional control valves are of the shear, plug, and poppet types. In most cases the plug type and poppet type are used where operating pressures are low. The shear type, shown in Fig. 15, shears the fluid flow as the rotor is shifted. This valve has an exceptionally low handle load even at extremely high operation pressure due to the use of large ball thrust bearings and the patented "Shear-Seal" design. The valve is designed so that the sealing surfaces are

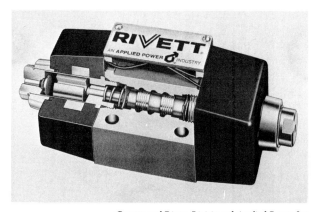

Courtesy of Rivett Division of Applied Power Inc.

Fig. 16. Cutaway view of a solenoid pilot valve.

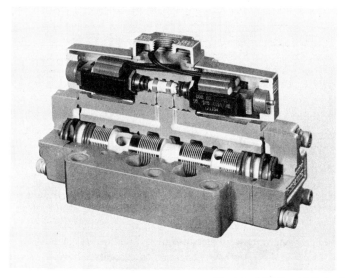

Courtesy of Continental Hydraulics
Fig. 17. Cutaway view of a double solenoid pilot valve mounted on a pilot operated valve.

kept clean to effect a positive seal. The flow takes place through the opening in the center of the "Shear-Seal," thus erosion or "wire drawing" does not become a factor when oil at high velocity is used. To prevent leakage, the flatness between the bottom of the rotor and the "Shear-Seal" must be held exceedingly close.

Three different types of connections are employed on valve bodies: flange type, threaded type and manifold type. With the manifold type, the complete valve can be removed from the system without disconnecting any piping.

Pilot Valve Applications. Pilot-operated master valves are favored in many circuit applications, especially the more complicated ones. There are several distinct advantages. In the first place, long runs of large piping can be eliminated. This is important when the operated cylinder may be fifty or sixty feet or even farther away from the operation station. It is much less expensive to run low-pressure small diameter tubing on long runs rather than the large high-pressure piping. The master control valve can be mounted next to the cylinder. Another advantage is that safety interlocks can easily be set up with pilot controls. These valves are ideally suited to provide the required safety features needed in fluid power-operated machines. Still another advantage is ease of operation. A large manually-operated valve requires considerable exertion to operate it during an

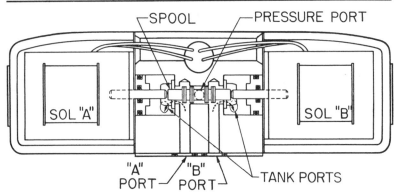

Fig. 18. Cross-sectional drawing of a small solenoid-actuated pilot valve.

eight-hour shift while the pilot valve can be operated by a mere flick of a finger.   Air-operated directional hydraulic valves, although not too commonly used, are advantageous in circuits where both air and oil or other liquid are used.   They are easily interlocked into the circuit, and safety circuits can be set up at a low cost.   These valves are capable of being cycled at a high rate of speed.

The pilot-operated master valves are controlled by very small pilot valves, usually with ⅛-inch or ¼-inch pipe capacity.   Often a small version of a large master valve, they are either of the three- or four-way type and are operated manually, mechanically or electrically. Figure 16 shows a small solenoid pilot valve and Fig. 17 shows a pilot valve mounted on a master valve.   Figure 18 shows a drawing of a small solenoid-actuated pilot valve.   The master valve is sub-plate mounted beneath the pilot valve as indicated by the location of the "A" and "B" ports.

Figure 19 shows a four-way electrohydraulic control that has a rated capacity of over 150 gpm at 3000 psi.   This valve incorporates design features which make the output flow insensitive to load pressure variations over a wide range of load pressures.   It utilizes a torque motor flapper-nozzle pilot stage.   The flapper-nozzle design insures high reliability and excellent resolution.   Sufficient null deadband is built into the valve to insure valve centering with zero input signal over a wide temperature range.   Centering springs return the valve to null in case of hydraulic and/or electrical power failure. The mounting pattern of the valve conforms to the D10 size given in the ANSI Standard B93.7-1968(R1973) [NFPA T3.5.65.1-1965] entitled "Dimensions for Mounting Surfaces of Sub-Plate Type Hydraulic Fluid Power Valves."

Figure 20 shows a 3-inch electrohydraulic pressure-compensated

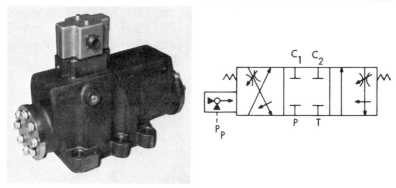

Courtesy of Sanders Associates, Inc.

Fig. 19. Four-way electrohydraulic control valve with subplate mounting.

control valve mounted on a large die-casting machine. In this valve, flow is proportional to input electric signal regardless of the load variation. This is a three-stage valve and it utilizes a force motor flapper-nozzle pilot stage for high reliability, excellent resolution, and fast response. The second and third stages are spool type with high force levels. Note the large flange connections and piping involved in this application.

Multiple Valves. Multiple valve units either of the three-way type, four-way type, or combination of the two find application in the industrial field mostly on mobile equipment. These units are of two styles.

Courtesy of Sanders Associates, Inc.

Fig. 20. Large electrohydraulic valve applied to die-casting machine.

One style has a common body for the several valve spools as shown in Fig. 21. In this body is also the relief valve. The other style is composed of a number of single valve sections which are held together with tie rods as shown in Fig. 22. Sections can be added to take care of additional cylinders. Figure 23 shows a large earth mover on which the valves in Fig. 22 are used.

Directional Control Valve Difficulties. Under proper operating conditions control valves should give long service with a minimum of maintenance. If the valve fails to function properly, some of the things to look for are:

1. *Heat.* Oftentimes heat will cause excessive leakage past the piston stem seal or the cover gasket. Extremely high temperatures

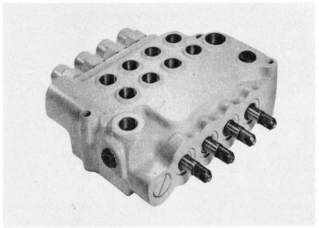

Courtesy of Parker-Hannifin Corp.

Fig. 21. Multiple valve units in one body such as this are commonly used in the industrial field on mobile equipment.

cause the packing seal to deteriorate. When valves are used around furnaces or if the oil temperature is 200°F or more (with some packings, less), it is necessary to use a high-temperature type packing that is built for temperatures up to 500°F.

2. *Dirt.* Dirt attacking the valve internally or externally can do considerable damage. Externally, certain abrasives and other hard foreign matter can score the valve stem so that it will cut the packing and cause a leak. Internally, dirt can score the valve body or piston and can cause internal leakage or may cause the piston to stick.

3. *Back pressure.* Back pressure due to a restricted exhaust will often cause the stem seals to freeze against the piston stem and make

the valve inoperative.   This is especially true with solenoid-operated valves.   With a manually-operated valve the operator usually has enough strength to push the piston from one extreme to the other, but with a solenoid there is only a limited amount of force available and a small amount of back pressure can quickly freeze the piston. This is especially true of spring-offset type solenoid valves.   The solenoid is usually just strong enough to overcome the spring pressure.

4. *Internal leakage.*   If the valve fails to allow the full pressure to go to the next component, check the valve for internal leaks.   Some of the pressure may be passing from the inlet port to the exhaust.   A wire drawing action may have taken place on the piston or the valve body may be scored.   In connection with internal leakage it should

Courtesy of Sundstrand Corporation

Fig. 22. Multiple valve units made up of a number of single valve sections which are tied together with rods.

be mentioned that a valve body usually wears much faster than the piston due to the softer material.   In replacing the piston enough grind stock should be available so that the piston can be ground and then lapped into the body.   A piston of standard size seldom fits into an old valve body.

5. *Broken springs.*   Broken return springs will not allow the piston to properly center when the piston is released on a spring-centered type valve.   A broken spring in a two-position spring-offset type valve will not allow the piston to shift to its normal position when released.

Broken packing-tension springs are a source of trouble in that leakage usually occurs around the valve stem whenever one breaks.   A broken tension spring may score the piston stem.

The Sequence or By-pass Valve

Sequence valves are used to set up the sequence of operations of the hydraulic system. While sequence valves are not the only method of setting up a sequence operation (see Chapter 12 for other methods), they do provide an inexpensive means. Figure 24 shows a sequence valve used in a simple circuit controlling two operations performed in sequence.

Sequence valves are of three general designs: the large direct-acting type with large spring, the pilot-operated type with small spring, and the remotely operated type which is a version of the pilot-operated type. Basically the sequence valve allows the oil to flow in one direction until a certain resistance is overcome and then the flow is sent in another direction. The resistance to the valve piston shift is set by the spring pressure. These springs have similar operating ranges to those in relief valves. Most sequence valves have built-in check valves so that the fluid can return freely through them.

Sequence valves of the remotely controlled type are ideal for use in "unloading" the pump pressure so as to reduce heat in applications calling for long stand-by periods during which the pump must continue to operate. This valve is also used on "high-low" circuits where the low-pressure, large-volume pump is unloaded when the high-pressure, low-volume pump goes into action. This action takes place automatically as the work cylinder meets resistance and calls for more pressure.

Figure 25 shows a cross section of a pressure control valve. One basic valve design performs sequence, counterbalance, and unloading functions. Only the pilot, drain, and return flow features differ depending upon the requirements of the specific function.

This normally closed, 2-way valve is used to block flow through a hydraulic circuit until pressure at the inlet port, or at the external pilot port, reaches the valve setting. The large-diameter seat and ball combined with the high rate disc spring in the control head provides control over the ball displacement producing exceptionally fast response to changing conditions with a high degree of repeatability. The valve has an adjustment range of 75 to 3000 psi.

Sequence Valve Difficulties. If any difficulty should be encountered with a sequence valve, here are some of the things which may be the cause:

1. *Back pressure.* Back pressure on drain connection will cause the valve to malfunction since a bank of oil under the piston will act as a solid block. This also happens when the user forgets to remove the drain plug.

Courtesy of Sundstrand Corporation.

Fig. 23. Earth mover with hydraulic system controlled by multiple valves made up of single valve sections fastened together with rods.

2. *Broken valve spring.* This will cause the valve to malfunction to the extent that it may become completely inoperative. A broken spring in a valve with a direct-acting spring is usually readily detected as the spring adjusting screw can easily be turned and screwed down further than normal.

3. *Dirt.* Dirt will cause the piston to stick and piston movement may become very erratic. As dirt passes through the valve, the piston may stick in one instance and in the next instance function properly. Not enough stress can be placed upon using clean oil at all times.

4. *Leakage.* Leakage between the piston and the valve body will cause oil to flow to the closed section of the valve defeating the purpose for which the valve is intended.

The Reducing Valve

The purpose of the reducing valve is exactly what its name implies—to reduce pressure. It is often desirable to use two pressures in a hydraulic circuit, one pressure to operate clamping devices, pilot controls, and auxiliary equipment, and another pressure for performing the operation. See Fig. 26 for basic circuit using a reducing valve. Oil flows from hydraulic power unit (1) to four-way, three-position, open-center valve (2) and on back to sump when the piston of valve (2) is in the neutral position. Operator shifts handle of valve to position *A* and oil flows to inlet port of reducing valve (3) and

SEQUENCE VALVE

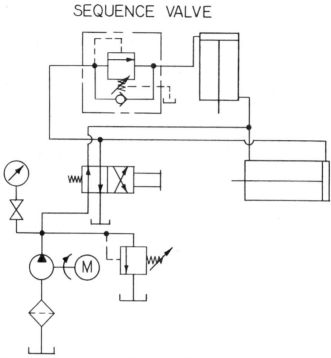

Fig. 24. Sequence valve in this hydraulic circuit controls two operations performed in sequence.

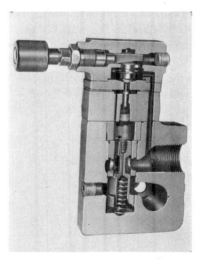

Fig. 25. Hydraulic pressure control valve.

on to blind end of cylinder (4) at reduced pressure and directly to blind end of cylinder (5) at full pressure. Piston rod of cylinder (4) moves forward locking workpiece in position at low pressure and piston rod of (5) moves forward performing the work operation. When operation has been completed, operator shifts handle of valve (2) to position *B* and oil flows to rod end of both cylinders (4) and (5) and full pressure returns their pistons to starting position. Operator then shifts handle to neutral position so that the oil will bypass back to sump at no pressure, thus reducing heat and electric power consumption.

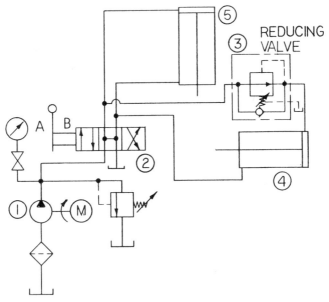

Fig. 26. Basic circuit of a system employing a reducing valve.

Figure 27 shows the cross section of a reducing valve of the direct-acting type. It is also made in the pilot type. Note the internal construction of this valve. The heavy alloy cast-iron body houses the working parts of the valve: the piston, seat and ball-check assembly. The valve cover contains the adjusting screw, spring support and spring. The valve is equipped with a drain to take care of any internal leakage. It is of the utmost importance when installing this valve that the drain plug be removed and the drain be connected to the reservoir. If this is not done, the valve will quickly malfunction.

The inlet or high-pressure port of this valve is to the left. Oil flows down around the piston (1), between the piston and the valve seat (2), underneath the piston, and on out through the low-pressure port. As the piston rises to close off the flow the main spring (3) is compressed. The load setting of the spring governs the differential between the high (inlet) pressure and the low (outlet) pressure. When the adjusting screw (4) relieves all of the load on the main spring, the inlet pressure and the outlet pressure are at their maximum differential. As the load on the spring (3) is increased, the differential begins to decrease. Pressure reductions of 10 to 1 may be effected. That is, the inlet pressure may be 1,000 lb per sq in. and the outlet

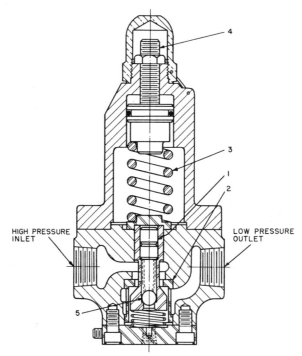

Fig. 27. Cross section of a reducing valve of the direct-acting type.

pressure 100 lb per sq in. If greater reduction in pressure is desired, then two reducing valves may be hooked up in series.

Pressure ranges for reducing valves run about the same as those of relief valves: 0 to 750 lb per sq in., 700 to 1500 lb per sq in., etc.

Note that in Fig. 27, the valve has a ball check (5) inside of the piston. If the line pressure is removed from the inlet port, the ball check opens, allowing the piston to be forced, by the large adjusting spring, to the bottom of its travel and oil can flow freely in the reverse direction. This eliminates an external check valve for producing free flow in one direction.

Reducing Valve Difficulties. Failure of reducing valves to function properly may be due to several causes:

1. *Dirt.* Again, as in other hydraulic valves, dirt is always a problem. Dirt wedging between the bore of the body and the piston will often cause the piston to stick and cause the valve to malfunction.

2. *Internal leakage.* Improper clearance between piston and valve body due to wear or leakage at the ball check seat will cause trouble.

3. *Broken springs.* Broken adjusting spring or other internal springs will disrupt the action of the valve.

4. *Back pressure.* As stated before, this valve cannot stand any appreciable amount of back pressure if suitable operation is expected. Noise in a reducing valve is often caused by the improper setting of the bleed-off needle. This needle should be adjusted as the tension on the adjusting spring is changed.

Flow Control Valves

Flow-control or speed-control valves, as they are commonly called, are used to control the volume of oil which passes a given point. The volume may be controlled as it approaches a component or as it leaves a component. For instance, in controlling the speed of the outstroke of the piston in a cylinder, some types of flow-control valves meter the oil as it leaves the rod-end cover of the cylinder while other types meter the oil before it enters the blind-end cover. Terms often used are "meter-in" or "meter-out."

Figure 28 shows a basic circuit using a flow-control valve. The circuit functions as follows: Operator places workpiece to be burnished into fixture and shifts handle of four-way, two-position valve (2) and oil flows to the blind end of burnishing cylinder (3); piston moves at speed set by flow-control valve (4) until end of cylinder stroke is reached. Operator then shifts handle of valve (2) to original position and oil flows freely through ball-check section in flow-control valve (4) and piston of cylinder retracts at rapid rate, completing the cycle. The speed of the outstroke of the cylinder can be changed by merely changing the setting of the adjusting needle in the flow-control valve.

Basically, a flow-control valve is a metering valve and a check valve built into one housing. This allows volume control in one direction and free flow in the other direction. The design of the metering valve may vary from a needle to a cam-shaped device.

Figure 29 shows the cross section of a flow-control valve of the

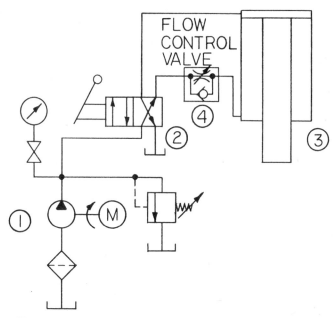

Fig. 28. Basic circuit of a system using a flow-control valve.

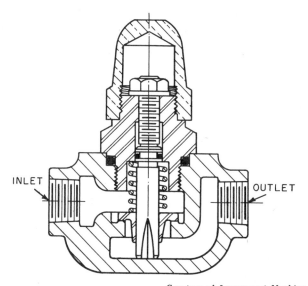

Courtesy of Logansport Machine Co., Inc.
Fig. 29. Cross-sectional view of a flow-control valve of the "meter" type.

"meter-out" type. Oil which is exhausted from the cylinder enters the left-hand port and flows to the center of the valve body, then through the V-shaped orifice in the feed needle and on out through the right-hand port, then on to the master control valve. The amount of oil which is metered through the valve is determined by how much of the V-shaped orifice extends above the top of the check valve. A very close fit is maintained between the feed needle and the check valve. The orifice is increased or decreased by screwing in or screwing out the feed needle.

When oil flow is reversed in this valve, the check valve rises on

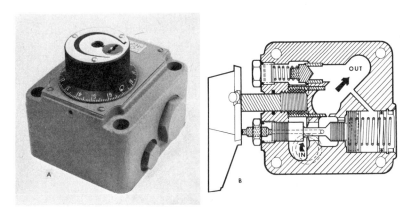

Courtesy of Continental Hydraulics
Fig. 30. Pressure compensated flow control.

the feed needle stem and oil is allowed to flow without restriction out of the left-hand port.

A mistake often made in using flow-control valves is to attempt to obtain a very minute flow. A recommended minimum flow is 5 cu in. per minute. This means that the minimum speed of a 3-inch bore cylinder will be approximately 0.7 inch per minute and on a 6-inch bore cylinder it will be 0.18 inch per minute. Flows less than that can become troublesome, especially if there is any lint or impurities in the oil which will clog the orifice.

Figure 30, view A, illustrates a hydraulic flow-control valve which is pressure compensated in order to maintain a constant flow out of the valve regardless of pressure changes to the inlet port. By using a unique, sharp-edged orifice design, the valve is made immune to temperature or fluid viscosity changes. (See approximated cross-section of valve in view B.) The dial on this valve is equipped with a lock in order to prevent tampering with the setting once it has been made.

Cam-operated, flow-control valves are of value where flow control is only required for a portion of the cylinder stroke. They can be used as a cushioning device to provide extra long cushioning for a cylinder or they can be used for slip feeding arrangements by placing multiple cams on a trip bar. Figure 31 shows the cross section of such a valve. When the cam roller is in the "up" position, oil flows freely through the valve, but when the cam roller is depressed by an

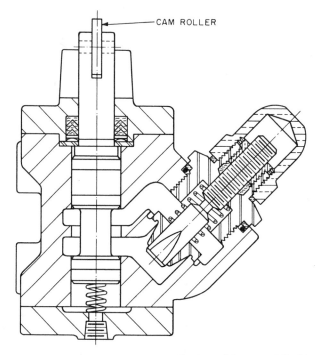

Fig. 31. Cross section of a cam-operated flow-control valve.

external mechanism, the piston closes off the inlet from the outlet and the oil must flow through the orifice in the needle. This produces the flow-control cycle.

When oil flows in the reverse direction while the cam roller is depressed, the check valve opens and allows full flow in this direction.

Special cam-operated, flow-control valves are built with metering grooves in the piston so that as the cam roller is depressed and the piston moves to close off the inlet passage, a gradual deceleration will take place. This eliminates any abrupt stoppage of flow as may be experienced in a valve where a needle arrangement is used. The one bad feature of this type of valve is that the orifice must be designed for each size of pump as there is no adjustment means available.

Flow-control valves are normally built in sizes from ¼-inch up to 2-inch pipe size. The majority are built for 1500 lb per sq in. maximum pressure but a number are available for much higher pressures.

Flow-control Valve Difficulties. When flow-control valves give trouble, it is usually due to one of the following reasons:

1. *Placed incorrectly in circuit.* One of the most common troubles is that the flow-control valve is placed backwards in the circuit. This often occurs in spite of the fact that they are usually marked with a large arrow showing the direction of the controlled flow.

2. *Dirt.* Again, dirt can be a big factor in causing this valve to malfunction. Dirt lodging in the orifice will cause it to clog up and produce very erratic feed. Dirt under the check valve seat will not allow the check to seal off thus producing considerable leakage.

3. *Change in viscosity.* A big change in oil viscosity will have a definite effect on the flow-setting, especially on a needle-type metering device.

4. *Excess leakage.* Excess internal leakage will produce poor flow control. If the leakage becomes too great, it is possible that even changing the setting on the needle from one extreme to the other will not change the flow rate appreciably.

The Air Release Valve

The air release valve as used in a hydraulic circuit is important although very small in size and simple in construction. It should be placed at the highest point in the hydraulic circuit. The valve is composed of a body and a bleeder screw which is seated in the body. When the bleeder screw is backed off from the seat and air is bled off as it passes the valve, a small amount of oil will also be expelled. When only oil—not a mixture of air and oil—bleeds off, then the bleeder screw should be seated. This valve is usually operated only when the system is put into operation or if air should get into the system after it is put into operation. Uneven feeds caused by air in the system can be eliminated by properly bleeding off the air.

Check Valves

Check valves, which are built to provide free flow in one direction and a check in the other direction, are used to prevent flow-back in hydraulic systems. For example, in a system in which the power unit is placed in the basement and the controls and cylinders are placed high above the unit, oil would drain back into the unit when the pump is shut off if a check valve were not placed in the circuit just beyond the relief valve. Oil from the pump flows freely through the check valve but the oil between it and the components cannot back through the valve.

Check valves are of several styles: swinging disc, ball, plunger, and poppet. Check valves of the pilot-operated type are also popu-

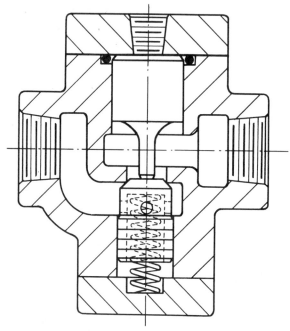

Courtesy of Double A Products Co.

Fig. 32. Pilot-operated check valve.

lar. Figure 32 shows a pilot-operated check valve. The connection at the bottom is connected to a remotely operated pilot valve. As in the ordinary check valve, free flow is permitted in one direction when the line pressure lifts the plunger thus opening the valve. The flow is checked in the opposite direction until pilot pressure is applied to the blind end of the plunger through the pilot connection. This causes the plunger to unseat the valve spool and oil is allowed to flow through the valve.

Leakage around the valve seat can be troublesome. To prevent

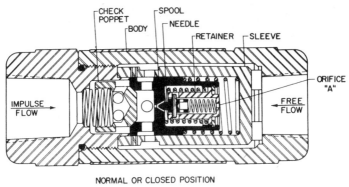

NORMAL OR CLOSED POSITION

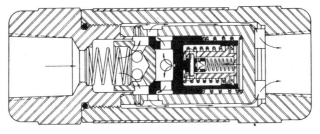

FLOW STARTING THROUGH VALVE

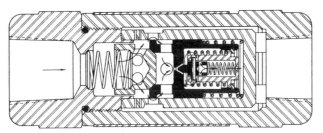

POSITION DURING RESETTING OF VALVE

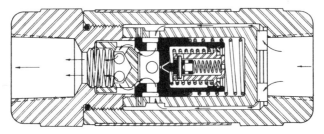

REVERSE FLOW POSITION

Courtesy of The Denison Div., Abex Corp.

Fig. 33. Four operating positions of a surge dampening valve.

leakage, many of the high-pressure valves have metal-to-metal contact at the seat and the parts are lapped to a perfect seal.

The Surge Dampening Valve

The vibrations caused by the shock of a hydraulic surge can be detrimental to the whole hydraulic system. As pressures are increased and oil flows are more quickly reversed, such shocks become more intense. The surge dampening valve is designed to prevent the damaging shock impulses at their formation. It allows only a gradually accelerated flow of fluid into the outlet line leading from the valve if hydraulic pressure is suddenly applied to the inlet of the valve, thus preventing shock or surge pressure in the outlet line and any device connected to the outlet line. The design of this valve is such that the rate of opening of the valve will be slower if a high pressure is applied to the inlet than if a low pressure is applied. After permitting the gradually accelerated flow from the inlet to the outlet line, the valve establishes a virtually free connection between the inlet and outlet which is maintained as long as the fluid continues to flow through the valve. Figure 33 shows four stages in the operation of a surge dampening valve.

Pressure Switches

While pressure switches could hardly be classed as valves, they do play an important role in the design of fluid power circuits. Pressure switches are used in conjunction with solenoid-operated valves and are very useful in setting up semi-automatic and automatic cycles. They eliminate the use of trip-dogs, limit switches, and other reversing devices. Figure 34 shows the external and internal views of a pressure switch. When the pressure of the oil in contact with the metal plunger is sufficient to overcome the pressure exerted by the spring, the electrical contacts are either open or closed—depending upon the type of switch. For instance, on a press circuit, the ram is started on its downward movement and when it makes solid contact with the work the pressure builds up causing the piston in the limit switch to rise, making the electrical contact. This in turn operates the master valve through a solenoid and the oil flow in the master valve is reversed so that it flows to the rod end of the ram and the ram retracts. The pressure at which the pressure switch functions depends upon the tension applied to the spring.

Combination type valves find many uses in hydraulic circuitry. A good example of such a valve is shown in Fig. 35. This is a directional control valve with automatic sequence control. The sequenc-

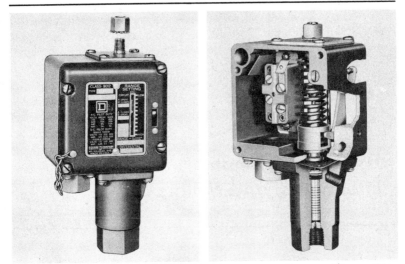

Courtesy of Square D Co.

Fig. 34. External and internal views of a pressure switch.

ing feature automatically causes reversal in cylinder piston travel at a pre-set pressure level.   Need for a pressure switch in the control circuit is thereby eliminated.   The diagram at right shows a circuit with this control.

Other valves which find applications in a hydraulic system are shuttle valves, gate valves, globe valves, in-line check valves, needle valves, and special valves of various types.   New special-purpose valves are being designed every day for some particular application.   Many of these would be of little value for general use.

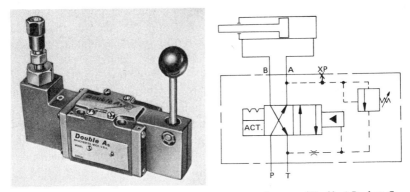

Courtesy of Double A Products Co.

Fig. 35. (*Left*) Directional control valve with automatic sequence control, and circuit (*Right*).

CHAPTER 7

Hydraulic Cylinders, Intensifiers, and Motors

Hydraulic cylinders receive the fluid pressure from the hydraulic lines and transmit it into lineal force. These cylinders are often referred to as jacks or rams and may be divided into two distinct types: the nonrotating type and the rotating type.

Nonrotating-Type Cylinders

The various designs of hydraulic cylinders cover such a wide range that it would be impractical to describe all of them. However; the basic elements of a nonrotating cylinder are a cylinder tube, piston, piston rod, covers, and packing. The cross section of a typical heavy-duty nonrotating hydraulic cylinder is shown in Fig. 1.

Cylinders of the nonrotating type may be classified into the *single-acting* type and the *double-acting* type. The single-acting type delivers force in only one direction, while the double-acting type delivers force in both directions. The single-acting cylinder may be either of the spring-return type or gravity-return type. Where a spring is used, it may be located either internally or externally.

In selecting a nonrotating cylinder several factors must be considered. One of these factors is design. Hydraulic cylinders range in price from a few dollars to many thousands of dollars depending on the bore size, stroke, and construction. It should always be remembered that an inexpensive cylinder may be extremely expensive to the user in the over-all picture. For example, a breakdown of a hydraulic cylinder on a high speed rolling mill in a steel mill can amount to several thousand dollars in lost production if repairs are extensive. A breakdown of a cylinder on a production machine can stall a whole production line. These examples are not given to confuse the reader by implying that a great amount of difficulty is experienced with hydraulic cylinders, but it is to make him aware that a proper selection must be made when choosing a cylinder for a particular application, especially those which are to be used in critical operations.

Another factor to be considered is the availability of replacement parts. It can be quite expensive to the user if replacement parts are not readily available. For instance, if a cylinder cover should break, which is not a frequent occurrence, and the manufacturer of the cylinder has none in stock, the user's equipment may be out of production for several weeks waiting for a replacement cover. Another example would be the inability of the user to obtain spare packings or other spare parts when he has a breakdown. Still another thing that should be watched is model obsolescence. If the user has designed his machine around a particular model of cylinder which sub-

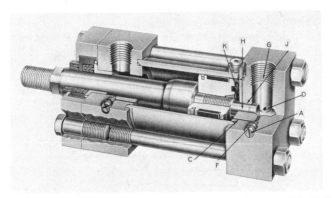

Courtesy of The Sheffer Corp.

Fig. 1. Cutaway section of a typical, heavy-duty non-rotating hydraulic cylinder which conforms with NFPA standards and specifications.

sequently becomes obsolete, the difficulty in obtaining repair parts may cause a very serious situation. It may also mean that the user must redesign a complicated machine for future operations.

The inability to obtain replacement parts, however small, can cost considerable production time, which may amount in dollars and cents to many times the price of the cylinder or its replacement parts.

Other factors to be taken into consideration are application factors such as frequency of operation, expected service life, temperature conditions, and other general operating conditions. The application factors are ones which should be given careful consideration. Users seem to be reluctant to give the manufacturer of cylinders all the facts regarding an application which often results in the manufacturer furnishing the incorrect cylinder and the user being dissatisfied with its performance.

Nonrotating Cylinder Tubes

The application dictates pretty much the design and material of the cylinder tube. Material available for cylinder tubes are centrifugal cast or plain cast iron, bronze, or steel; cast aluminum; and drawn seamless steel, stainless steel, brass, or aluminum. Steel tubes are often chromium plated in order to provide resistance to wear and corrosion.

In order to resist internal leakage the bore of the cylinder tube must be honed or ground to a smooth finish of from 10 to 50 micro-inches, rms, depending upon packing used, and it must be held to close tolerances from one end to the other. For example, thin-walled tubing often becomes egg-shaped after honing, which is not true of the heavy-walled cast tubes.

A new cold-forming process for cold shaping steel under compression may open a whole new field for manufacturers of hydraulic

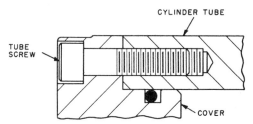

Fig. 2. Method of fastening tube to a cylinder cover employing the use of tapped holes and machine screws.

cylinders. These tubes can be formed with one cover integral with the tube, itself. The finish of the bore and the finish on the outside diameter are in the order of 50 micro-inches, rms. Concentricity is held to within 0.005 inch.

At present this process is designed for manufacturers who make large quantities of cylinders of the same size. Prices on small quantities would be prohibitive due to die costs.

There are several methods of fastening the tube to the cylinder covers. Figures 2, 3, and 4 (left) are typical methods. Figure 2 shows a tube which is tapped for the cover screws. While this method helps streamline the appearance of the cylinder, it has the disadvantage of being a costly operation on long cylinder tubes. It is often necessary to construct a pit in the floor beneath the machine which

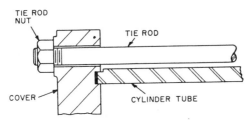

Fig. 3. Method of fastening tube to a cylinder cover using tie rods.

drills and taps the cover screw holes so that the long tubes can be properly handled. Secondly, if a tube screw should break during disassembly, due to corrosion, or, should the thread in the tapped hole be stripped, considerable down time may be involved.

Figure 3 shows the use of tie rods, which is one of the oldest and least expensive methods. When using tie rods on high-pressure cylinders, elongation must be taken into consideration and the sealing means should be designed accordingly. From an appearance standpoint, it will be noted that tie rods do not add to the present-day streamlining of equipment.

Figure 4 (left) shows the tube-ring construction, which is more or less standard on heavy-duty mill-type cylinders. The cover screw holes in the tube ring are drilled after they are welded to the tube. These holes are clearance holes for bolts or studs. The advantage of the tube-ring construction is that if the nuts and bolts should become rusted due to their location in a mill, the cover can be removed by using a sledge hammer and cold chisel to cut off the bolts, yet the cylinder cover or tube will not be damaged.

When using gasket-type seals between the tube and cover, the tube ring is set back a short distance from the end of the tube, as shown in Fig. 4 (left), so that the end of the tube is seating against the

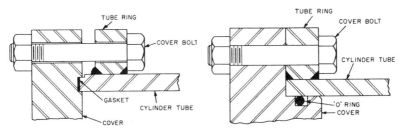

Fig. 4. Tube-ring method of fastening tube to a cylinder cover. (*Left*) Standard arrangement for heavy-duty mill-type cylinders. (*Right*) Variation employing an "O" ring which is commonly used on high-pressure cylinders.

gasket. Care must be taken not to tighten the cover screws excessively as the leverage exerted on the end of the tube may tend to bow it in. On high pressure cylinders, the end of the tube is usually flush with the tube ring, as shown in Fig. 4 (right), and the static seal is made with an "O" ring. With this arrangement, pressure from within or without will not cause the tube to flex. Tube rings certainly do not help streamline a cylinder; however, this type is usually placed in locations where function is more important than appearance.

Nonrotating Cylinder Piston and Piston Packing

The design of the cylinder piston is dependent upon the type of packing to be used. The piston material of an oil hydraulic cylinder

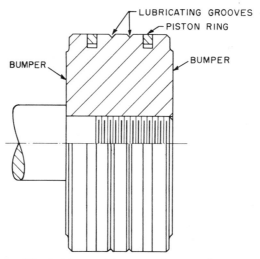

Fig. 5. Piston with automotive-type rings.

is usually cast iron or steel with a bronze facing, while in a water hydraulic cylinder it will be bronze. The bearing qualities of cast iron make it an ideal piston material, especially where piston rings are employed. Figure 5 shows a typical piston with automotive-type rings. Notice the depth of the ring grooves and also the lubricating grooves. Pistons may be designed with as few as two rings and as many as six or eight rings depending upon the length of piston and the operating pressure. Some pistons are designed to be exceptionally long so as to act as a bearing to counteract side thrust on the end of the piston rod.

The bore of the piston is precision machined so that it closely fits the pilot on the piston rod. The piston should be finish-ground after it has been assembled to the rod so that it will be concentric with the rod.

Note the extended section of reduced diameter on each end of the piston. These are called *bumpers* and are used to keep the piston from squeezing and binding the rings as a result of the piston making repeated contact with the cylinder covers at each end of its stroke.

The extended bumper sections also help to keep any small particles of impurities from lodging under the ends of the piston at either end of the stroke. The outside diameter of the bumper is slightly less than the bottom diameter of the ring groove so that the bumper acts against a solid section all the way through the piston.

Pistons employing piston rings cannot be considered leakproof; however, by proper fitting, leakage can be held to a minimum. Some of the advantages of the ring-type piston are: extremely long life, ease

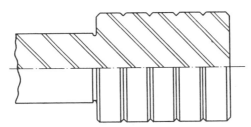

Fig. 6. Piston without any packing or seal.

of servicing, suitability for high temperatures, long bearing surface, and low cost. Pistons of this type are standard with nearly all manufacturers.

Figure 6 shows a piston without any packing or seal. Pistons of this type are honed and lapped to fit the cylinder tube and are usually hardened to a Rockwell C of 55 to 60 as any slight wear would allow excessive leakage. Because of its design, this piston operates in its cylinder with very little friction. Although there are not too many applications for this type of piston, it should be considered where high pressures are to be handled.

Figure 7 shows a piston design which is used on many low-pressure hydraulic applications and, with proper design and packings, can also be used on high-pressure applications. The piston is made up of the center section, two metal retainer rings, two cup packings, and the retainer screws. The shoulder on each end of the center section around which the cup packing fits should be about ten per cent less in axial height than the thickness of the back of the cup. This permits precompression to be applied to each cup by its retainer ring to prevent leakage between the center section and the retainer but prevents the cup from being squeezed so tightly that the lips will turn in. The retainer rings absorb the shock as the piston assembly strikes the covers. Single-acting pistons require one cup packing and one piston ring.

The cup-packing material may be leather, synthetic rubber, or a fabricated composition. (See Chapter 13 on packing.)

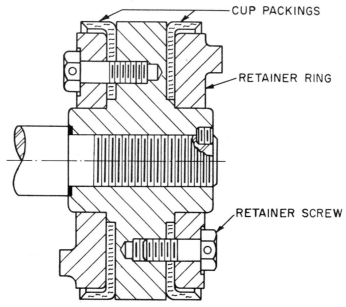

Fig. 7. Piston design which is used on many low-pressure hydraulic applications. It can also be used on high-pressure applications with proper design and packing.

In a cup-packing type piston the bearing between the piston assembly and tube is in the center section. The length of this section depends on the bearing requirements.

Factors which contribute to failure of this type of packing are clearance between the center section of the piston and the cylinder tube, improper support or backup, heat, and poor finish on the cylinder tube.

While Fig. 7 only shows one basic design of a cup-packing type piston, many designs have been established.

Figure 8 shows a piston of the chevron or "V"-type packing design. This principle is, in effect, similar to that of the cup-packing type except there are a number of sealing lips instead of only one. Chevron-type pistons are usually designed for high-pressure applications where leakage is a critical factor. Again, as in the cup packing, the bearing is provided for in the center section and the sets of packing are held in place by retainers. In order to use this type of packing, tolerances on the tube internal diameter must be closely held. The number of chevrons is determined by the operating pressure. (See Chapter 13 on packing for details.)

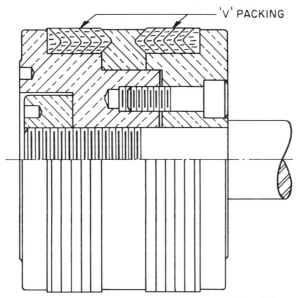

Fig. 8. Piston of the chevron or "V"-type packing design.

The center section of the piston should be designed so that when the retainer rings are pulled into place, there will be a slight compression against the chevrons. Too much pressure, however, destroys the effectiveness of the multi-lip seal and causes it to act as a one-piece solid seal.

Figure 9 shows a piston in which an "O" ring is used as a packing seal. The "O" ring is one of the newer seals, having come into prominence during World War II. The advantages of the use of "O" rings are:

1. The piston is inexpensive; only one groove is required.

2. The seal is effective in both directions, which allows a thinner piston to be used.

3. The working pressure can be up to 3000 lb per sq in. (when using backup washers).

4. The seal is easy to replace. There is no need to disassemble the piston as is required on most other seals.

Caution must be used in providing proper clearance between tube and piston, otherwise extrusion of the seal into the clearance space may result. (For details on "O" rings see Chapter 13 on packings and seals.)

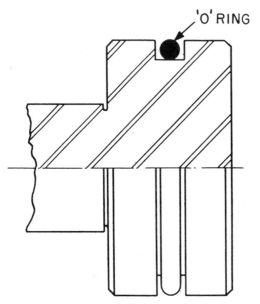

'O' RING

Fig. 9. Piston with an "O" ring as a packing seal.

Pistons are also designed with other types of packing such as "U" rings, packaged sets of metallic seals, and other formed packings. New developments to provide a better seal between the piston and cylinder tube are constantly under way.

Piston Rods

The piston rod is generally thought of as a section of ground and polished steel shafting threaded at both ends. Actually, in many applications there are factors which necessitate a wide variation in the design of piston rods. For example, in applications where moisture or acid fumes may be present, a ground or polished plain steel piston rod may become corroded in a very short time, thus ruining the rod packings. By using a stainless steel piston rod or a chromium-plated plain steel rod, the corrosive atmosphere will have little or no effect.

There are many applications where large quantities of abrasives are present. If a plain steel rod were used, the abrasives would soon ruin it. A hardened and ground rod with a hardness of, say, 55 to 60 Rockwell C would, in all probability, be a solution to the problem. Another solution which has found considerable use is the placing of a protective covering over the rod in the form of a "boot." This "boot" is made of either a soft leather, a duck material impregnated with synthetic rubber, or of synthetic rubber. The first and second types are fabricated in a form similar to that shown in Fig. 10 and

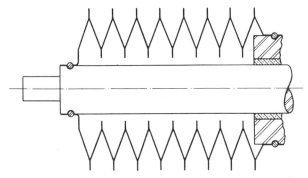

Fig. 10. Cross-sectional sketch of a "boot" which protects a moving piston rod from foreign abrasive matter.

are very short when compressed but when extended can cover quite a range. Since these are fabricated, there is really no limit as to the number of convolutions available. The third type is molded and is limited in length to the size of the molding press available. When using "boots" it is necessary to allow enough rod extension on the cylinder to compensate for the collapsed length of the "boot." Another method of keeping the piston rod clean is to use a wiper, either of metal or of synthetic rubber. The function of the wiper is to remove the foreign materials from the rod before they enter the packing.

Whatever material is used in making piston rods, it should have both high tensile and compressive strengths. It should also be able to withstand sudden shock loads.

Diameters of piston rods of cataloged cylinders are pretty much standardized by the various cylinder manufacturers. Manufacturers list a standard size rod and a "2-to-1" rod. The latter is approximately one-half the area of the bore size of the cylinder. This type rod is used on applications where added strength is required or where a rapid return stroke is desired, such as on large presses.

Piston rods may be either solid or hollow. Solid rods are usually more easily machined due to the stability of the material but its weight may be bothersome. For example a piston rod, eight inches in diameter by twelve feet long weighs approximately 2050 pounds. In a cylinder mounted horizontally this weight may cause trouble in the packings and bearings unless given careful consideration when designing the remaining parts of the cylinder. The end of the rod must be properly supported at all times. If a hollow rod is to be used for a double-acting cylinder, both ends are usually closed by first inserting threaded plugs and then welding around the periphery of each. One end will have provision for connecting the piston and the other end will have an adaption for the equipment to which it is to be attached. The rod will be ground and polished after the welding operation.

On applications which require the piston of the cylinder to come to a stop without a shock, piston rods are designed with cushioning sections. The length of these sections depends upon the amount of load to be stopped. Often cushion sections are six or eight inches long, but on normal applications they are usually approximately an inch long. These sections must be precision finished to a close fit with the cover bore in order to hold leakage to a minimum.

Cylinder Covers

Standard cylinder covers are designed with mountings which can be adapted to nearly all fluid requirements. Standard mountings available are rabbeted, foot, trunnion, centerline, front flange, blind flange, and clevis. Combinations are also available such as a cylinder with a flange mounting on both ends or one with foot mounting on one end and flange mounting on the other. Some cylinders are designed with one cover welded to the cylinder tube while still others have both covers welded to the tube. However, neither of these two types have wide application in industrial plants. Most of these are used on road machinery, farm equipment, and other outdoor applications.

Cylinder cover material is usually either a high tensile strength cast iron or steel. If steel is used, the area that is in contact with the piston rod should be of a good bearing material, such as bronze, cast iron, etc.

The sealing section on the cover may be one of two types. The cover may be recessed as shown in Fig. 3 and contain a gasket, either of a paper or a synthetic or metallic material. This surface must be square with the bore and cannot be very wide unless it is scored. The other means of sealing is shown in Fig. 2 in which an "O" ring is used as a static seal. The big advantage of this type is that the cylinder cover screws need only to be pulled down snugly to complete the sealing. When using the first type, considerable stress must be set up between the cover and tube just to effect a suitable seal. This method is usually used for low-pressure sealing while the "O" ring type is used for high pressures. Figure 1 illustrates another method.

Cylinder covers are either of the cushion type or non-cushion type, and are designed accordingly. The cushion type contains the cushion needle assembly and the ball check assembly. The bores in the covers may be the same for either the cushion or non-cushion covers. Figure 1 shows a cylinder which is cushioned at both ends. As the cushion nose H enters cushion recess J, oil is trapped in section K and its only means of escape is through orifices C and F. The rate of escape is determined by the setting of cushion needle A. This slows down the movement of the piston and cushions it at the end of the piston's retracted stroke.

When oil is directed from the line to port G it moves quickly through orifice D, unseats ball check (not shown) and moves into section K. The oil under pressure contacts the end area of the piston causing a full-force start of the piston on the forward stroke. If it were not for the ball check, the starting area would only be on the end of the cushion nose and if the piston were pushing a heavy load, the pressure placed on the end of the cushion nose would not be sufficient to make the piston move, thus causing a stall. There are other designs which also allow for a full-force startup.

Pipe ports in the cylinder covers should be large enough to allow passage of oil to the piston with a minimum of frictional loss. Care must be exercised especially on cylinders with "2-to-1" rods. Only a small amount of oil is required on the rod end to start the piston moving at a rapid rate. However, if the pipe port on the blind end is not of sufficient size to exhaust the large amount of oil behind the piston, considerable back pressure will be built up slowing down the piston movement. This is a fact that is often overlooked in a fluid power system layout.

The rod-end covers are designed with built-in stuffing boxes for retaining the rod packings. The stuffing box design depends on the type of rod packing used. Figure 1 shows a modified V-type packing. Note that the packing gland cap is pulled against the front cover so that no external take-up on the packing is possible. V-type packings do not generally require gland pressure since, due to their design, they are under slight radial compression when mounted properly. For details on rod packing, see Chapter 13.

The applications for nonrotating hydraulic cylinders are practically unlimited. Their sizes are only limited by the manufacturer's ability to produce them. Large cylinders are often made with the cylinder tubes and piston rods in sections. In order to manufacture large cylinders, large capacity boring, grinding, turning, and honing machines are required. Consider the type equipment which would be required to manufacture an 18-inch bore by 24-foot stroke hydraulic cylinder having a 12-inch piston rod.

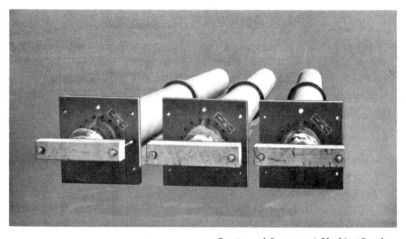

Courtesy of Logansport Machine Co., Inc.

Fig. 11. Large low-pressure hydraulic cylinders of 14-inch bore by 237-inch stroke which have a 7-inch diameter piston rod. Note the protective covers placed over the ends for shipping purposes.

Figure 11 shows large low-pressure hydraulic cylinders of 14-inch bore by 237-inch stroke and having a 7-inch diameter piston rod. Note the protective covers over the ends for shipping purposes.

Figure 12 shows heavy-duty cylinders for steel mill use. Note bolts through covers for connecting covers to tube. Also note center support which helps keep heavy tube from sagging. These cylinders have a 12-inch bore by 18-foot stroke. Their operating pressure is 1500 lb per sq in.

Courtesy of Logansport Machine Co., Inc.

Fig. 12. Heavy-duty cylinders for steel mill use. Bolts connect covers to tubes. Note center support which helps to keep the heavy tube from sagging.

Figure 13 shows a heavy-duty cylinder with fabricated steel covers. Note the provision for pipe flange connection at the cover ports instead of threaded pipe connections. These cylinders are used for high operating pressures.

Figure 14 shows a specially built hydraulic cylinder having a male type trunnion so that the cylinder can be mounted to oscillate as required. Note that the pipe connections are in the trunnion which affords easier piping.

A long stroke double-acting hydraulic cylinder with a flange mounting on the piston rod end is shown in Fig. 15. Note how the

Fig. 13. Heavy-duty cylinder with fabricated steel covers which is used for high operating pressures.

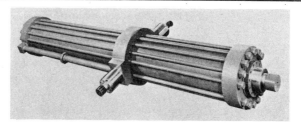

Courtesy of Lynair, Inc.
Fig. 14. Specially built hydraulic cylinder having a 16-inch bore and 133-inch stroke with trunnion mounting and heavy-duty tie-rods.

piping is run from the blind cover to the front cover flange. A setup of this type is often employed if the cylinder is to be mounted in a pit where it is difficult to make a pipe connection to the blind cover after the cylinder is installed. This cylinder has a 10-inch bore by 240-inch stroke.

A hydraulic cylinder with tie-rod construction is shown in Fig. 16. This type of cylinder is used for pressures up to 5000 psi, non-shock. Note the wrench flats on the piston rod. Cylinders of this type are mass produced at a competitive price. See Charts 1 and 2, pages 7-18 and 7-19, for determining cylinder size. See Table 2, in Chapter 16, for piston-rod-diameter recommendations.

Rotating Cylinders

There are not nearly as many applications for rotating cylinders as there are for nonrotating cylinders. Their applications, however, are very important in industrial processes. They are used on metal-removing machines such as lathes, grinders, and honing machines, on coilers and uncoilers for feeding strip stock through mills, and on any other heavy-duty application requiring a cylinder with a stationary distributor housing and a rotating distributor shaft.

Basically, the design of a nonrotating cylinder and that of a rotating cylinder are about the same except that the rotating cylinder has a distributor. Figure 17 shows a distributor for a rotating hydrau-

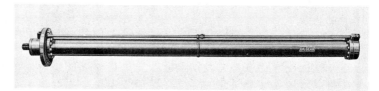

Courtesy of The Oilgear Co.
Fig. 15. A heavy-duty double-acting hydraulic cylinder with rod end flange mounting.

lic cylinder. The oil shaft stem (3) is made of hardened steel and is drilled for the oil passages which feed both ends of the cylinder. The oil shaft body (2) is often referred to as the distributor housing and contains the pressure ports as well as the drain port. The seal between the oil shaft and the oil shaft body is effected by quad rings (12). Bearing (4) allows the oil shaft stem (3) to rotate

Courtesy of Logansport Machine Co., Inc.

Fig. 16. Hydraulic cylinder with tie rod construction which is used for higher operating pressures.

freely within body (2). An oil film is also present to provide lubrication between (2) and (3). "O" rings (8) provide a seal between the oil shaft stem and cover of cylinder. Drain port must be open at all times so as not to place any back pressure on distributor assembly. The distributor assembly is suitable for speeds up to 3600 rpm and pressures up to 500 lb per sq in. Special distributors can be designed for much higher operating pressures.

Difficulty is often encountered when designing rotating cylinders with a hole through the center for bar feed applications, the difficulty being that the packings available do not readily lend themselves to

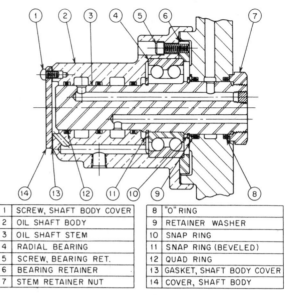

1	SCREW, SHAFT BODY COVER	8	"O" RING
2	OIL SHAFT BODY	9	RETAINER WASHER
3	OIL SHAFT STEM	10	SNAP RING
4	RADIAL BEARING	11	SNAP RING (BEVELED)
5	SCREW, BEARING RET.	12	QUAD RING
6	BEARING RETAINER	13	GASKET, SHAFT BODY COVER
7	STEM RETAINER NUT	14	COVER, SHAFT BODY

Courtesy of Logansport Machine Co., Inc.

Fig. 17. A rotating cylinder and distributor.

Courtesy of Jones & Lamson Machine Co.

Fig. 18. Rotating hydraulic cylinder mounted on a lathe. Note flexible connections to distributor on rotating cylinder.

CHART 1

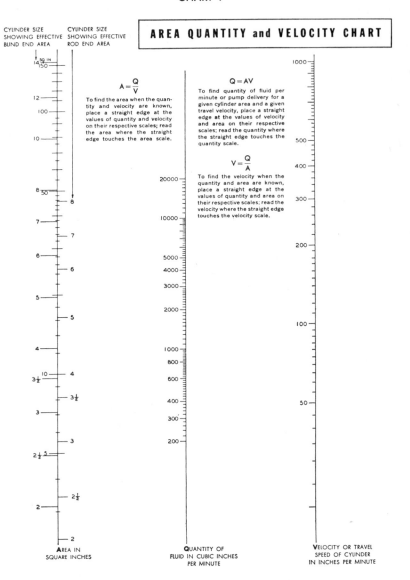

CYLINDER SIZE
SHOWING EFFECTIVE
BLIND END AREA

CYLINDER SIZE
SHOWING EFFECTIVE
ROD END AREA

AREA QUANTITY and VELOCITY CHART

$$A = \frac{Q}{V}$$

To find the area when the quantity and velocity are known, place a straight edge at the values of quantity and velocity on their respective scales; read the area where the straight edge touches the area scale.

$$Q = AV$$

To find quantity of fluid per minute or pump delivery for a given cylinder area and a given travel velocity, place a straight edge at the values of velocity and area on their respective scales; read the quantity where the straight edge touches the quantity scale.

$$V = \frac{Q}{A}$$

To find the velocity when the quantity and area are known, place a straight edge at the values of quantity and area on their respective scales; read the velocity where the straight edge touches the velocity scale.

AREA IN
SQUARE INCHES

QUANTITY OF
FLUID IN CUBIC INCHES
PER MINUTE

VELOCITY OR TRAVEL
SPEED OF CYLINDER
IN INCHES PER MINUTE

sealing a revolving shaft under fairly high pressures. What happens is that the friction caused by the seals, particularly at high speeds, creates so much heat that the seals will break down. A water cooling jacket built into the distributor housing is helpful but is quite an

CHART 2

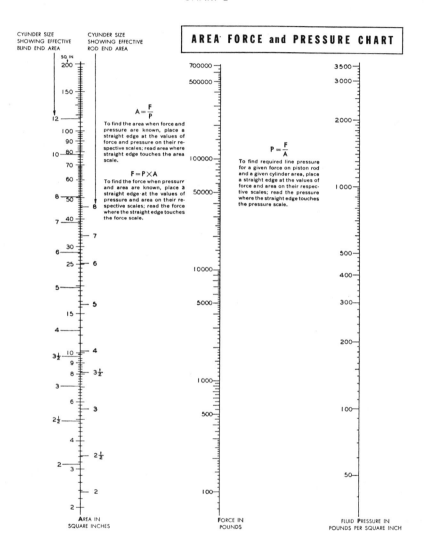

added expense and also increases the weight. Formed metallic packing appears to be the best approach to the problem.

Figure 18 shows a rotating hydraulic cylinder mounted on a lathe for actuating a chuck. In Fig. 19 a rotating hydraulic cylinder is shown mounted on the spindle of a roll straightener for actuating the power chuck.

Courtesy of Industrial Metal Products Co.

Fig. 19. Rotating hydraulic cylinder *A* actuates the power chuck *B* on a roll straightener which processes shafts and tubes of various configurations. Many hydraulic components are employed on this machine.

Intensifiers

As shown in Fig. 20, an intensifier consists of two cylinders of different sizes with a common piston, assembled as a unit. Oil is admitted under pressure to the larger cylinder and exerts a force on the large end of the piston. Neglecting losses due to friction, the smaller end of the piston exerts the same force on the hydraulic fluid in the smaller cylinder or intensifier chamber. Since the area of the end of the piston in the intensifier chamber is smaller than that end in the larger cylinder the pressure exerted by the smaller piston is increased or "intensified."

If P_o = pressure exerted on large or "operating" end of piston; A_o = area of this end of the piston; P_i = pressure exerted by small or "intensifying" end of the piston; and A_i = area of this end of the piston, then

$$P_o \times A_o = P_i \times A_i$$

or

$$P_i = \frac{A_o}{A_i} \times P_o$$

In other words, the pressures in the two chambers vary inversely as the areas of the piston ends operating in them, neglecting frictional losses, and the intensifying pressure ratio is the ratio of the area of the large end to that of the small end. The volume of oil which can be furnished at the intensified pressure is equal to the area of the small end of the piston times the stroke. The operating end of the intensifier must, of course, be a double-acting cylinder in order that the piston can be retracted after each stroke.

Intensifiers often have pressure ratios ranging up to 50 to 1, so that they find most useful applications where the pressures required are

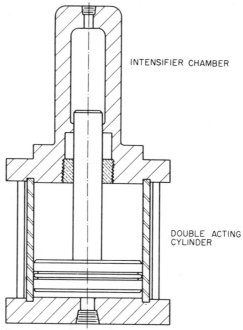

INTENSIFIER CHAMBER

DOUBLE ACTING
CYLINDER

Fig. 20. Intensifier unit consisting of two cylinders of different sizes with a common piston.

much higher than are obtainable from a pump, as in high-pressure testing machines, for example.

Intensifiers may be either hydraulically or pneumatically operated, the advantage of the latter being that no power unit is required since the air can be taken from the general supply or storage tank.

Figure 21 shows a basic circuit with an intensifier in a punching operation. The circuit functions as follows:

1. Operator places workpiece in fixture and shifts handle of four-way, two-position valve (1) and oil flows to blind end of clamp cylinder (2) and its piston moves forward locking workpiece in fixture. Pressure then builds up opening sequence valve (3) and oil flows to blind end of intensifier cylinder (5) and oil pressure forces the piston forward directing intensified pressure to single-acting spring-return punch cylinders (6), (7), (8), (9), and (10) causing the punches to perform operation on workpiece.

2. Operator then shifts handle of valve (1) back to original position and oil flows to rod end of intensifier cylinder (5) returning

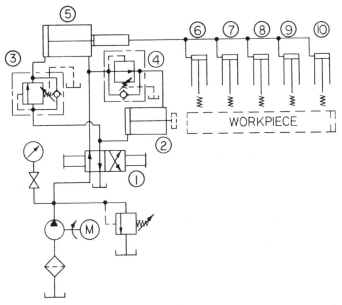

Fig. 21. Basic circuit for use in a punching operation which employs an intensifier.

piston to original position. Springs return the punch cylinder pistons. Oil pressure builds up opening sequence valve (4) and oil flows to rod end of clamp cylinder, opening clamp.

3. Operator unloads finished workpiece and reloads for next cycle.

In this circuit the punch cylinder pistons must be leakproof or as nearly so as possible. Also the packing in the rod end of the intensifier cylinder must keep the intensified pressure from leaking to the low-pressure side. All air in the lines between the intensifier and the punch cylinders should be bled out as quickly and completely as possible. For air-oil boosters, see Chapter 16.

Hydraulic Motors

Hydraulic motors are now playing an important role in industrial applications. They convert hydraulic power from the hydraulic lines into reversible, variable-speed, and rotary mechanical power. Applications for hydraulic motors are found on pottery machines, winding machines, machine tools, drilling rigs, mining equipment and on many other types of machinery.

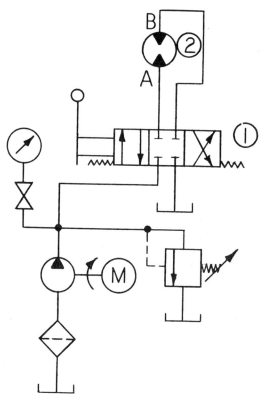

Fig. 22. Hydraulic motor circuit for unwinding and rewinding a hose.

The advantages of this type of motivation are numerous:

1. Stepless variations in speeds are available from zero to that obtainable from the maximum output of the pumping system. A very simple adjustment controls the complete range.

2. This type of drive is explosion proof. It requires no electrical connections—only two hydraulic lines.

3. It is capable of sudden stops without harm to the mechanism. By closing the control valve, the motor immediately stops.

4. Frequent starting, stopping, and reversing are all controlled by a single four-way, three-position control valve. The operator has complete control at his fingertips.

5. It is suitable for drives in machines in which controlled variable torque is required, such as on paper winding machines.

6. It is advantageous for drives on machines requiring controlled maximum torque output such as thread-tapping equipment. It has also proved especially suitable for index table drives, since any outside resistance will immediately cause the hydraulic motor to stall.

The design of hydraulic motors is closely related to hydraulic pumps. In fact some pumps can be used as motors. The three principal types are vane, gear, and piston. Hydraulic motors are very smooth in operation.

Figure 22 (Circuit No. 2) shows a basic layout of a hydraulic motor circuit, which operates as follows:

1. Operator shifts handle of four-way, three-position valve (1) to first position and oil flows to port A of hydraulic motor (2) and motor turns hose reel, unwinding hose. When enough hose is unwound, operator releases valve handle and spool in valve returns to the center position, blocking the oil flow and immediately stopping the hydraulic motor.

2. When operation is finished, the operator shifts handle of valve (1) to second position and oil flows to port B of motor (2) which now runs in the reverse direction and hose reel winds up hose. Operator releases valve handle when hose is rewound and stops motor, completing cycle. By addition of flow control valves, he can control speed of unwinding and winding.

Figure 23 shows a hydraulic hoisting winch for use on mobile cranes, ship cranes, lifting machines, and special handling applications. The winch has a heavy-duty, radial piston type motor which is close-coupled to the drum. The winch has a safety valve which is shown in the lower-right-hand corner of the photograph. This valve controls the speed of lowering (the safety of the load), limits the maximum pressure (safety of the system), and provides automatic actuation of the fail-safe brake.

Figure 24 shows a hydraulic winch circuit with a winch safety valve, and the hydraulic power source which includes a directional control valve.

Figure 25 shows the circuit for a hydraulic winch without a winch safety valve. The winch safety valve is incorporated in the direc-

Courtesy of Poclain Hydraulics

Fig. 23. Hydraulic winch equipped with winch safety valve.

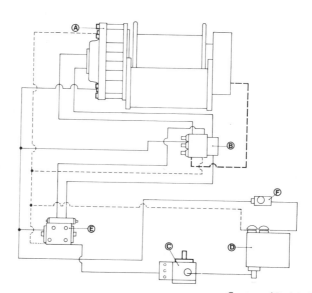

Courtesy of Poclain Hydraulics

Fig. 24. Hydraulic winch circuit with a winch safety valve at winch. *Winch and Valve Assembly*: Winch *A*; Winch safety valve *B*. *Hydraulic Power Source*: Pump *C*; Reservoir *D*; Directional valve (to control raise, lower, and stop) *E*; Relief valve *F*.

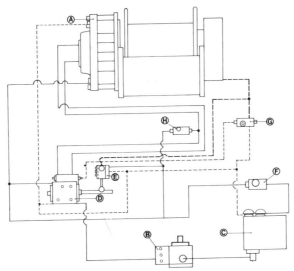

Courtesy of Poclain Hydraulics

Fig. 25. Hydraulic winch circuit with winch alone. *Winch Alone*: Winch *A*. *Hydraulic Power Source*: Pump *B*; Reservoir *C*; Directional valve incorporating winch safety valve *D*; Brake selector *E*; Relief valve *F*; Safety check valve *G*; Overload relief valve *H*.

tional control package of the hydraulic power source. Also in the system are: a safety check valve, an overload relief valve, and a brake protector.

As designers become more aware of the many advantages of hydraulic motors, their applications and use will greatly increase in all types of machines.

CHAPTER 8

Heat Exchangers
for Hydraulic Systems

Heat and dirt are the two trouble makers in a hydraulic system. A question which often arises is "What causes heat in a hydraulic system?" Heat in the system may be caused by two things—friction and external temperature. A large volume of high-pressure oil spilling through the relief valve or passing through long lengths of piping is bound to create excessive heat. Very high external temperatures will transfer considerable heat into a hydraulic system. For example, a hydraulic system which is used for feeding and controlling the opening and closing of furnace doors is subjected to extensive heat. Heat transferred to the piston rod of the pusher cylinder for a furnace will, in turn, be transferred to the hydraulic oil. Other hydraulic installations where high external temperatures are present are on blast furnaces, rolling mills, forging equipment, test chambers, etc.

Heat causes hydraulic oil to become less viscous and also causes it to break down. Heat will cause packings to leak, valves to malfunction and pumping equipment to lose its efficiency. Packings will become brittle, especially if they are made of synthetic material or leather. The close fitting precision parts of valves often seize when excessive temperatures are present. Oil with very low viscosity will cause considerable slippage in the pump with consequent reduction in efficiency and maximum output pressure.

It is generally agreed that the operating temperature (temperature of oil in the reservoir) should not exceed 140°F. Even somewhat lower temperatures are desirable. It is easily seen that if the external temperature were 130°F, the internal temperature of the unit would quickly go beyond a safe operating temperature.

Various methods have been devised to reduce the temperature of the hydraulic oil. One method is to provide larger oil reservoirs for the power unit. Even with larger reservoirs, it may not be possible to dissipate enough heat to provide satisfactory cooling. Another method is to place a coil of copper tubing in the oil reservoir and circulate cold water through this coil. This method is not too practical since it consumes space in the reservoir, and condensation, which often forms, mixes with the oil.

A most satisfactory method of oil cooling is the use of a heat exchanger in the hydraulic system. Some of the hydraulically operated machines which make use of heat exchangers are presses, gear shapers, die-casting machines, spot welding machines, molding machines and broaching machines.

The location of the heat exchanger in the system depends upon the conditions which create heat. If heat is caused by a large volume of high-pressure oil spilling through the relief valve, then the heat exchanger should be placed in the exhaust line which runs from the exhaust port of the relief valve to the oil reservoir. This will dissipate the excessive heat in the oil before the oil returns to the reservoir. A common error is not to place the heat exchanger in this location when

Courtesy of Kewanee-Ross Corp.

Fig. 1. Heat exchanger mounted on the top plate of a hydraulic power device. The arrow points to the heat exchanger which is in the line between the relief valve and the oil reservoir.

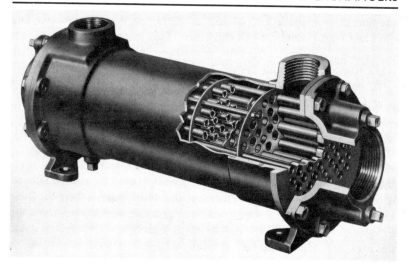

Courtesy of Kewanee-Ross Corp.
Fig. 2. Cutaway view of a single-pass type of heat exchanger.

heat is due to the above cause but to place it in the return line of the system. Then, the large volume of hot oil spilling through the relief valve is not relieved of its heat and the temperature in the reservoir quickly builds up. A heat exchanger is of value, however, when placed in the return line of the system if heat is caused by friction in the piping, restrictions, or by external heat. The complete system should be carefully studied before locating the heat exchanger.

Figure 1 shows a heat exchanger mounted on the top plate of a hydraulic power device. In this installation the heat exchanger is in the line between the relief valve and the oil reservoir.

Commercially standard heat exchangers are of several designs and types. Figure 2 shows the interior of a single-pass type exchanger. Oil enters through the right-hand top port and leaves by the left-hand top port. Cooling water enters the port on the right-hand cap and flows through the interior of the tubes and leaves by the port on the end of the left-hand cover. Heat is extracted from the oil as it circulates around the water tubes during its passage through the exchanger. Exchangers must be built so that there is no possibility of leakage between the oil section and the water section. Non-ferrous materials are used to reduce possibilities of corrosion. Note that in this design the end cover is removable so that if there is any need to clean the small tubes they are easily accessible.

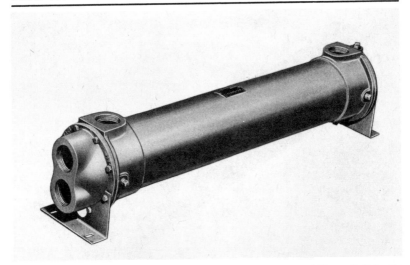

Fig. 3. Two-pass type of heat exchanger.

Heat exchangers are also built in two-pass and four-pass types. Figure 3 illustrates a two-pass type while Fig. 3A shows an exploded and cutaway view of the same exchanger. Note the separation between the top and bottom sections of tubes of the exchanger. In

Fig. 3A. Exploded and cutaway view of the two-pass heat exchanger shown in Fig. 3.

Courtesy of Young Radiator Co.

Fig. 4. Four-pass type of heat exchanger.

this model the cooling water enters the upper left end port, flows through the top section of tubes, returns through the bottom section of tubes and exhausts out the lower port *A,* on the left end. Oil enters the left-hand top port *B* and leaves by the right-hand top port. Figure 4 shows a design of a four-pass type heat exchanger. In this exchanger, the same as in the two-pass, the cooling water and the oil should enter their respective ports at the same end of the exchanger— water on the end port and oil on the top port. In this type, however, the cooling water passes to the far end, back to the inlet end, once again back to the far end and then comes back to exhaust at the out- let (adjacent to the inlet) at the inlet end.

A somewhat different type of heat exchanger is shown in Fig. 5.

Courtesy of The Heat-X-Changer Co., Inc.

Fig. 5. Heat exchanger that can readily be disassembled for cleaning or replace- ment of parts which is suitable for oil pressures up to 250 pounds per square inch and water pressures up to 150 pounds per square inch.

In this exchanger the oil flows through the copper inner-fins while the water flows around these inner-fins. The unit can readily be disassembled for cleaning or replacement of parts. The seal between the oil section and water section is effected by "O" rings.

As auxiliary equipment to a cooler, a thermostatic control (see Fig. 6) may be placed in the water inlet line to the cooler. This will automatically start and stop the flow of water and keep the oil at the desired temperature. When the oil becomes too warm, the water is

Courtesy of Young Radiator Co.

Fig. 6. Two-pass heat exchanger shown on large hydraulic power unit equipped with a 60 horsepower electric motor. A thermostatic control is connected to the water inlet port of the exchanger.

allowed to flow and when it cools off the water flow is stopped. On some applications it is desirable to use several heat exchangers to provide ample cooling as shown in Fig. 7.

The capacity of a shell-and-tube type heat exchanger depends upon several factors such as: the properties of the fluids flowing through it, their mass rates of flow, the temperatures of the fluids, the tube diameter and the surface area. The highest temperature fluid should generally be circulated through the shell side of the exchanger to minimize the difference in expansion of shell and tubes.

In order for the heat exchanger manufacturer or his representative to select a suitable heat exchanger for a particular installation, he must be supplied with certain information. This would include: the specifications of the oil to be cooled, the maximum operating

Courtesy of Young Radiator Co.

Fig. 7. Five two-pass heat exchangers are used on this large 1500-ton press to cool the hydraulic fluid.

temperature of the oil, the allowable pressure drop of the oil and the cooling water as they pass through the heat exchanger, the possible temperature range of the inlet cooling water and the amount of space available for the installation of the heat exchanger. As far as the oil specifications are concerned the heat exchanger manufacturer would like to know its specific gravity, viscosity, specific heat, thermal conductivity and its fouling tendencies at the mean fluid temperature. It is more than likely that the manufacturer knows these specifications if some standard oil is being used.

In making his selection the heat exchanger manufacturer equates the heat loss of the oil to the heat gain of the cooling water keeping in mind the thermal efficiency of the heat exchanger (which is determined by the thermal properties of the heat exchanger materials), the fouling tendencies of the oil, and the limitations placed on the size of the unit by the space available.

Liquid-to-Air Coolers

Air cooling of hydraulic oil offers a number of advantages over the water cooling method. Among these are: (1) elimination of feedwater supply which is often expensive, sometimes unreliable, and may be inconvenient to pipe to the cooler due to the location of the hy-

Courtesy of Modine Manufacturing Co.

Fig. 8. Air-cooled oil cooler designed specifically for use on hydrostatic, hydraulic, and other oil-cooling systems.

draulic equipment; (2) elimination of antifreeze solutions where hydraulic systems would set over a weekend at very low temperatures such as in a steel-processing mill; and (3) elimination of the water lines and water controls.

In Fig. 8, a liquid-to-air cooler of brazed aluminum construction is shown. This provides a bond between fin and tube which is equivalent to the strength of the parent metal and offers optimum heat transfer between primary and secondary surfaces. The use of aluminum greatly increases the cooling-capacity-to-weight ratio.

CHAPTER 9

Synchronizing the Movement
of Fluid Power Rams

One of the big problems in fluid mechanics, which has many approaches but few satisfactory solutions, is synchronizing the movement of two or more fluid power rams. There are many industrial applications which require nearly perfect synchronization of movement of two or more rams in order to complete some phase of operation. This usually comes about where exact movement is required at both ends of a span, such as on rolling equipment, power shears, power brakes, large presses, etc. If satisfactory synchronizing effects were available, many mechanical design problems could be simplified.

At first glance it might appear that if a medium oil is forced against two rams of the same size in the same circuit (as shown, for example, in Fig. 1) both rams will move forward at the same rate. However, this happens only in theory, not in actual practice. There are too many variables in the manufacture of commercial valves and cylinders which, in the final installation itself, can easily disrupt perfect coordination. Let us consider some of these problems or variables which present themselves in setting up such a layout.

Factors Affecting Synchronization

1. *Friction.* Probably the greatest problem is friction, both internal and external. Since the fluid will follow the path of least resistance, the ram having the smallest amount of restraint will move to the end of its stroke. Then the second ram will follow. Now what will cause more internal friction in one cylinder than in another, both being of the same bore and stroke? The rod packing may vary slightly from one cylinder to another. The packing gland on one cylinder may be pulled up more tightly on the cylinder rod than on the other. The packing recess may be slightly larger on one cylinder than on the other. The finish on one piston rod may be better than on

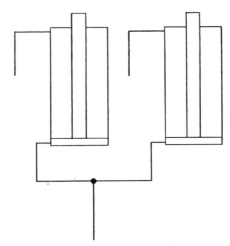

Fig. 1. Portion of a circuit containing two identical fluid power rams.

the other. Either the piston ring fit or the cup packing fit, whichever
is used, may be tighter on one cylinder than on the other. These are
some of the many variables which must be taken into consideration
on available commercial cylinders. Obviously, if the cylinders were
laboratory-made, the cost would be so exorbitant as to make them
impractical for industrial use.

External friction also presents a problem due to the fact that the
ram having the least resistance will move first. Gibs, tracks, rollers,
etc., must be designed so as not to allow either cylinder to be at a
disadvantage.

2. *Leakage.* Leakage can be another difficult problem in the
synchronization of two cylinders. Since we are dealing with very
minute amounts of fluid, any loss of this fluid will change the amount
of flow. All external leakage in pipe fittings, valve cover gaskets,
valve packings, cylinder cover gaskets and cylinder rod packings must
be eliminated. Internal leakage must also be minimized. "O" rings
or some other type of leak-proof packing must be used in the piston
so as to prevent oil from slipping past. When equalizing valves or
similar arrangements are used in the system, minimum leakage is a
must. Valve leakage must be watched after the system has been in
operation for a time since metal-to-metal sliding fits between valve
body and valve piston finally wear to a point where the pistons should
be replaced.

3. *Cylinder bore size.* The nominal bore of two hydraulic cylinders may be the same, yet the actual bore may vary by several

thousandths. Even though the exact amount of oil could be metered to each cylinder, the ram speed would be somewhat different on a long-stroke application.

4. *Cleanliness of fluid.* Where flow-control valves are being employed to meter the fluid from two cylinders, any impurities in the oil can clog up the small orifices in these valves resulting in functional failure. Clean oil is a must in any hydraulic application.

5. *Eccentric loading.* This is another factor which may affect synchronization. There are numerous applications where the loading must be eccentric and will therefore require that the proper circuit be chosen to combat the problem. Where one system may work well on concentric loading it may not be at all suitable for eccentric loading.

Other Factors Affecting Use of Synchronizing Circuit

It will be seen from a perusal of the above difficulties, that the designer should use caution in the selection of synchronizing circuits.

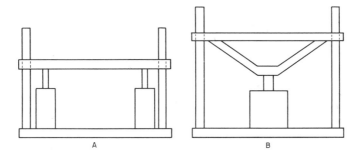

Fig. 2. The design of the ram in B which has the same cylinder working area as the device shown in A is superior because it does not necessitate synchronizing the movement of two fluid power rams.

As shown schematically in Fig. 2, a second cylinder can oftentimes be eliminated by properly designing the structure of the machine since the use of one large cylinder having the same working area as two small cylinders will impart the same pressure and eliminate the necessity of synchronizing. Although used in the majority of press applications, this design has its over-all limitations since it could not be used in extremely long spans.

Which system to choose will depend upon how closely the rams of the cylinders must be synchronized. It may also depend upon the cost of the system. Where one system may run into considerable

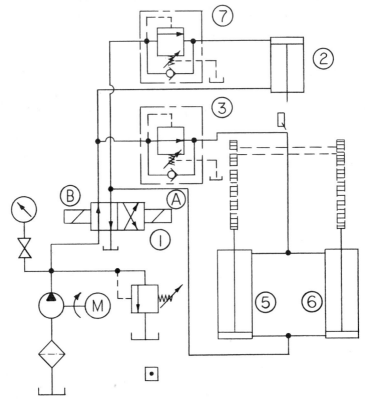

Fig. 3. Circuit diagram of one of the most positive in performance of all synchronized cylinder layouts. This design allows for the synchronization of movement in both directions.

cost, it may be that the selling price of the machine or the use for which it is intended will not bear such a high cost. Space limitations may also dictate which system is the better for a certain application.

The following basic circuits represent various ways and means of tackling the problem of synchronized movement of fluid power cylinders.

Use of Mechanically Linked Pistons

Circuit No. 1 (Fig. 3) shows one of the most positive in performance of all synchronized cylinder layouts. The use of a rack and gear mechanism as a mechanical linkage between the two pistons provides perfect timing so long as wear is kept to a minimum in the rack and gear teeth. This design allows for synchronization of movement in both directions. One disadvantage is that considerable space is required for installing such a mechanism, particularly if the cylinder strokes are especially long.

The circuit as used on a large drilling fixture functions thus:

1. Large workpiece moves down power conveyor until a positive stop is contacted. This stop is actuated by a small air cylinder (not shown) and is retracted after operation has been performed so that finished part may proceed down conveyor.

2. Operator depresses electric pushbutton which energizes solenoid (A) of valve (1), allowing oil to flow from pump to blind end of transfer-clamp cylinders (5) and (6). Oil pressure moves pistons forward while oil in rod ends passes to sump through valve (1). The end of each piston rod is fastened to a rack which meshes with a gear. The two gears are connected by a common shaft. Thus, the piston rod on cylinder (5) cannot move ahead of piston rod on cylinder (6) because of the gear and rack mechanism. As the pistons reach the end of the outstroke the workpiece is clamped tightly into the drill fixture.

3. Pressure builds up in line between valve (1) and blind end of cylinders (5) and (6), and sequence valve (7) opens allowing oil from pump to flow to blind end of drill cylinder (2) while oil in rod end of cylinder (2) flows out to sump through valve (1). Drill cylinder piston moves drill head forward and hole is drilled in workpiece. When drill cylinder reaches the end of the stroke a limit switch is contacted energizing solenoid (B) of valve (1).

4. Valve (1) shifts, allowing oil from pump to flow to rod end of drill cylinder while oil in blind end returns to sump through valve (1). Drill retracts to original position and drill stops rotating. Pressure builds up in line and sequence valve (3) opens, allowing oil from pump to flow to rod ends of transfer clamp cylinders (5) and (6) while oil in blind ends returns to sump through valve (1). Piston rods then retract uniformly bringing back workpiece onto conveyor. Positive stop is retracted until workpiece has moved on down conveyor, then is returned to lock position by the tripping of a limit switch as workpiece leaves this section of conveyor.

Use of Hydraulic Motors as Metering Devices

Circuit No. 2 (Fig. 4) shows the synchronization of three hydraulic cylinders by the use of hydraulic motors coupled together, to act as metering devices. Figure 5 shows three such motors coupled together by a silent chain. By use of this scheme, which proves effective in many cases, the fluid is divided equally as it passes through the three hydraulic motors and is then forced to the cylinders. In such a setup, eccentric loading creates no problem as it often does in other layouts. Friction, also, is not a problem; but leakage, both internal and external, can be critical.

Although the use of fluid motors for synchronization is one of the best solutions, it is also somewhat expensive. The greater the oil volumes and the higher the operating pressures, the more expensive

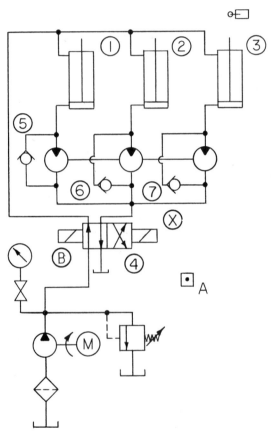

Fig. 4. Synchronization of three hydraulic cylinders is being accomplished by using hydraulic motors coupled together. These motors act as metering devices.

is the fluid motor. The use of fluid motors for synchronization has the advantage that the cylinders can be placed at various positions and need not be parallel like those shown in Circuit No. 1.

The circuit is shown for the operation of a precision drilling fixture and functions as follows:

1. Operator places workpiece in drill fixture and depresses electric starting button (A).

2. Solenoid (X) of valve (4) is energized and oil from pump flows to the three fluid motors which are coupled together with a silent chain drive. Equal volumes of oil are metered through the three motors and flow to the blind ends of the three precision drill cylinders (1), (2), and (3) while oil from the rod ends flows out to sump through valve (4). The piston rods feed the drilling spindles forward and the drilling operations are performed.

3. A limit switch is contacted at the end of the drilling stroke, solenoid (B) of valve (4) is energized, and the oil flow is shifted to the rod ends of cylinders (1), (2), and (3) by master valve (4) and

Fig. 5. Three hydraulic motors positively coupled together by means of a silent chain.

oil in the blind ends of these cylinders flows out through unseated ball checks in valves (5), (6), and (7), bypassing the fluid motors, and on to the sump through valve (4). Pistons of drill cylinders retract, although not in perfect unison. Operator then unloads finished workpiece completing the cycle.

One thing that must be remembered when using fluid motors for synchronization is that the flow of oil to each cylinder will be the volume of the pump output divided by the number of cylinders (or

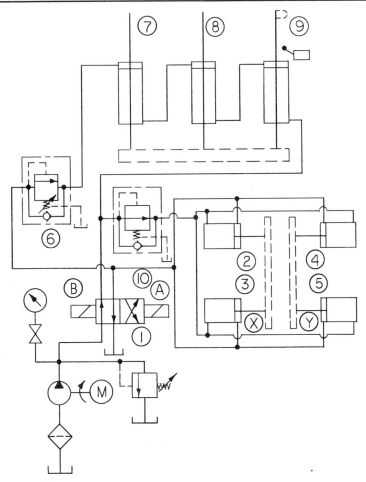

Fig. 6. Circuit which employs hydraulic cylinders of the double-end type for synchronizing the movement of fluid power rams.

fluid motors) in the circuit. For example, if the output of the pump is 30 gpm and three cylinders are used, then the volume to each cylinder will be 10 gpm.

Use of Double-end Cylinders in Series

As shown in Circuit No. 3 (Fig. 6), hydraulic cylinders of the double-end type can be used in series where several cylinders are involved, but this often presents certain problems. The cylinders must be free of both internal and external leakage for if there is any escape of fluid, the cylinders will quickly get out of phase. Also, the lines between the cylinders must be full of oil and free of air. These conditions are sometimes difficult of one-hundred-per-cent achievement.

One objection to the use of double-end cylinders is that they may not fit into a machine due to the space needed for the rod extensions. The double-end cylinder, however, does provide better rod support.

The circuit as applied to a dipping fixture functions as follows:

1. Several operators load workpieces onto pins on large rack which is traveling on a conveyor. Operator then allows rack to roll into position over dip tank. He then depresses button on momentary contact electric push button switch and current energizes solenoid (A) of valve (1). This shifts the piston of valve (1) and oil flows to rod ends of conveyor rail cylinders (2), (3), (4), and (5) while oil in blind ends flows out to sump through valve (1). The pistons retract causing conveyor rails (X) and (Y) to move clear of rack.

2. Pressure builds up, opening sequence valve (6), allowing oil to flow to top end of rack cylinder (7) which in turn transmits pressure to rack cylinder (8), which in turn transmits pressure to rack cylinder (9). Piston rods of all three cylinders move the large rack down into the dip tank. When the cylinders reach the end of their strokes, a limit switch is contacted by the tail rod on cylinder (9), an adjustable timer cuts in and the parts are held in the dip tank until the time cycle runs out, then solenoid (B) of valve (1) is energized and oil then flows to bottom end of rack cylinder (9).

3. Oil pressure is now applied to the bottom end of cylinder (9) and by transmission of pressure, all three rack cylinder piston rods (7), (8), and (9) move up and the rack and workpieces move up to top position.

4. Sequence valve (10) opens and oil from pump then flows to blind ends of conveyor rail cylinders (2), (3), (4), and (5) while oil in rod ends passes out to sump through valve (1). Conveyor rails move back into place under rollers of rack. The conveyor rail cylinders are mounted on a slight incline so that the weight of the rack is taken off of the ends of the rack piston rods when the track is in place. This allows ample clearance between the couplings on the end of the rods and the coupling on the rack so that the racks can be easily moved on and off of the dipping section.

5. Operator pushes completed rack off of dipping section and brings up next rack for dipping. Operators load one rack while other is in dipping cycle. Several dipping stations can be set up where dippings of various substances are involved.

Use of Flow Control Valves

In Circuit No. 4 (Fig. 7), flow-control valves are used for equalizing the speed of two pistons. This method is used in many instances when "close to synchronization" is desired and is probably the most inaccurate, but there are many cases where the speed of the two

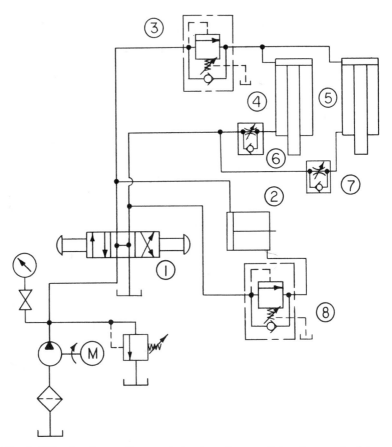

Fig. 7. This circuit uses flow control valves for equalizing the speed of two pistons.

pistons need not be exact. It is by far the least expensive, inasmuch as flow-control valves usually cost less than fifty dollars each depending upon the size used, whereas fluid motors usually cost several hundred dollars each. Piston leakage should be held to a minimum in such a circuit, although it is not as critical as in other types of flow-proportioning circuits. Variation in oil temperature will change the speed of the rams and may call for a readjustment of the flow-control needle. Extremely slow feeds may cause trouble as the orifices may become clogged if any dirt or lint is present in the system. The inlet lines to the cylinders should be the same size and the lines from the tee connection to the cylinders should be the same length.

This circuit as applied to the operation of a bending fixture functions as follows:

1. Operator places workpiece on bending fixture and shifts handle of four-way three-position valve (1). Oil flows from pump to blind end of clamp cylinder (2) while oil in rod end passes out to sump through valve (1) and workpiece is securely clamped in fixture.

2. Oil pressure builds up opening sequence valve (3) and oil from pump flows to blind ends of forming press cylinders (4) and (5). Pistons in cylinders (4) and (5) move forward at a speed set by flow-control valves (6) and (7) as oil in rod ends of cylinders (4) and (5) passes out to sump through valve (1) and workpiece is formed.

3. Operator shifts handle of valve (1) to opposite position and oil from pump flows to rod ends of forming press cylinders (4) and (5) through the free-flow section of valves (6) and (7) while oil in blind ends returns to sump through valve (1). Pistons of cylinders (4) and (5) retract at a rapid rate. Oil pressure builds up opening sequence valve (8) and oil from pump flows to rod end of clamp cylinder (2) while oil from blind end flows out through valve (1), unclamping workpiece. Operator then shifts valve (1) handle to neutral position which allows oil to flow directly to sump eliminating over-heating of oil while he loads and unloads workpieces.

Use of Air-hydraulic Cylinders in Series

By the use of two or more air-hydraulic cylinders hooked up in series as shown in Circuit No. 5 (Fig. 8), very satisfactory synchronized movement is assured. It is also an economical solution to the problem. Air is the motivating power and is contained in the back section of each cylinder. Oil is the controlling means and is contained in the front section of each cylinder. The front sections are connected in series by a closed circuit. In each cylinder the piston in the oil section is equipped with an "O" ring packing in order to eliminate leakage. The piston rod on the back side of the oil piston is integral with the air piston.

In using the air-hydraulic cylinder the connecting lines between the oil sides must be absolutely free of leaks and void of air. Either

will throw the system out of phase.   Air must also be kept out of
the oil chambers at all times.

The possible disadvantages of this layout are (1) that it requires
considerable space in order to mount the cylinders and (2) that the
total force available is only that of the air line pressure multiplied by
the area of the air cylinder bore, less whatever friction is set up by
the packings and metering devices.   In cases where such conditions
are not detrimental factors, this is a very economical layout for many
difficult applications.   Since very little heat is created due to the elimination of the hydraulic power device, there need be no further adjust-

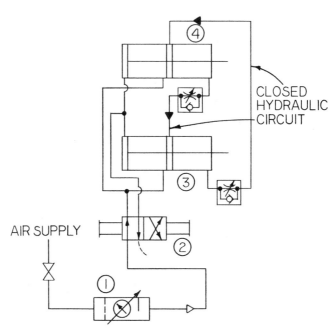

Fig. 8. Circuit employing the use of two air-hydraulic cylinders hooked up in
series to provide very satisfactory synchronized movement.

ment of the flow-control needle once it is set at the correct speed.

This circuit for a rolling mill functions as follows:

1. Operator loads work onto large reel and feeds end of work into
rolls. Cylinders (3) and (4) are located at either end of rolls.

2. Operator shifts handle of four-way air valve (2) allowing air
to flow to blind end of air-hydraulic cylinders (3) and (4). (Since
oil cylinders (3) and (4) are connected in series in a closed circuit,
the flow of oil into and out of each is the same and the movement

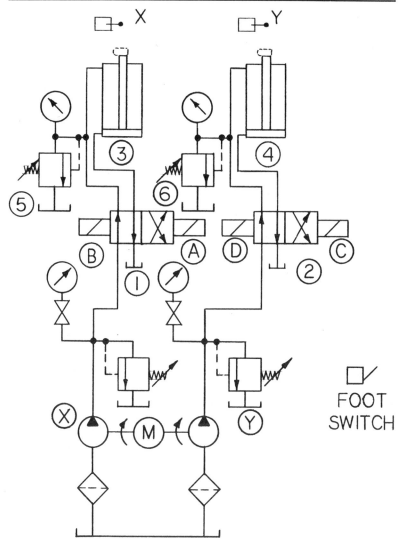

Fig. 9. Two pumps of the same capacity are used to synchronize piston movements of two hydraulic cylinders of the same bore and stroke in this circuit.

of one oil piston is synchronized with that of the other, the speed of movement being controlled by the rate of oil flow through the flow control valves.) Pistons of both cylinders move out simultaneously closing one roll against the other, setting the correct amount of pressure on workpiece. Pressure is adjusted by regulating valve (1) which sets the operating pressure.

3. After material is processed, operator shifts handle of valve (2) to original position allowing air to flow to rod end of air-piston side of cylinders (3) and (4) and rods retract opening the space between rolls thus completing cycle.

Use of Equal Capacity Pumps

In Circuit No. 6 (Fig. 9), two pumps of the same capacity are used to synchronize the piston movements of two hydraulic cylinders of the same bore and stroke. Leakage must be held to a minimum to assure satisfactory results. The variance in pump delivery will also be the variance in the piston movement of the cylinders.

The circuit for a pressing fixture functions as follows:

1. Operator swings large casting onto fixture and locates over pins in fixture bed.

2. Operator depresses and releases pedal as foot-operated electric switch momentarily energizes solenoids (A) and (C) of four-way valves (1) and (2) respectively. Oil flows from pump (X) to blind end of bushing cylinder (3) and from pump (Y) to blind end of bushing cylinder (4) while oil in rod ends is exhausted to sump through valves (1) and (2). Pistons move out and the pins carrying bushings move into holes press-fitting bushings into place. At end of stroke, cams on end of piston rods of cylinders (3) and (4) contact limit switches (X) and (Y) (connected in series) which energize solenoid (B) and (D) of valves (1) and (2) respectively. Pistons of valves (1) and (2) shift allowing oil from pumps to flow to rod ends of cylinders (3) and (4) and the pistons retract at low pressure as set by relief valves (5) and (6), while oil from blind ends passes out to sump through valves (1) and (2). Pressure remains low during stand-by periods due to the oil dumping through low-pressure relief valves (5) and (6).

3. Operator unloads workpiece and is ready for next cycle.

Flow-equalizing valves are used in some synchronization applications but in many industrial applications their capacity is too small. The equalizing valve divides the flow into two equal parts and is placed in the circuit following the master control valve.

Dual Pressure Hydraulic Systems

In some hydraulic systems if only one operating pressure were available, some of the components would be very much out of proportion with respect to the job they have to do. For example, on a hydraulic press the workpiece may be positioned or clamped by a light-weight hydraulic cylinder but the work to be performed must be done by a heavy-duty hydraulic cylinder. If the operating pressures of both cylinders are the same, the positioning cylinder may be too small to be practical or the work cylinder may be too large in diameter to fit into the allotted space. Also, the volume of oil required by a cylinder of very large bore may be out of the question due to the cost of large pump units.

The elimination of such conditions can be accomplished by operating one portion of the system at one pressure and another portion at another pressure. Often it is desirable to use several different pressures at once or several pressures in steps during a complete cycle. By using a dual-pressure system, such factors as heat, component wear, leakage, and power consumption can be greatly reduced. This chapter will deal with the many methods employed to obtain two or more pressures from one system. These various methods make use of reducing valves, combination pumping units, cam-operated relief valves, individual pumps, and intensifiers.

Use of Pressure-reducing Valve

In one of the most common and least expensive dual-pressure systems a pressure-reducing valve, such as that shown in Fig. 1, is employed. The operating pressure of the oil is reduced as it passes through the valve, the regulation of the pressure being obtained by the adjustment of the spring tension. As the spring load is increased, the pressure on the low-pressure side of the valve is increased. The pressure differential from inlet to outlet may range up to a ratio of 10 to 1. When greater pressure differentials are desired, two valves may be connected in series. Some hydraulic reducing valves are constructed with a built-in check valve which affords free flow return of the oil. All drain lines in reducing valves must be connected to the sump if satisfactory results are expected, since back pressure in the drainage system will cause the valve to malfunction.

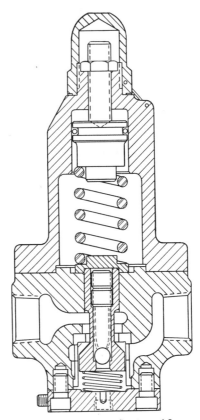

Courtesy of Logansport Machine Co., Inc.

Fig. 1. Pressure-reducing valve of the type commonly used on the less expensive dual-pressure systems.

Circuit No. 1 (Fig. 2), used on a large hydraulic press, functions as follows:

1. Operator loads work into fixture and shifts handle of master valve (1) to position (A). Oil at low pressure flows out of reducing valve (2) to blind end of lock-pin cylinder (3) locking workpiece in fixture while oil in rod end flows back to sump through valve (1).

Oil at high pressure flows to blind end of ram cylinder (4) while oil in rod end flows back to sump through valve (1) and ram descends and performs pressing operation.

2. Operator shifts handle of valve (1) to position (B) and oil flows to rod end of ram cylinder, while oil in blind end flows to sump through valve (1) and ram moves to up position. Then sequence valve (5) opens and oil flows to rod end of lock-pin cylinder, while oil in blind end flows to sump through valve (1), and lock pin is retracted. Operator then shifts handle of valve (1) to position (N), for stand-by period while he unloads and reloads press.

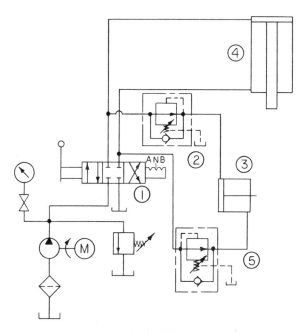

Fig. 2. Large hydraulic press circuit.

Use of a Hydraulic Intensifier

A dual-pressure system designed around a hydraulic intensifier not only has many applications but it also has many advantages. Intensifier circuits are used on tubing fixtures for testing bursting pressures, on high-pressure punching fixtures, high-pressure holding fixtures and in many other installations. The intensifier may be operated by air pressure or oil pressure depending upon the ultimate pressure desired. Both are similar in design. The advantage of the air-operated type is that the hydraulic power device can be eliminated. This intensifier is used with many air-operated hydraulic presses to reduce construction cost. Another advantage of air operation is the rapid response of the medium.

The pressure exerted by an intensifier is governed by the area of the intensifier cylinder, the operating pressure of the system and the ram area of the intensifier. The theoretical intensified pressure, neglecting the effect of friction, may be found by the formula:

$$P_I = \frac{A_C}{A_R} \times P_O$$

where P_I　= intensified pressure
A_C　= area of operating cylinder
A_R　= area of intensifier ram
P_O　= operating pressure

For example, a twelve-inch intensifier cylinder operating at 900 lb per sq in. and having a 3-inch diameter ram will create a theoretical intensified pressure of 14,400 lb per sq in.

Extremely large intensifiers have been developed for certain applications which require delivery of a large volume of hydraulic fluid at intensified pressure. Rod packing may then become an important problem, since it is difficult to provide suitable packing for very high pressures.

When using a pump system to create high pressures, excessive heat may become a detrimental factor. This is not the case when using intensifiers, as high pressures can be held for long periods without causing heat in the intensifier system.

If a single-acting intensifier is used, the operations must be completed before the intensifier ram reaches the end of the stroke or the intensified pressure will not be maintained. There has been some experimental work on double-acting type intensifiers but pulsations are created which may be undesirable.

The use of a hydraulically operated intensifier to provide a different operating pressure is illustrated by Circuit No. 2 (Fig. 3). This circuit actually has three different operating pressures, one controlled by the relief valve at the power unit, one in the return line of the clamping cylinders, and one created in the intensified part of the system. The circuit, as applied for actuating punches, operates as follows:

1. Operator places workpiece in clamping fixture and shifts handle of four-way valve (1) to allow oil from pump to flow to blind ends

of clamping cylinders (2) and (3) while oil in rod ends flows out to sump through valve (1). Clamp cylinders securely lock work in fixture.

2. Operator then shifts handle of four-way valve (4), allowing oil from pump to flow to blind end of intensifier cylinder (5), while oil in rod end (Y) of intensifier cylinder flows back to sump through

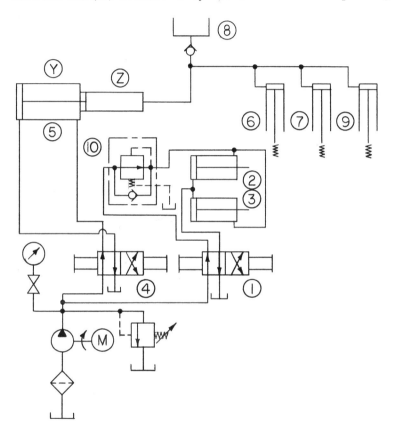

Fig. 3. Circuit provided with a hydraulically operated intensifier which provides an increased operating pressure.

valve (4), and piston starts to move forward. Oil at intensified pressure flows from chamber (Z) to blind end of single-acting piercing cylinders (6), (7), and (9). Their pistons move forward as piston of intensifier (5) moves forward.

3. When piercing of workpiece is completed, operator shifts handle of valve (4), oil from pump flows to rod end (Y) of intensifier

cylinder while oil in blind end passes out to sump, and piston of intensifier cylinder retracts to original position. Springs in piercing cylinders retract pistons in these cylinders forcing oil back into chamber (Z). Make-up unit (8) takes care of any oil loss in the intensifier chamber due to leakage in the closed system.

4. Operator shifts handle of valve (1) directing oil from pump through reducing valve (10) and into rod ends of clamp cylinders (2) and (3), while oil in blind ends is exhausted to sump. Clamp cylinders release the workpiece and cycle is then completed. Spring-loaded check valves may be placed just before each of the four-way valves in order not to have any pressure drop-off in the system when one of the four-way valves is operated. By using a pressure switch and solenoid master valve, a semi-automatic circuit may be set up so that when the punch pistons "bottom" they will automatically return.

Use of Two Pumps

A hydraulic power device consisting of two pumps attached to a double-shaft electric motor is ideal for many applications. There are also double pumps designed which only require a single-shaft electric motor. In either case, one pump is operated at one pressure and the other pump is operated at another. Both designs afford the same possibilities in the end use. The relief valve connected to each pump sets the operating pressure for each portion of the system. The one pump may have a very small volume while the other may be large or both pumps may be of the same volume, depending upon the application. A small volume at high pressure may be used for clamping; a large volume at low pressure may be used for the work cycle of the machine. By using two pumps in one system: (1) heat will be reduced; (2) there will be a reduction in packing leaks and oil break-down; and (3) because of the wide selection of volumes and pressures which are possible, cylinders can be readily selected to fit into almost any installation.

Circuit No. 3 (Fig. 4), used on a piercing fixture, shows a typical application of the use of two pumps to meet one of the more common applications of a dual-pressure system.

1. Operator places casting in clamping device and shifts handle of valve (1). Oil from pump (A) flows to the rod end of the sleeve-clamping cylinder (2), while oil from blind end flows out to sump, and piston retracts, expanding sleeve and securely clamping casting in fixture. The operating pressure from pump (A) which is set by relief valve (X) keeps constant pressure on fixture during work cycle.

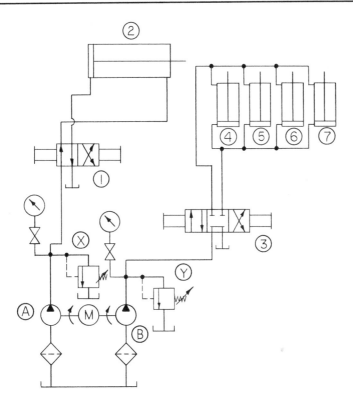

Fig. 4. The use of two pumps in this dual-pressure system is used for operating a piercing fixture.

2. Operator then shifts handle of four-way, three-position valve (3) to forward position and piercing cylinders (4), (5), (6), and (7) move in and perform operation.

3. Operator then returns handle of valve (3) to opposite position and pistons of cylinders (4), (5), (6), and (7) retract. Valve handle is then returned to neutral position.

4. Operator shifts handle of valve (1) to original position and piston rod of cylinder (2) moves forward, collapsing sleeve, which releases workpiece, thus completing cycle. Pump (B) is a high-pressure, large-volume type which is required in order to move the punches into the workpiece quickly at high pressure. Its pressure is controlled by relief valve (Y).

Use of Two Relief Valves

A two-pressure circuit designed with two relief valves is of great value on press circuits where there is a long stand-by between operations. This is usually brought about by complicated loading, the time for which may be several times as long as that required by the actual operating cycle. One relief valve is located on the power unit while the other is placed in the line between the four-way valve and the rod-end port of the operating cylinder. The relief valve on the power unit is set so that the operating pressure is sufficient to perform the required work and the other relief valve is set so that the pressure in its line is just enough to return the ram carrying the die fixture. This eliminates over-heating of the oil during the long stand-by periods when the stand-by pressure is usually 50 to 60 lb per sq in.

Where only one relief valve is used and the master valve is of the two-position type, a large volume of oil must be pumped at high pressure and excess heat is the result. The author has viewed installations that have become so hot that the oil actually boiled over the top of the power unit. By installing the second relief valve, the situation was immediately cleared up.

While some press circuits are designed with an open-center valve to relieve the pressure during the stand-by periods, a heavy fixture on the end of the ram may cause it to drift after the internal parts of the valves and cylinders become worn and this may cause considerable trouble.

A typical example of the use of two relief valves is shown in Circuit No. 4 (Fig. 5).

1. Operator places work under press ram and shifts handle of four-way, two-position valve (4). High pressure, as set by relief valve (2), is applied to blind end of press cylinder ram (5) and ram descends, performing the operation.

2. Operator shifts handle of valve (4) to original position and oil from pump flows to rod end of ram cylinder (5), returning ram at low pressure, as controlled by the setting of relief valve (6), while oil in blind end is exhausted to sump.

Use of a High-low Pump

In circuits which require a low starting pressure and a high finishing pressure, such as press circuits where the ram approaches the work at high speed and low pressure and then performs the operation at high pressure and low speed, the use of a high-low pump with automatic change-over is a satisfactory solution. The high-low pumping unit may be two separate pumps with a change-over system or it may be two pumps built into one compact unit. The big advantage of the two individual pumps is that a greater variety of pressures

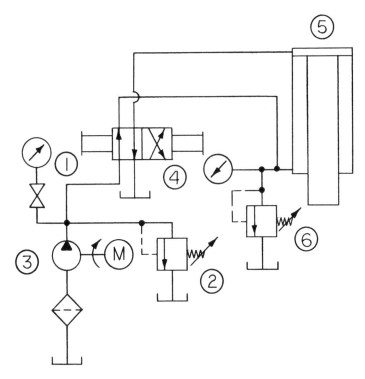

Fig. 5. Typical example of the use of two relief valves is illustrated in this circuit.

and volumes is available. The advantage of this high-low pumping unit is the saving in electric power consumed and the reduction in heat. For example, a high-low pump delivering 12.6 gpm at 300 lb per sq in. and 1.8 gpm at 1000 lb per sq in. requires a 3-horsepower motor. To get both the large volume and high pressure from a single pump, it would require approximately 7.5 horsepower. If the high-pressure holding period is long, any excessive internal leakage in the valve or cylinder must be eliminated or the change-over valve will be actuated and the low-pressure system will be cut back in. This may result in damage to the workpiece since the holding pressure is reduced. Excess heat in the system may also result due to the large volume of oil spilling through the relief valve.

Circuit No. 5 (Fig. 6) shows the use of two pumps with change-over valving. The circuit is used to operate a clamping fixture for large ingots.

1. Operator swings large ingot from conveyor into clamping device. He then shifts handle of valve (2) and oil flows from low-pressure,

large-volume pump (4) to blind end of clamping cylinders (A) and (B) and their rams move at rapid speed under low pressure, while oil in rod ends is passed through valve (2) to sump. As soon as ingot is contacted, pressure builds up until pilot pressure (through dashed pipe line shown in Fig. 6) causes unloading valve (1) to open, dumping the output of low-pressure pump (4) back into the reservoir. Ingot is then clamped under high pressure and low volume supplied by pump (5). Operator brings in mechanically operated saw to perform work.

2. Operator then shifts handle of valve (2) to original position. Valve (1) closes and rams of clamping cylinders (A) and (B) retract at high speed under low pressure until end of stroke is reached and pressure again builds up, opening valve (1). High-pressure,

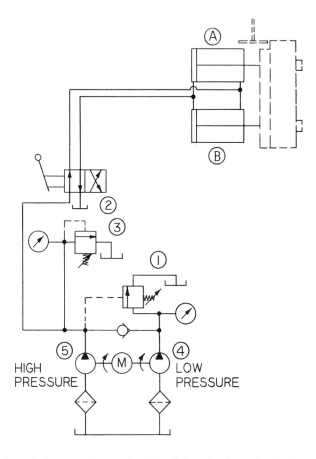

Fig. 6. Circuit for operating a clamping fixture for large ingots incorporates the use of two pumps with change-over valving.

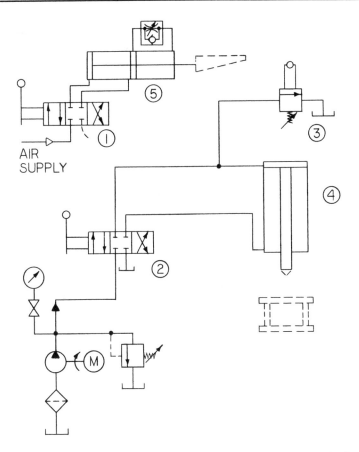

Fig. 7. An air, air-hydraulic, and hydraulic combination type of circuit employing a cam-operated relief valve.

low-volume pump (5) now, alone, delivers oil to hold rams in retracted position.

Use of Cam-operated Relief Valve

Another effective multiple-pressure system is designed to use air, a combination of air and hydraulic fluid, and hydraulic fluid. The design has proved effective for testing applications where pressure must be varied many times during each cycle. The basis of the design is a cam-operated hydraulic relief valve which is capable of handling a large range of operating pressures. A long cam, with a very little rise, connected to the rod of an air-hydraulic cylinder gradually depresses the roller of the relief valve, so that the pressure in the line is increased until the maximum is reached. The cylinder and the cam can both be stopped at any spot by shifting a three-position air valve controlling the cylinder movement back into the neutral position.   This blocks the cylinder ports.   If a very small hydraulic pump is used as the power device, a slight fluctuation shows up on the pressure gage and appears to be caused by the pulsation of the pump.   The addition of a bladder-type accumulator tends to overcome this condition.

Circuits No. 6 and 7 (Figs. 7 and 8) are examples showing the use of a cam-operated relief valve.   Circuit No. 6 is an air, air-hydraulic and hydraulic combination and Circuit No. 7 is all-hydraulic.   Figure 9 shows Circuit No. 7 applied to a testing machine.

Circuit No. 6 functions as follows:

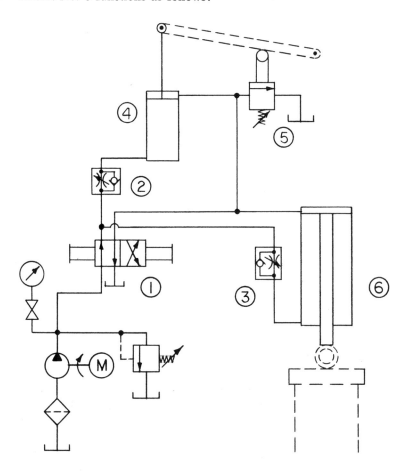

Fig. 8. An all-hydraulic circuit incorporating the use of a cam-operated relief valve.

1. Operator places test bar on fixtures and brings hydraulic ram (4) down at very low pressure by shifting handle of four-way, three-position hydraulic valve (2). When ram contacts the work, operator then shifts handle of four-way, three-position air valve (1) to forward position and air flows to blind end of air-hydraulic cylinder (5) so that cam is moved out depressing roller of hydraulic relief valve (3). Operator stops cam by shifting valve (1) into neutral and checks deflection on test bar. The cam has several graduations so that the operator can stop the cam at each and take a deflection reading. The farther the roller on the cam valve is depressed, the higher the operating pressure on the ram.

2. After operator has taken readings at the various settings, he shifts handles of both valves (1) and (2) to return the pistons of

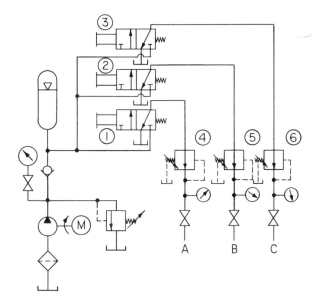

Fig. 9. Several different pieces can be subjected to a hydraulic pressure test at the same time using this circuit.

both cylinders, then shifts handles back into neutral position during the stand-by period.

In Circuit No. 7 (Fig. 8) the small hydraulic cylinder (4) changes the setting of the relief valve (5) by cam action, thus increasing the operating pressure after the ram has met the work.

By using a dead-weight accumulator and several reducing valves, a multiple-pressure circuit may be set up for pressure-testing purposes. Advantages of this circuit are that very little heat is generated even though the operating pressure is high, and where a small pump is used to charge the accumulator, the electrical output is reduced.

In Circuit No. 8 (Fig. 9) the operator can subject several different pieces to a hydraulic pressure test at the same time. This circuit functions as follows:

1. Operator places workpieces on testing stations (A), (B), and (C). He then shifts three-way valves (1), (2), and (3) and oil under pressure flows from accumulator through these valves, on through the reducing valves (4), (5), and (6), respectively, and on to stations (A), (B), and (C), where it enters the workpieces. The pressure is regulated by the reducing valves and is registered on the pressure gages. The pressure will be steady since the dead-weight accumulator acts as a large storage tank in which the pressure is always the same even though the volume changes. Operator may test three pieces at one time or only one piece at a time, depending on the requirements.

2. When the testing time is completed, operator shifts handle of three-way valves back to original position and then removes workpieces as fluid in them is exhausted to atmosphere.

The accumulator is automatically recharged, since a limit switch is contacted when the oil volume drops to a predetermined point and cuts in the hydraulic pump. The pump operates until another limit switch is contacted, when the accumulator is full and the motor stops. A check valve must be placed between the pump and the accumulator in order to keep the accumulator from discharging through the pump when the pump is idle.

Combinations of the basic circuits described in this chapter may be developed to solve many fluid power problems.

CHAPTER II

Safety Controls for Hydraulic Circuits

It has not been too many years since one could go into an industrial plant and see workmen with one, two, or three fingers missing due to accidents while operating punch presses or other types of machines. Such accidents were accepted as part of the risks of the job. In contrast, safety is the keynote in today's plant operations.

This change has come about as a result of several factors:

1. Employers now recognize the economic liability of careless operators.

2. Workmen today are safety conscious and are reluctant to work on unsafe machines. In fact, work stoppages have often resulted from the use of such machines.

3. Employers have come to realize that safe working conditions promote good employer-employee relations since they are important to the welfare of the workmen and contribute substantially to better workmanship and maximum production.

4. Furthermore, insurance companies are exacting in their demands for improvements in industrial working conditions to make them as near accident proof as is humanly possible.

As a result, both management and labor are continually striving to improve safety conditions and have taken steps to impress the facts upon legislative bodies to such a degree that improved safety codes are regularly on legislative agendas for study and possible action. Many "must" specifications have been set up for machines in order to assure maximum safety to the workman. In short, it is now clearly recognized that regardless of whether accidents are the result of carelessness on the part of the employee or are due to the lack of safety factors on the machine, steps must be taken to prevent or eliminate those conditions that make them possible.

Advantages of Hydraulic Controls

The use of hydraulic components affords many safety advantages such as operator protection in dangerous locations where there are explosive atmospheres, high or low temperatures, toxic or deadly fumes and in other similar circumstances. In explosive atmospheres, the machine operated by hydraulic cylinders may be in one room. the power unit and solenoid control valves in another room and the operating pilot controls in another. The operator may only see the machine indirectly but still have perfect control of all of its functions through the pilot controls. The master valves are usually placed as near the cylinders as possible in order to get instant response.

An operator sitting in a high pulpit in a steel mill has the complete control of certain phases of the process at his fingertips by manipulating hydraulic devices. This keeps him away from the heat of the mill and also gives him a chance to see exactly what is taking place all during the processing period. Hydraulic controls in this instance not only safeguard the operator but also protect the expensive mechanisms of the mill.

Around heat-treating furnaces and other hot spots it is not necessary for the operator to be near the heat when controlling these various devices. He can be located at a safe distance yet have perfect control by using hydraulic components. The same is true around toxic or deadly fumes. (See Chapter 7 on hydraulic cylinders with regard to protective devices for cylinders which come into direct contact with heat, abrasives, etc.)

Control Circuits Designed for Safety

In addition to the safety advantages offered by hydraulic components, the control circuit itself can and should be so designed that it will safeguard the operator, the machine, and the workpiece. This chapter will cover some of the basic hydraulic safety circuits and will point out their various features.

Hydraulic presses should be equipped with safety controls to protect the operator should he, for any reason, attempt to place his hands under the ram when it is descending. Some, such as the protective cage-type enclosure shown in Fig. 1, prevent the employee from approaching the press ram while it is in operation. The front and rear sliding doors of this enclosure must be closed before the ram will operate. Other types of safety controls require that the operator have both hands on control buttons during the pressing stroke. If he should remove either or both hands from the controls, the press ram will either return or stop, depending upon the type of master control valve designed into the circuit. Whether these controls be electric push buttons or pilot control valves with a spring return feature, will depend upon the preference of the user. Some

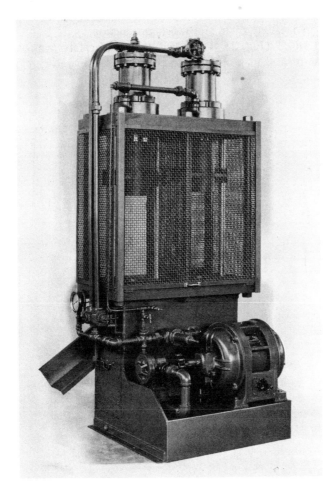

Fig. 1. Hydraulic press is provided with a protective cage-type enclosure which prevents the employee from approaching the press ram while it is in operation.

safety engineers have frowned upon the use of electrically-operated safety control equipment due to the possibility of control part failures; but with present-day electrical controls, most of the objections have been overcome.

Two-hand Safety Control Circuit

Often it is a battle of wits between the circuit designer and the machine operator, each trying to see who can outsmart the other. Even though the designer works hard to create a fool-proof circuit, some operators will still try to figure out a way to tie down a valve in order to increase their piecework production.

Circuit No. 1 (Fig. 2) is a typical two-hand safety control circuit which has proved its worth in safeguarding the operator. The cost

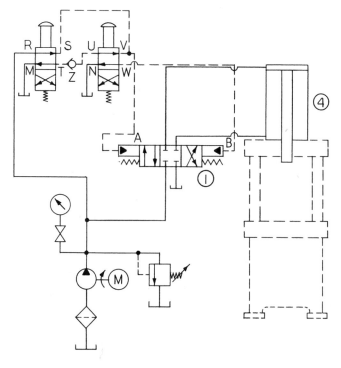

Fig. 2. Typical two-hand safety control circuit.

of this safety feature is nominal in comparison to the protection that it provides.

In this circuit, the operator would gain nothing by trying to tie down either of the operating levers of the two small pilot valves, because the press cannot be recycled if either is tied down. This is an extremely important feature. There are several types of controls available to operate the master valve, but the setup shown is positive, inexpensive, and responds quickly.

The circuit functions as follows:

1. Operator places mixture in compression chamber and then depresses handles of pilot valves (2) and (3). (Flow diagram of valves (2) and (3) in Circuit No. 1 shows flow when both valve handles are depressed.) Oil flows in port R of valve (2) and out port S, then through port V of valve (3) and out port U, but check valve (Z) stops flow. Oil also flows to pilot connection A of valve (1) causing the piston to shift, connecting the pressure line from the pump to the blind end of cylinder (4) and piston rod starts to descend with oil leaving rod end of cylinder (4) and returning to sump through valve (1). Should the operator remove his hand from the handle of valve (3) during the descent of the ram, the pilot pressure instead of flowing to pilot connection A of valve (1) flows into port V of valve (3) and out port N to sump releasing pressure to pilot connection A. This allows the four-way spring-centered valve to return to neutral causing the ram descent to be immediately stopped. If operator should keep handle of valve (3) depressed but releases handle of valve (2), oil flows in port R of valve (2) and out port T, then unseats ball in check in valve Z, then on to port U of valve (3), and out port V. Oil follows the path of least resistance so it passes in port S of valve (2), out port M and into sump. It does not build up enough pressure to keep pilot pressure on pilot connection (A) so valve (1) shifts back to neutral, stopping the ram.

When the operator depresses the handles of valves (2) and (3) as he should, the ram of cylinder (4) completes its down stroke.

2. Operator then releases handles of both valves (2) and (3). Oil flows in port R of valve (2) and out port T, then through check valve Z and into port U of valve (3). Oil flows out port W into pilot connection B of valve (1) shifting piston of valve (1) and pressure line from pump is connected to rod end of cylinder 4. Piston starts to retract and oil in blind end of cylinder flows through valve (1) and back into the sump.

3. Operator unloads press and is ready for next cycle.

Four-hand Safety Control

Four-hand safety controls are often necessary on large presses or machines where two operators are working. The safety problem arising from the inability of one operator to see the other or, because of noise, to contact him before starting the press, is usually solved inexpensively with four safety push buttons operating electric controls connected in series. The release of any one push button by either operator will cause the press ram to return to its original position.

By using electrical controls, one station may be located at quite a distance from the other, yet when both operators depress the push buttons, the master valve responds at once.

If the demand is for pilot valve controls, then either air pilots or hydraulic pilots may be used to actuate the master valve. Considerably more piping will be involved if either of these two means is used over what would be required if electric controls are used, so that the two operating stations should be as close together as possible.

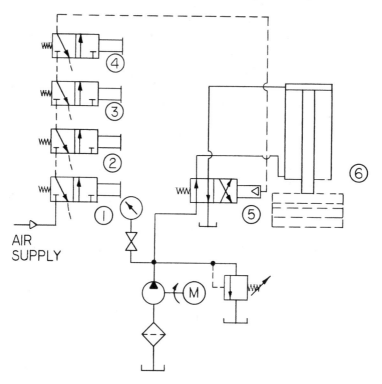

Fig. 3. Four-hand safety control circuit using air-pilot operated valves.

By using the series hookup, other stations may be added if the application so dictates. However, each additional station will usually slow up the end results due to the time lag created by the operators in depressing or shifting their controls.

A four-hand safety control circuit using air-pilot operated valves is shown in Circuit No. 2 (Fig. 3) and functions as follows:

1. Operator on one side of press slides large sheet of material onto press platen. Operator on opposite side of press helps him adjust

this material and then each operator places additional material in slots in original sheet. When each completes his operation, he depresses the palm buttons on the pilot controls. Operator "A" depresses palm buttons on air valves (1) and (2) and operator "B" depresses palm buttons on valves (3) and (4). Air flows from the regulator, filter, and lubricator units (not shown) through valves (1), (2), (3), and (4) on to pilot connection of air-operated, spring-return, four-way hydraulic control valve (5). Even though the two control stations are quite a distance from valve (5), the air travels very quickly to it and shifts the spool of this valve allowing oil to flow from the pump, through the valve, and then to the blind end of the large ram cylinder (6). The ram moves down toward the work as oil in the rod end passes to reservoir through valve (5). If either operator sees that all is not going well, he can release either palm button and the press ram will immediately return.

2. After the ram has descended and has completed the pressing operation, either operator can return the ram by releasing his controls. On electrical four-hand safety-control circuits it is often desirable to locate a switch at a position where the ram will trip it just before the ram contacts the work. This will connect the power directly to the solenoid master valve and allow both operators to remove their hands from the controls, since the ram has traveled to a point where it would be impossible for either to get his hands under the ram. This gives both operators a chance to perform other duties while the press is completing its pressing cycle. A timer tied in with the switch will allow the master valve to be shifted automatically at the end of the time cycle thus returning the press ram to the "up" position.

Use of Safety Interlocks to Protect Machine

Machine safety or protection of the machine is a very important factor especially on those machines which perform multiple operations. The sequencing of the machine must be such as to eliminate any possible mishaps. This is done hydraulically through safety interlocks such as are shown in Circuit No. 3 (Fig. 4). This circuit as applied to a spinning and staking machine in a large auto parts manufacturing plant has several features which make it outstanding. *First*, by shifting the handle of valve (12) to "emergency stop," the machine cycle stops immediately. *Second*, by the interlocking of valves (10) and (11), the turntable cannot index until the spinning and staking cylinders are in the "up" position. *Third*, if the turntable should meet undue resistance caused by a jam, it will stall because

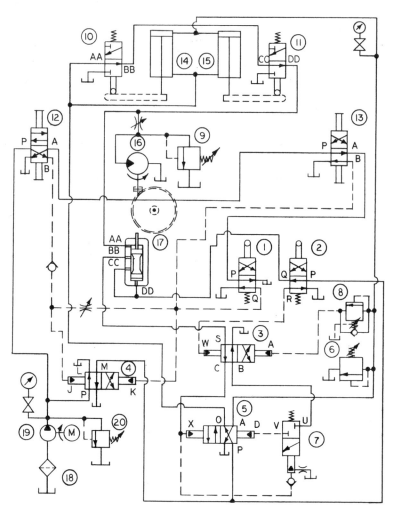

Fig. 4. Circuit of a spinning and staking machine in a large auto parts manufacturing plant which incorporates the use of safety interlocks for proper
sequencing.

relief valve (9) is set so that the pressure to the fluid motor will just
be enough to properly index the table under normal conditions.
Fourth, lock plunger (17) definitely locates the table position and
locks it. *Fifth,* this circuit is completely hydraulic with no electrical
controls, other than on the power unit, which is remotely located, and
can be readily used around explosive atmospheres.

This circuit operates as follows:

1. Operator shifts handle of valve (12) to "run" position and latches handle of valve (13) in "start" position if he wishes continuous cycling of machine or, if he only wants the machine to go through one cycle and then stop, he momentarily shifts handle of valve (13) to "start" position and then releases it. This directs pump supply through "out" port (B) of valve (13) to pilot chamber (K) of valve (4). Piston of valve (4) moves to left, directing pump supply through "out" port (M) to "in" port P of valve (5). Oil also passes on to valve (2), entering through port P and leaving by port Q and then goes on to port DD of valve (17). This causes piston of valve (17) to move into the locking position. With the piston of valve (17) in this position, oil enters port CC, leaves port BB and passes through valve (3) by ports S and C to pilot port X of valve (5). This causes the piston of this valve to shift to the left so that oil can leave through port A. Oil pressure is now applied to the blind end of the spinning and staking cylinders (14) and (15), the pressure on them being controlled by the setting of relief valve (6).

At the end of the down stroke of the spinning and staking rams, the pressure build-up in the line is sufficient to open valve (8), directing pilot pressure to port A of valve (3). Piston of valve (3) moves to the right, port C being connected to the sump and port B to port S. Oil pressure is no longer applied to port X of valve (5) which is now opened to the exhaust line and pressure which has been applied through the check valve Y to hold time-delay valve (7) closed is also removed. After a delay of from 0 to 10 seconds, depending upon the setting, valve (7) opens permitting oil from outlet B of valve (3) to enter port U and leave port V of valve (7) so that pressure is applied to the pilot port D of valve (5). This causes the piston of valve (5) to shift to the right. Oil now passes from inlet port P out through port O and on to the rod ends of the spinning and staking cylinders and to port AA of valve (10).

As the spinning and staking cylinder rams reach the ends of their upstrokes, the pistons of valves (10) and (11) are shifted by cam action, permitting oil to flow out port BB of valve (10) and through ports CC and DD of valve (11) to fluid motor (16) and to port AA of valve (17). Oil pressure on the piston of valve (17) causes the lock plunger to be withdrawn from the table which is now indexed by the fluid motor rotating an index cam. As the table is indexed, a cam shaft is rotated, momentarily shifting the pistons of valves (1) and (2). The piston in valve (2) is shifted to connect port P to port R so that oil flows momentarily to pilot chamber W of valve (3), shifting its piston to the left, which is the position shown. Port S of valve (3) has no pressure on it as it is supplied from port BB of valve (17) which is blocked when the locking plunger is retracted. Valve (1), which has been shifted momentarily by cam action when the table was indexed, directs oil momentarily from port F through port G to pilot chamber J of valve (4), shifting its piston, completing the operating cycle and by-passing the oil from the pump back to the supply tank through port L. If valve (13) is latched in the

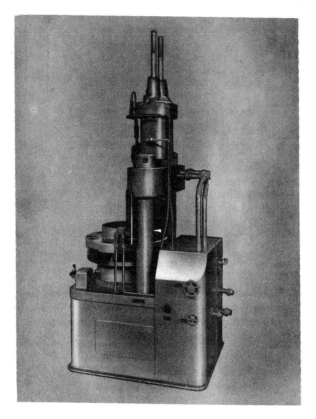

Fig. 5. Side view of a twin-ram powder press for compressing high explosives.

"start" position, oil passes out through port B and the operating cycle is repeated.

Figures 5 and 6 show a twin-ram powder press for compressing high explosives in which the same basic circuit is used.

Use of Emergency Cut-off Valves

Another approach to machine safety when using hydraulic controls is to install an emergency cut-off valve between the pump and the master control valve. If an accident should occur, the operator can stop all hydraulic action by actuating the emergency stop valve. This not only stops damage to the machine but also allows the sequence of operation to continue when the emergency stop valve is released. The single emergency stop works exceptionally well when cylinders are mounted in the horizontal position. If they are mounted in a

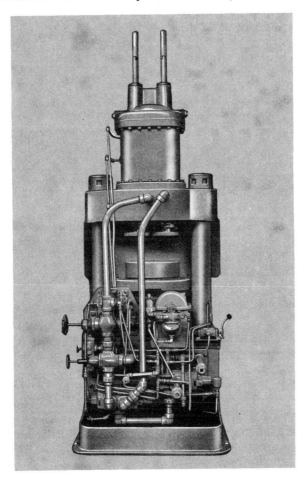

Fig. 6. Front view of the twin-ram powder press shown in Fig. 5.

vertical position with heavy loads on the end of the rods, a single cut-off valve usually is not suitable since the weight would cause the cylinders to coast after the shut-off. Cut-off valves of the two-way style may be necessary in several of the lines if the sequence is to be continued after the shutdown.

Circuit No. 4 (Fig. 7) shows the use of the cut-off valve in a safety circuit for controlling a push-off device. This circuit functions as follows:

1. Workpieces move down conveyor line. As they reach station A, operator momentarily depresses push button marked "forward" and valve (2) is shifted to position shown, allowing oil pressure to flow to blind end of cylinder (3) while oil is being forced out of rod end through orifice of flow control valve (4), through valve (2) and on to sump. If operator sees a jam occurring he presses "Emergency" button and oil pressure is blocked when solenoid valve (1)

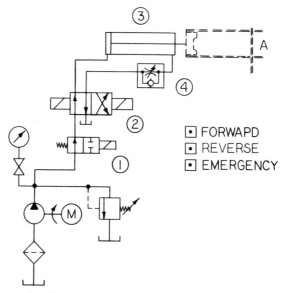

Fig. 7. A cut-off valve is used to control the push-off device in this safety circuit.

shifts, stopping movement of ram. When block is cleared he releases "Emergency" button and ram continues forward movement.

2. To reverse the ram, operator depresses "Reverse" button, shifting solenoid valve (2) allowing oil to flow to rod end of cylinder (3) through valve (2) on to sump while piston of cylinder (3) retracts to starting position.

3. Operator is then ready to eject the next work piece.

Safeguarding the Cutting Tool

Under machine safety should be listed tool safety. Even though tremendous pressure is being exerted hydraulically, it can be applied smoothly and evenly so that tool breakage will be kept to a minimum.

Circuit No. 5 (Fig. 8) is a good example of what can be accomplished with a smooth, accurate hydraulic feed. Before this circuit was set up, the production of the part was accomplished on two lathes

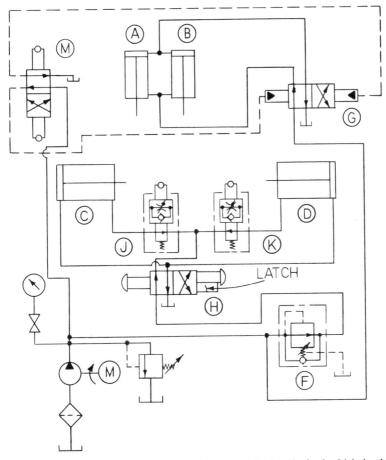

Fig. 8. Circuit which provides a smooth accurate hydraulic feed which in this instance safeguards a cutting tool. This helps to increase the tool life through decreased breakage.

being operated "around the clock." By the installation of the hydraulic circuit on one double-end machine, the same production schedule is met in one-fourth the time with far greater machining accuracy, less operator fatigue, and no down time for tool breakage. The pressure is quickly adjusted to meet the requirements of different types of metals being cut, thus eliminating tool breakage.

The circuit functions as follows:

During the stand-by period the pistons of cylinders (A), (B), (C), and (D) are in retracted position as shown. Valve (H) is then in position No. 2, as shown, with latch released and oil pressure is holding sequence valve (F) open.

Operator shifts valve (H) to position No. 1, where it is latched, thus directing pressure through valve (F), which is held open by the oil pressure, to blind end of boring bar cylinders (C) and (D), causing their pistons to advance. Almost at once, a trip dog actuates valve (M) to reverse valve (G) from the position shown, thus directing pressure to the blind ends of clamp cylinders (A) and (B) causing them to advance and clamp workpiece. While clamp cylinder pistons are advancing, those of cylinders (C) and (D) are at rest, since these cylinders are "starved" of oil by action of sequence valve (F) which has closed due to the drop in pressure as the clamp cylinders begin to move.

Spring in sequence valve (F) holds it shut until clamp pistons reach end of stroke, then, when workpiece is firmly clamped, build-up pressure opens sequence valve (F), permitting oil to flow again through valve (H). Cylinders (C) and (D) then continue their "out-stroke," rapidly, at first, until straight-line cams actuate flow-control valves (J) and (K) and then slowly at a "feed" rate determined by the setting of valves (J) and (K). Cylinder (C) should be somewhat "later" than cylinder (D), thus permitting cylinder (D) to complete its "out-stroke" first.

Then when cylinder (C) completes its "out-stroke," the latch on valve (H) is released, reversing the valve so that oil is directed to rod ends of cylinders (C) and (D), causing them to retract rapidly.

Near end of "in-stroke," a trip dog actuates valve (M) which in turn operates to reverse valve (G), causing it to return to position No. 2 in which oil is fed to the rod ends of cylinders (A) and (B) to retract and release workpiece. All action then ceases.

Use of Hand Pump for Emergency

The insertion of a hydraulic hand pump into some circuits is often a useful safety feature in event of a power failure as heavy loads may then be lifted or moved without undue effort. This added feature is extremely important in many instances and can save life and property. To serve a similar function, an accumulator is often inserted in the circuit as a pressure storage reservoir to take care of emergency power breakdowns. This is discussed in Chapter 4 on accumulators.

Circuit No. 6 (Fig. 9) shows the use of the hand pump in a hydraulic circuit controlling a penstock gate. This circuit functions as follows:

Let us assume that gate-raising piston (G) is at the end of its stroke, as shown, with gate in lowered position. To raise gate, operator momentarily depresses "raise" push button station which closes motor starter, starting pump motor. Oil pressure is then pumped into cylinder for raising gate. Gate rises to a predetermined partially open position, when limit switch (LS-2) is tripped and breaks holding current to motor starter coil. Motor starter opens and stops pump motor. Gate remains in this position as hydraulic oil is trapped in cylinder by check valves and normally closed solenoid valve (C). When penstock into which gate opens has filled, a pressure switch closes and again starts motor and gate is raised to full open position.

To lower gate by its own weight without hydraulic power, operator momentarily depresses "lower" button. This energizes relay R-1 which energizes solenoid of valve (C), opening this valve, and permits oil to exhaust from rod end of cylinder through open valves (E_1) and (B). Gate lowers at speed governed by setting of speed control valve (I).

Should gate jam and it is desired to cut in hydraulic pressure, operator depresses the "power to lower" push button. This energizes the solenoid in valve (B), closing same, and closes motor starter, starting pump motor. Hydraulic pressure is then directed to both sides of cylinder piston and gate lowers due to differential of areas between rod side and blind end of piston. Hydraulic pressure is governed by setting of relief valve (D). After jam is broken, operator can shut off motor by depressing "stop" button, which will break holding current to motor starter coil. Gate will, however, continue to lower by its own weight. When gate is fully closed, limit switch (LS-3) is opened, breaking holding current to relay (R-1) which de-energizes all electrical controls to circuit.

The two hydraulic pumps have capacities of 1.7 gpm and 15.4 gpm delivery, respectively. The combined delivery of both pumps is available to the hydraulic system up to a pressure of 880 lb per sq in. If system requirements are more than this, the unloading valve (F) will open and "dump" the 15.4 gpm pump. The hydraulic system then operates at high pressure and low speed from the 1.7 gpm pump until

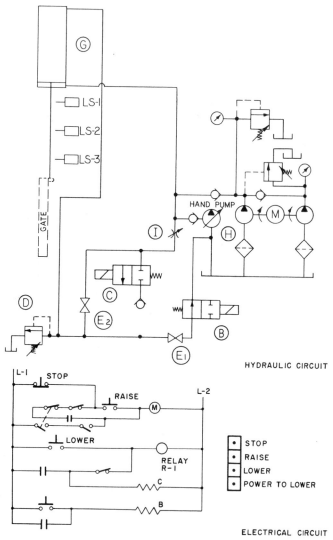

HYDRAULIC CIRCUIT

ELECTRICAL CIRCUIT

Fig. 9. The hydraulic hand pump in this hydraulic circuit which controls a pen-stock gate is a useful safety feature in that the heavy gate may be opened or closed without undue manual effort in the event of a power failure.

the system requirements drop back to 880 lb per sq in. or lower.

Hand pump (H) and shut-off valves E_1 and E_2 are incorporated in the hydraulic circuit so that the gate may be raised or lowered in event of electrical power failure. To raise gate in emergency, it is only necessary for operator to operate hand pump. However, to lower gate by its own weight in case of electric power failure, hand valve (E_2) must be opened. This permits oil to exhaust from rod end of cylinder into blind end or to sump through valves E_1 and B so that gate can lower. If a jam occurs, hand-operated valve (E_1) is then closed and operator can break jam by use of hand pump.

Use of an Accumulator as a Reserve Safety Device

There are applications where it might be necessary to provide a reserve power source which would function automatically, if necessary, to complete at least one cycle of a cylinder even if the main power supply should fail. In Fig. 10, cylinder (G) could be used to operate a butterfly valve, a machine control, a clamping device, a furnace feed device, etc. This circuit functions as follows: Pump (J) directs oil through check valves (C) and (B) to accumulator

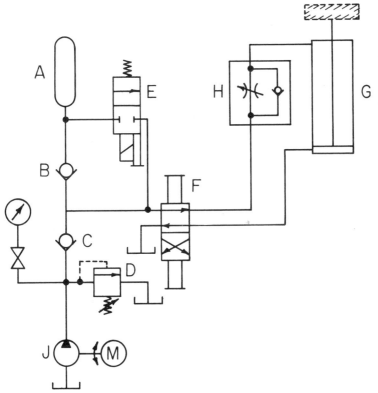

Fig. 10. Circuit showing use of an accumulator as a reserve source of power for emergency, short-interval use.

(A). Oil is also directed through four-way valve (F) and through flow-control valve (H) to the head end of cylinder (G). Oil is trapped in accumulator (A) by two-way valve (E) and check valve (B). Pressure for the system is set by relief valve (D).

To advance the piston of cylinder (G), operator shifts spool of valve (F) and oil, under pressure, is directed to cap end of cylinder (G). Piston of (G) advances at a speed controlled by valve (H). To reverse the motion of the piston of cylinder (G), operator shifts spool of valve (F) to original position, and oil, under pressure, is directed to head end of cylinder (G) and piston of (G) then retracts at a rapid rate.

Should there be an electric-power failure or a failure of the pump-spring offset, the two-way valve (E) could be actuated either automatically or through a push button, depending upon the condition of the failure. This would cause oil to be directed from accumulator (A) to valve (F). Oil could now flow back to relief valve (D) as it would be stopped by check valve (C). Operator could cause cylinder (G) to function as long as there was a sufficient supply of oil, under pressure, coming from accumulator (A). The size of the accumulator would be determined by the cylinder requirements.

Application of Hydraulic Safety Valve

On applications where a workpiece is being revolved at a relatively high speed and work is performed, such as on a lathe, grinder, machining center, etc., and where power-operated chucking equipment is holding the workpiece, provision should be made to protect the operator, the machine, and the tooling in case of power failure. Power failure could be caused by a ruptured fluid line, a pump failure, a faulty relief valve, etc. Figure 11 shows a hydraulic safety valve used in conjunction with a rotating hydraulic cylinder. This device is attached to the hydraulic distributor housing and contains a double pilot-operated check valve which is by-passed, in normal operation, by another pilot-operated check valve having a restricted flow. In normal operation, any pressure buildup due to frictional heat can relieve itself through the by-pass check valve, which is piloted open, regardless of which port is being pressurized. There must be provision in the circuitry to allow for this pressure relief further upstream.

In the event of oil-line rupture, the by-pass check valve loses its pilot pressure and closes. The check valve in the line that was pressurized closed when the cylinder was pressurized, so there are now no escape paths for the pressure within the cylinder. If the spindle continues to rotate after hydraulic power loss, pressure

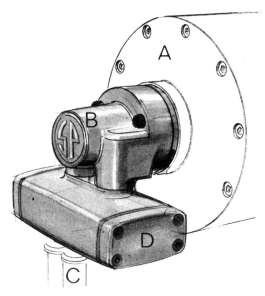

Courtesy of the SP Manufacturing Corp.

Fig. 11. Hydraulic safety valve used in conjunction with rotating hydraulic cylinder. (A) Cylinder, (B) Hydraulic distributor, (C) Hydraulic lines, and (D) Safety valve.

buildup—due to frictional heat—cannot relieve itself, and could continue to rise; therefore, a pressure switch capable of sensing the loss of pressure should be incorporated in the circuitry.

Ports are provided on the valve for relieving trapped pressure when desired. These ports can also be used to attach small accumulators which can provide an additional volume of oil, for added safety.

CHAPTER 12

Sequencing of Hydraulic Cylinder Motion

Nearly every hydraulic circuit having two or more cylinders has some kind of sequence of operation. This sequencing of hydraulic cylinder motion may be accomplished in a number of ways such as: (1) with sequence valves; (2) with a cam valve; (3) with a dual-pressure pump; (4) with a camshaft drive; (5) by using pilot operated controls; and (6) with solenoid controls. Whatever means are utilized, it is important to remember in designing a sequential circuit that the least number of components required to do the job satisfactorily should be employed. Circuits overloaded with valves may be low in efficiency and high in maintenance cost. The various ways in which sequence operations may be controlled will now be described.

Sequence Valves

Sequence valves are used extensively for sequencing operations when two or three cylinders are involved. These valves, which are usually spring loaded, have been described in Chapter 6. Since a sequence valve is set to remain closed at one pressure and to open at a higher pressure, it is not applicable where the operating pressure of the second cylinder must be lower than that of the first one.

Circuit No. 1 (Fig. 1) shows the use of a sequence valve in a two-cylinder circuit.

1. Operator loads casting into a vise which is operated by cylinder (4). He then shifts handle of two-position, four-way control valve (2) and oil flows from pump to blind end of cylinder (4), while oil in rod end flows out to sump through valve (3), and casting is securely clamped in vise. Oil pressure then builds up, opening sequence valve (6) and oil flows to blind end of drill cylinder (5), while oil in rod end flows to sump through valve (2). Piston rod of drill cylinder brings drill in to the workpiece, performing the drilling operation.

2. Operator shifts handle of valve (2) back to original position. Oil flows to rod end of drill cylinder (5) while oil in blind end flows to sump through valve (6) and drill retracts. Oil pressure then builds up opening sequence valve (3) and oil flows to rod end of vise

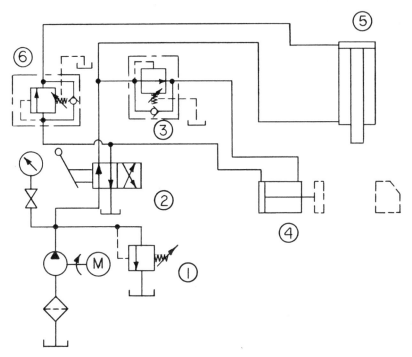

Fig. 1. Two-cylinder hydraulic circuit employing a sequence valve.

cylinder, while oil in blind end flows to sump through valve (2), opening jaws of vise.

3. Operator removes workpiece and is ready for next cycle.

Sequence Operation by Use of Cam Valve

A cam-operated, two-position, four-way control valve used in conjunction with a manually operated, four-way control valve gives a very positive sequence of operations with minimum time lag between one operation and the next. The sequence of operations in this type of circuit is controlled by the position of the piston rod rather than by pressure through a sequence valve.

Circuit No. 2 (Fig. 2) operates as follows (piston rods of both cylinders are retracted in starting position):

1. Operator moves workpiece in tray onto elevating mechanism. He then shifts handle of four-way two-position valve (3) allowing oil to flow from pump to blind end of elevating cylinder (4) while oil in rod end flows to sump through valves (5) and (3). The elevator raises the load at a rate set by flow control valve (5). As

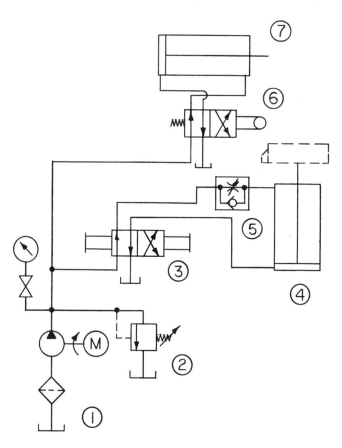

Fig. 2. Hydraulic circuit using a cam-operated two-position four-way control valve in conjunction with a manually operated four-way control valve to give a very positive sequence of operations with a minimum time lag between one operation and the next.

elevator nears top of stroke, a cam on the side of the platform contacts roller of four-way cam valve (6) which shifts, allowing oil to flow to blind end of ejection cylinder (7), while oil in rod end flows to sump through valve (6), and end of piston rod moves into contact with tray and shoves it onto conveyor.

2. Operator then shifts handle of valve (3) to original position and oil flows to rod end of elevating cylinder while oil in blind end flows out to sump through valve (3) and elevator begins to retract. As cam on platform of elevator rides off of cam roller on valve (6) the valve is shifted by spring action and oil flows to rod end of ejection cylinder (7) while oil in blind end passes directly to sump through valve (6). Piston rods of both cylinders retract to their original position. Operator is then ready to move next tray onto elevator.

It must be remembered that when both the cylinders are in motion, their speeds will be considerably less than when only one cylinder is in motion since the pump delivers the same amount of oil in both cases. In this circuit, the two cylinders are in motion together on the retracting stroke but the area of their piston rods sufficiently reduces the working area of the pistons so that they move at a suitable retracting speed.

Dual Pressure Pump in Sequential Operation

Circuit No. 3 (Fig. 3), while somewhat novel in design, is very simple in principle and can be used to good advantage in a number of applications. In order to make this circuit perform properly there must be a considerable difference in the two loads involved and the light duty must be performed by the cylinder which operates first.

This circuit as applied to a machine for performing a burnishing operation functions as follows:

1. Operator places workpiece in fixture and then depresses electric push button (A) which energizes solenoid X of four-way valve (2). Oil flows from large-volume pump (1) and small-volume pump (5) at low pressure, through valve (2) on to the blind end of cylinder (3). Oil also flows to blind end of cylinder (4) but the force it provides at low pressure is not great enough to move piston of cylinder (4). As piston of cylinder (3) moves forward, the oil in the rod end of this cylinder flows to the sump through valve (2). As locking device which is connected to end of piston rod locks the workpiece in the fixture, the pressure builds up and shifts piston in unloading valve (6) so that large-volume pump (1) is unloaded back to sump. Pump (5) continues to supply oil at high pressure which is controlled by relief valve (7). This oil cannot flow toward pump (1) as it is stopped by check valve (8). The high pressure as created by pump (5) moves piston of burnishing cylinder (4) forward and oil in rod end of this cylinder flows back through valve (2) to sump.

When cylinder (4) reaches end of stroke, operator depresses push-button (B) energizing solenoid (Y) of valve (2). The pressure in the supply line is reduced, valve (6) closes and oil from pumps (1) and (5) flows to the rod ends of cylinders (3) and (4) and oil in their blind ends flows through valve (2) to sump. Rams of both cylinders retract under low pressure since the force required to return them is very small. When both rams reach the retracted position, pressure in the supply line again builds up, valve (6) reopens and large-volume pump (1) is spilled to the sump during the stand-by period at no pressure while the rams are held retracted at high pressure by pump (5). Oil at high pressure is spilling through relief valve (7) during this period, but since it is only in low volume, not much heat is created. This high-low pump arrangement saves considerable horsepower.

Camshaft Drive for Sequential Operation

By the use of a camshaft driven by a speed reducer connected to a small electric motor, a sequence of operations may be set up for any number of cylinders. There are several types of controls available for such an operation. The camshaft lobes may contact limit switches, three-way hydraulic pilot valves, or three-way air-operated pilot valves. The limit switches, when contacted, allow electric current to energize the solenoids connected to the ends of the master valve thus shifting the spool of the master valve. If the master valve is of the single-solenoid type with spring-return feature, only one limit switch is required per master valve. As the lobe of the cam rides off of the limit switch roller, the spring shifts the valve back to original position and the contact is broken. When using the single-solenoid master valve, the cam must be of relatively large diameter and the camshaft must turn very slowly so that other valves can be cut in to set up the sequential operation of several cylinders. Instead of using several cams on a camshaft, one cam plate of larger diameter may be used to operate several limit switches in one revolution of the cam plate. This is a more practical arrangement where double-solenoid valves are used. Figure 4 is a sketch showing such a cam plate and limit switch layout.

When using three-way hydraulic pilot valves to operate the master valve, they work on the same principle as limit switches except oil is the operating medium to the master valve instead of electric current.

Circuit No. 4 (Fig. 5) shows the use of a cam shaft operating three-way air pilot valves. The circuit functions as follows:

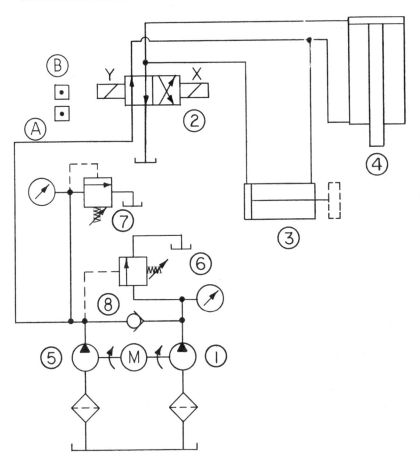

Fig. 3. Hydraulic circuit which performs properly when there is a considerable difference in the two loads involved, the light duty being performed by the cylinder which operates first.

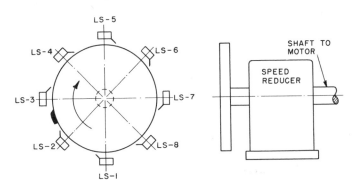

Fig. 4. Sketch of a cam plate and limit switch layout.

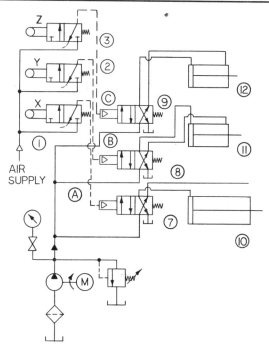

Fig. 5. Circuit showing the use of a cam shaft in operating three-way air-pilot valves.

1. Operator places workpiece in fixture, then starts electric motor. As camshaft starts rotation, cam depresses cam roller of three-way valve (1) allowing air to flow to air pilot chamber (A) of four-way, two-position, spring-return hydraulic valve (7) shifting valve piston and allowing oil to flow to blind end of cylinder (10) while oil in rod end flows out to sump through valve (7). Piston moves out performing first operation. As cam rides off of roller of valve (1), air is exhausted from chamber (A) and piston of valve (7) returns to original position and piston of cylinder (10) retracts. Cam (Y) then depresses the roller on valve (2) as the cam shaft continues on the first revolution and air flows to chamber (B) of valve (8) and piston is shifted allowing oil to flow to blind end of cylinder (11). Piston then moves out performing second operation. As cam (Y) rides off of roller of valve (2) and air exhausts from chamber (B) of valve (8), piston of cylinder (11) retracts to original position. Cam (Z) then contacts roller of valve (3) and air flows to chamber (C) of valve (9) shifting valve piston and allowing oil to flow to blind end of cylinder (12). Piston rod moves out performing third opera-

tion. As cam (Z) rides off of roller of valve (3), air is exhausted from chamber (C) of valve (9) and valve shifts to original position allowing oil to flow to rod end of cylinder (12) and piston retracts.

2. Operator stops motor at end of one revolution and removes workpiece from fixture.

By changing the cam design and repositioning the pilot valves, other sequences can be worked out. The use of this scheme allows flexible design in sequencing circuits at a very reasonable cost. All of the units are simple in construction and require little maintenance.

Pilot-operated Control Used for Reciprocating Action in Sequence Circuit

Circuit No. 5 (Fig. 6) makes use of pilot control for high-speed reciprocation of a hydraulic cylinder after the part is securely held. The circuit as applied to a planing operation functions as follows:

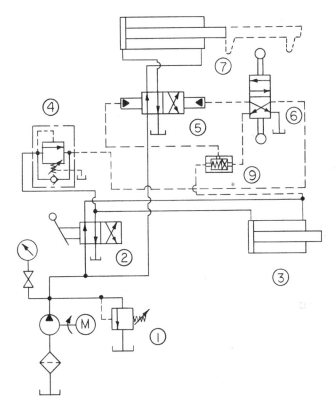

Fig. 6. Circuit making use of a pilot control for high-speed reciprocation of hydraulic cylinder after the part is securely held.

1. Operator shifts handle of four-way, two-position valve (2) and oil flows from pump to blind end of clamp cylinder (3) while oil in rod end exhausts to sump through valve (2). Piston rod of clamp cylinder (3) moves forward securely clamping workpiece which has moved into fixture from conveyor. Line pressure now builds up opening sequence valve (4) allowing oil to flow to pilot valve (6) which controls the flow of oil to the pilot cylinders of valve (5). Valve (6) has a trip control and there are trip dogs on the piston rod of cylinder (7). As long as oil pressure is applied to valve (6), valve (5) will be actuated and the piston of cylinder (7) will reciprocate. After operation is completed, operator shifts handle of valve (2) to original position. Piston of cylinder (7) will return to starting position due to positioning of spool in valve (5) caused by the oil flowing directly from valve (2) through spring offset shuttle valve (9) to the oil pilot connection. This feature assures positive retraction of the piston of cylinder (7) even though valve (2) is shifted when piston of cylinder (7) is moving on the outstroke. Piston of cylinder (3) returns to original position, allowing workpiece to drop down to outbound conveyor thus completing the cycle.

Another use for this sequence cycle is on hydraulic tamping operations where cylinder (3) does the clamping and cylinder (7) does the tamping. Fast heavy blows can be administered with very little valving. The speed of reciprocation may reach several hundred cycles per minute depending upon the length of stroke, the pressure required, and the size of the power unit.

Solenoid Controls Used in Sequence

Circuit No. 6 in Fig. 7 shows one of the ways in which solenoid controls can be arranged to set up a sequential operation. In this case, a flywheel (7) is to be operated at various speeds and oil pressures in the same way that the crankshaft of an automobile engine is driven by the movement of the pistons and connecting rods.

Solenoid-operated control valves (1), (2), and (3) are used to control the flow of oil to cylinders (4), (5), and (6), respectively, the relief valve (8) serving to prevent build-up of supply line pressure beyond the setting of the valve. In operation, the trip on the flywheel contacts a series of limit switches as it rotates and causes solenoids of valves (1), (2), and (3) to open and close these valves in sequence. The resulting flow of oil to each cylinder in turn provides the power to turn the flywheel. Each cylinder has pressure on the blind end for 180 degrees of the cycle, followed by pressure on the rod end for the remaining 180 degrees.

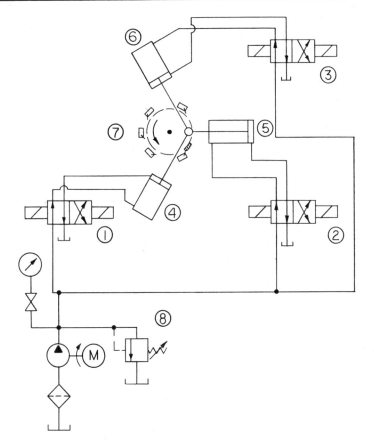

Fig. 7. One of the ways in which solenoid controls can be arranged to set up a sequential operation is illustrated in this circuit.

To eliminate any side thrust or bending action on the piston rods, each cylinder is pivot mounted, and flexible hose connections are provided to permit free movement as the cylinders oscillate.

CHAPTER 13

Packings and Seals

The important role that packings and seals play in fluid power systems cannot be overemphasized. As mentioned in a previous chapter, it was not until Joseph Bramah's invention of the cup packing for hydraulic pistons, late in the eighteenth century, that the application of hydraulic power to industry was made possible. Today, packings and seals of a variety of designs and materials are essential elements of hydraulic devices. But even though a great deal has been accomplished in solving the problem of packing or sealing pistons and piston rods of cylinders, stems of valves, shafts of pumps, rotary joints of couplings and distributors, and other fluid power components, there is still much to be done before completely satisfactory sealing means for all applications are developed.

Applications of Packings and Seals

The wide use of packings and seals in fluid power devices is indicated by the following tabulation:

Fluid Power Device	Type of Seal or Packing
Pumps	Shaft Seals Gasket Seals Internal Seals
Oil Reservoirs	Gasket Seals
Cylinders	Rod Packings Piston Packings Cover Gaskets or Seals Cushion Needle Packings Rotary Joint Packings for Rotating Cylinders

Valves	Cover Gaskets
	Stem Packings
	Needle Packings
	Piston Packings

Materials Used for Packings and Seals

In the early days of fluid power, packings were made of such materials as rope, sawdust, rags, etc., jammed tightly into a stuffing box by means of a packing gland. Their use led inevitably to extrusion of the packing through clearance spaces, rapid wear, and almost continual leakage. As a result, they demanded almost constant attention and adjustment. Today, packings and seals are made of materials which have been carefully chosen or developed for the job they have to do. These include: leather; leather impregnated with synthetic rubber or wax; homogeneous synthetic rubbers such as "Neoprene," "butyl," various types of "Bunas," and "Viton"; plastics such as "Teflon," "Kel-F," and "Polyurethane"; fabricated synthetics such as woven cotten duck, canvas, or asbestos cloth impregnated or bound with synthetics; asbestos; felt; bronze or cast iron, as used in piston rings; and copper, stainless steel, "Monel," and other materials which are used for gasket seals.

Fabricated versus Homogeneous Rubber Packings

Two classes of packings which are widely used in hydraulic system components are the fabricated and the homogeneous synthetic rubber packings.

The fabricated packings, which are made by moulding woven cotton duck or fine-weave asbestos cloth impregnated with rubber or synthetic rubber of various types, are by nature of their inherent make-up capable of operating over a wide temperature range, from 0° to 600°F.

In general terms, fabricated packings, as compared with homogeneous packings, are adapted to the following conditions:

1. Ordinary metal finishes
2. Greater machine clearances
3. Higher temperatures
4. Pressures up to 10,000 lb per sq in.
5. Alkali and acid conditions

Unlike the fabricated class, the homogeneous packings are elastic in character. Their use is adapted to the following conditions:

1. Fine metal finishes
2. Minimum machine clearances
3. Minimum eccentricity
4. Temperatures of sub-zero to 600°F for short periods
5. Pressures up to 5,000 lb per sq in. (motion)
6. Pressures up to 15,000 lb per sq in. (static)

Although there are a few applications where either fabricated or homogeneous packings can be applied, as a rule each class and type has its own particular application and place. It depends entirely upon the type of equipment and operating conditions as to which will give the best results.

There are three basic factors which determine the suitability of homogeneous synthetic rubber packings for a given application:

1. Diametral clearances must be held to from 0.005 to 0.008 inch, depending on size.

2. Eccentricity must be held to from 0.002 to 0.005 inch, depending on size.

3. All surface finishes through which the packing moves must be ground and honed to a maximum roughness of 16 microinches rms, as measured with a profilometer.

Unless the above qualifications are adhered to, it is not advisable to use the homogeneous class of packings, especially "O" rings.

Types of Packings and Seals

There are many types of packings and seals: "O" rings, "V" packings, "U" packings, "U"-cup packings, flange or hat packings, cup packings, diaphragms, gaskets, and special shapes. Each has its preferred place in fluid power equipment, although some types are suitable for several kinds of applications. Also available are materials in the form of sheets and ropes which are used for many general applications.

Seven general types of seals and packings used in hydraulic systems are:

1. *The "O"-ring packing and gasket.* As shown in Fig. 1A, this type is in the shape of a toroid or ring with a round cross section and is made of one of the homogeneous synthetic rubber compounds. "O"-ring *packings* are used where there is either reciprocating or rotating motion. On rotating shafts they are limited to applications where the speed is low—200 surface feet per minute, or less. "O"-ring packings are generally not used for pressures above 1500 lb per sq in., although when installed with back-up rings they have

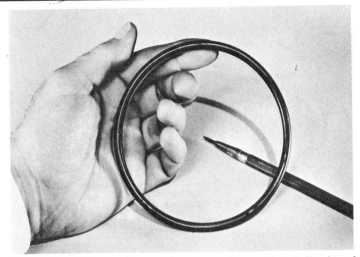

Courtesy of E. F. Houghton & Co.

Fig. 1A. A synthetic rubber "O"-ring packing.

been used for pressures as high as 5000 lb per sq in. 3000 lb per sq in. is the usual limit. "O"-ring packings can also be used as static seals.

"O"-ring *gaskets* are used only as static seals where there is no movement, as for sealing cylinder heads, and have been successfully employed for pressures ranging up to 15,000 lb per sq in. An

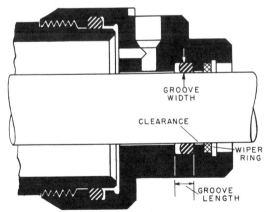

Courtesy of Mobil Oil Corp.

Fig. 1B. "O"-ring seal. The clearance that exists between the rod and the outside edge of the groove is an important factor in seal performance. When the rod and groove are properly dimensioned, the ring will be squeezed slightly as installed.

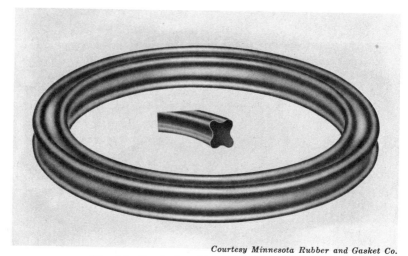

Fig. 2A. Synthetic rubber "Quad-Ring" seal.

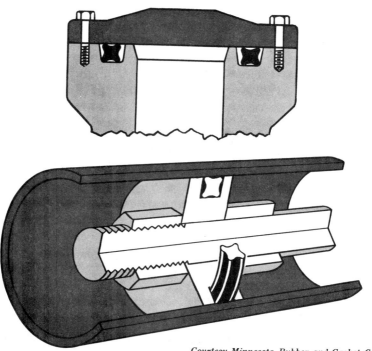

Fig. 2B. (*Upper*) "Quad-Ring" being used as a cover gasket.   (*Lower*) "Quad-Ring" being used as a piston packing.

"O"-ring piston rod packing and an "O"-ring cylinder gasket are shown installed in Fig. 1B.

2. *The "Quad-Ring" packings and gaskets.* This is a relatively new type of industrial seal which is made up in the form of a homogeneous synthetic rubber ring with a modified square type of cross section, as shown in Fig. 2A. It can be used as a packing for reciprocating or rotary motion and also as a static seal. Figure 2B shows a "Quad-Ring" used as a piston packing and also a "Quad-Ring" used as a cover gasket. "Quad-Rings" are applicable for use over a wide range of pressures. They are especially suitable where pressures are low, particularly below eight lbs per sq in., and also for extremely high pressures. They are also useful where lubrication is difficult yet friction must be kept to a minimum.

3. *The "V" packings.* The *fabricated* type of "V" packing is furnished in the form of a single ring (not in coils or spirals) and is usually split on an angle unless a solid ring is requested. These rings are also known as pressure rings. Complete sets are furnished, their shape permitting a number of them to be stacked together to form a composite packing, as shown in Fig. 3. The male and female adapters shown at either end of the stack are usually made of the same material as the rings but are harder in texture. Sometimes metal male and female rings which are commonly referred to as *support* rings may be used instead. "V" rings and support rings are shown installed in Fig. 4.

Fabricated "V"-ring packings are capable of holding almost any hydraulic or pneumatic pressure that may be experienced in industrial applications. Leather "V"s, such as that shown in Fig. 5, are able to withstand even higher pressures than can the fabricated "V"s.

The *homogeneous* type is furnished in solid ring form only and is supplied as individual units. These are not recommended for pressures over 5,000 lb per sq in.

4. *The cup packing.* Leather cup packings have long been accepted as standard for high-pressure service. Recent developments include a rubber-impregnated leather packing (Fig. 6) combining the best features of both materials, also a new design with a slight flare on the side walls and square shoulders—a "pre-loaded" packing.

Cup packings are also available in fabricated and homogeneous rubber compositions. The *fabricated* cup packing is shown in Fig. 7. A homogeneous cup packing is also available but is not widely used. "O"-ring packings are preferred in place of these, for hydraulic service, unless friction is a critical factor. Packings of the cup type are commonly used on the ends of pistons or plungers.

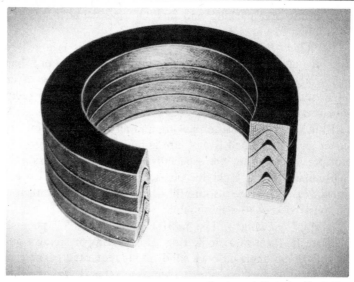

Courtesy of Raybestos-Manhattan, Inc.

Fig. 3. A composite packing of "V" rings.

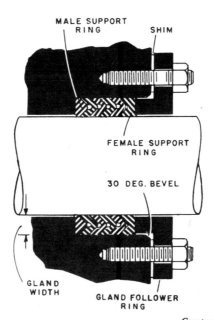

Courtesy of Mobil Oil Corp.

Fig. 4. "V"-ring seal. Gland follower ring is bolted metal-to-metal with a shim inserted to prevent the follower ring from exerting pressure on the sealing rings. Note the chamfer on the edge of the gland which greatly aids installation of the rings.

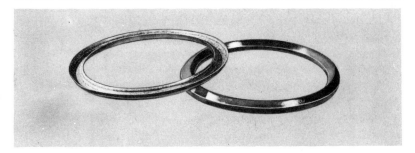

Courtesy of E. F. Houghton & Co.

Fig. 5. Top and bottom views of leather "V"-rings.

Courtesy of E. F. Houghton & Co.

Fig. 6. Rubber-impregnated leather cup packing.

Courtesy of E. F. Houghton & Co.

Fig. 7. A fabricated cup packing.

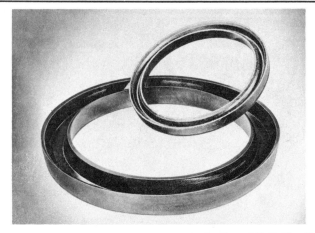

Courtesy of E. F. Houghton & Co.

Fig. 8. A wax-impregnated leather "U" packing.

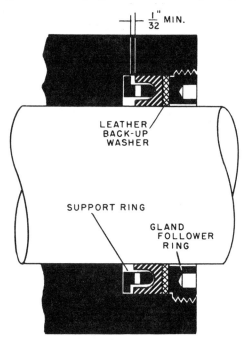

Courtesy of Mobil Oil Corp.

Fig. 9. "U"-ring seal. The support ring should be machined so it will center in the gland and the follower ring should be adjusted so that no pressure is exerted between the sealing ring and the support ring. The clearance between the rod and the follower ring should be as small as consistent with good shop practice. When it is excessive, a leather back-up ring should be used to prevent extrusion.

Courtesy of E. F. Houghton & Co.

Fig. 10. A "U"-cup packing.

5. *The "U" packing.* The fabricated "U" packing is shown in Fig. 8. This packing is also made from wax-impregnated leather. The "U" packing is never used in sets, one nested on top of another, but is, strictly speaking, a unit seal. A "U" ring with support ring and leather back-up washer is shown installed in Fig. 9.

6. *The "U"-cup packing.* The "U"-cup packing is shown in Fig. 10 and is furnished in homogeneous synthetic rubber of 70 Shore durometer hardness for pressures of less than 1000 lb per sq in. and

Courtesy of E. F. Houghton & Co.

Fig. 11. A fabricated flange packing sometimes referred to as a hat packing.

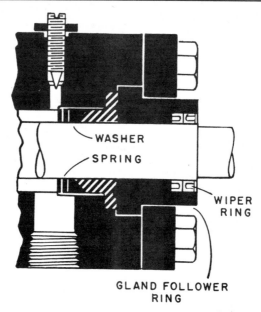

WASHER

SPRING

WIPER
RING

GLAND FOLLOWER
RING

Courtesy of Mobil Oil Corp.

Fig. 12. Flange seal. In this design, initial pressure is applied to the lip of the seal by a washer and spring. "U"-rings are also used for dirt and oil wiper rings.

in synthetic rubber of 90 Shore durometer hardness for pressures of 1000 to 2000 lb per sq in.

7. *The flange packing.* The *fabricated* flange packing, sometimes referred to as the *hat* packing, Fig. 11, is recommended for use only where a "V" packing or a "U" packing cannot be used due to lack of space, especially for rotary motion. It is also available in leather and in homogeneous rubber composition but is the least popular of the homogeneous packings. A flange seal used with washer and spring is shown installed in Fig. 12.

In all of these packings, the sealing material is flexible and generally elastic. The pressure of the fluid forces the sealing material against the members being sealed. None of these packings produce their sealing effect as a result of pressure applied by a gland follower.

In the later sections of this chapter, the applications of each of these seven general types of packings and seals will be discussed.

Packing Standards

A move which began with the standardizing of "O" ring sizes in Army-Navy specifications has spread to the establishment of standards for all types of hydraulic and pneumatic packings. This was prompted by the Joint Industry Conference called by automotive engineers in 1949, which was endeavoring to set up standards for all types of replacement parts, including hydraulic mechanisms.

As a result, the packing industry is gradually adopting standards for the various types of packings so that users can stock by code number, or "Dash" number, as it is termed, to indicate exact size. For example, a Dash 30 "O" ring is $\frac{3}{16}$-inch thick, $1\frac{3}{4}$-inch inside diameter and $2\frac{1}{8}$-inch outside diameter. Plates containing Dash numbers can be attached to hydraulic equipment showing the code. For instance, a marking of LV-40-.62 would mean Leather "V" packing, Dash 40, stack height .62 inch.

Adoption of standards means that all "V" packings will be interchangeable, whether they are rubber, homogeneous or fabricated, or leather. The same is true of "U" packings or cups of varied materials. And the dimensions will not vary as between manufacturers, which will be of great aid to packing buyers. Standards are rapidly being adopted by manufacturers using packings in their assemblies. The packing suppliers have tables and drawings available to users upon request.

Factors Affecting Selection of Packings and Seals

There is such a wide variance in operating conditions, component design, and system design that it is almost obvious that no one type of packing yet developed will take care of all conditions. Under operating conditions must be considered: pressure—whether low or high; temperature—maximum and minimum, as well as the range; type of medium—water, oil, air, etc.; speeds; metal clearances; metal finishes; motion—rotary or reciprocating; allowable space—"V" seals, for example, because of the number of "V" rings required (never less than three) need glands of greater depth than other types of packing; alignment; atmosphere—clean or dirty; and life expectancy.

Probably the two most common enemies of hydraulic packings are dirt and heat. To cite one example, by reducing the operating temperature of a hydraulic system by from 40° to 60°F, it was found that a system which leaked at every packing at the higher temperature became free of leaks at the lower temperature. In many cases the effects of dirt can be avoided by proper location of the equipment and careful maintenance; in other cases the condition of the surrounding atmosphere may be such that the best that can be done is to select the kind of packing which will stand up longest under the severe conditions to which it will be subjected.

Alignment may seem like a trivial matter in reference to packings, but if a piston rod of a long-stroke cylinder is not properly guided or aligned with a bearing support, the side thrust against one type of packing will cause leakage much quicker than with one of a different type.

The following table may be used as a general guide to the applicability of leather and synthetic rubber packings for various fluids and operating conditions. The manufacturer should be consulted whenever there are conflicting requirements and a clear-cut selection cannot be made.

TABLE I

APPLICABILITY OF LEATHER AND SYNTHETIC RUBBER PACKINGS
FOR VARIOUS FLUIDS AND OPERATING CONDITIONS

Key: G = Good; F = Fair; P = Poor; and X = not recommended

Fluids and Operating Conditions	Type of Packing		
	Leather	Synthetic Rubber	
		Homogeneous	Fabricated
Oil	G	G	G
Air	G	G	G
Water	G	G	G
Oxygen	G	X	X
Steam	X	X	G
Some Solvents	X	G	G
Mild Acids	X	X	F
Mild Caustic	X	F	F
Resistance to Heat	+180°F (Max)	+250°F (Max)	+700°F (Max)
Resistance to Cold	−65°F (Min)	−65°F (Min)	−20°F (Min)
Rotary Motion	G	P	F
Reciprocating Motion	G	G	G
Metal Finishes (Micro Inches) Max.)	40	20	40
Types of Metal	Ferrous and non-Ferrous	*Ferrous, except cast iron	*Ferrous
Back Support Clearances (Tolerance)	Medium	Very close	Close
Resistance to Cold Flow (Extrusion)	G	P	F
Friction (Coefficient)	Low	High	Medium
Resistance to Abrasion	G	P	F
Pressures, p.s.i. (Max.)	125,000	5,000	10,000

* Chromium-plated steel and only those non-ferrous alloys which have desired hard smooth surface.

Life of Sealing Rings

When correctly selected sealing rings are properly installed in a gland that is properly designed and finished for the particular seal selected, there should be only enough leakage to assure lubrication of the packing. Under normal conditions a seal should last at least one year. Under severe conditions a seal may last only six months; in light service, it may last one and a half to two years or longer. The service rendered by any seal obviously will be influenced by: (1) its original design; (2) installation procedures; (3) operating conditions; and (4) mechanical maintenance.

"O"-ring Packings and Gaskets

Simple, rugged, and dependable, the "O"-ring packing has grown in popularity by leaps and bounds in recent years. It makes an eminently satisfactory seal between reciprocating pistons and cylinders at pressures ranging up to 1500 lb per sq in. and when used with leather or synthetic back-up washers, it is suitable for applications ranging up to 3000 lb per sq in. The "O"-ring packing can also be used where there is oscillating or rotating motion, providing speeds are slow (200 ft per min, max) and pressures are low.

Use of Back-up Washers. The action of the "O"-ring packing in the rectangular groove, in the "V" groove, and in the rectangular groove with back-up washers at various pressures is shown in Fig. 13. It can readily be seen that without back-up washers, as the pressure increases, the ring is forced or extruded into the clearance space. When the back-up washers are used, however, this is prevented and packing life is thus extended. Leather back-up washers are favored by some engineers for pressures under 1500 lb per sq in. Although not needed to prevent extrusion at these pressures, they do absorb oil and provide a means of lubrication for the "O" ring.

Leather back-up washers should not be cut. Results have shown that when using split washers the "O" rings and washers have a tendency to fail at the cut section. Synthetic spiral back-up rings are used for pressures to 3000 psi.

In the case of thin-walled cylinders, even though leather back-up washers are used for pressures over 1500 lb per sq in., the diametral expansion of the cylinders under pressure, called "breathing," may cause the metal clearances to increase, resulting in some extrusion of the packing.

Either one or two washers can be used with "O"-ring gaskets. However, if only one washer is used, it should be placed on the side

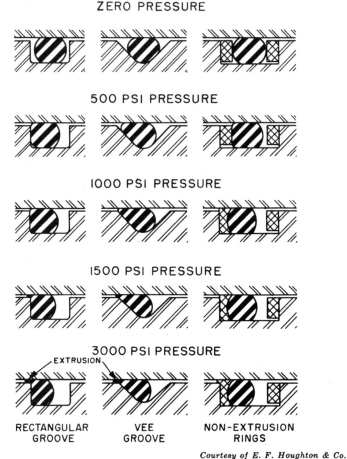

Courtesy of E. F. Houghton & Co.

Fig. 13. Relative positions of "O"-ring packings in different grooves at increasing pressures. Where the packing is under pressure, the pressure is being applied from the right-hand side of the groove.

of the ring away from the pressure, as this is the side where extrusion of the ring would otherwise take place.

The same principles which apply for "O"-ring packings and washers also apply for "O"-ring and back-up gaskets.

Pneumatic Service Applications. When "O"-ring packings are used in pneumatic service, the packings require lubrication. Otherwise they will abrade, twist, and fail very rapidly. Under such conditions, friction will also become excessive. At one time grease was used as a lubricant, but oil is now generally recommended.

One approved method for lubricating pneumatic systems is to feed an oil fog into the air stream. This method not only lubricates the "O" rings, but also the entire internal system. Such an automatic air line lubricator is shown in Chapter 14. It can be installed in the air line where it regulates the air pressure and oil flow and at the same time filters the air and traps and drains off moisture. The flow rate is always visible to the operator and oil is injected only when air is flowing.

Use of "O" Rings as Static Seals or Gaskets. "O"-ring gaskets are intended for use as static seals, where no movement takes place. They are being used rather extensively for cylinder heads and other similar types of equipment where severe compression of the gasket is not desirable. It eliminates the tendency—as when flat gaskets are used—to submit cylinder covers and cover bolts to undue stress caused by excessive tightening just to make certain that no leak will occur. When "O"-ring gaskets are used, it is only necessary to pull the cover screws up fingertight to effect a perfect seal.

Where "O"-ring gaskets are used to seal internal pressure, the outside diameter of the groove should equal the outside diameter of the "O" ring. If, however, an external pressure or internal vacuum is being sealed, the internal diameter of the groove should equal the internal diameter of the "O" ring.

"Quad-Ring" Packings and Gaskets

Because of their cross-sectional design, "Quad-Rings" are particularly useful as piston seals in those cases where "O"-rings have a tendency to "spiral twisting" with accompanying slight leakage of oil. (This twisting action results when the piston thrust tends to roll one section of an "O"-ring while another section is sliding.) Another important application is for packings at low pressures—below eight pounds per sq in. When so used, they eliminate "torque leak".

"Quad-Rings" are also suitable as seals on rotating devices. In such applications it is usually important to avoid excessive seal pressure on the shaft. This requirement is easily met because of the small amount of squeeze needed—one to two per cent of the cross section is recommended. An added advantage is the low break-away friction drag.

When used as a static seal or gasket, the design of the "Quad-Ring" prevents excessive distortion under pressure.

"Quad-Ring" sizes are comparable to those of "O"-rings in wall thicknesses and diameters and all grooved dimensions exactly correspond to those recommended for "O"-ring installations.

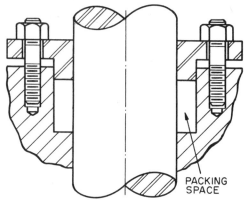

Courtesy of E. F. Houghton & Co.
Fig. 14. Stuffing box with a square-nosed gland ring.

"V" Packings

Of the fabricated class of packings, the "V" type is the most popular. As previously stated, it is furnished in single rings which are split at an angle, unless solid rings are requested. Solid rings are recommended wherever they are feasible. In either case, the rings are assembled in sets to form a complete packing. When split packings are used, the joints on successive packings are staggered alternately 180 degrees and 90 degrees apart.

"V" packings are made slightly oversize at the heel and slightly undersize at the lips to the nominal inside diameter and slightly undersize at the heel and oversize at the lip to the nominal outside diameter dimensions. Hence, when the rings are used on a shaft or rod of

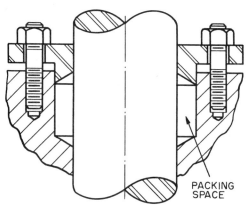

Courtesy of E. F. Houghton & Co.
Fig. 15. Beveled stuffing box.

correct diameter and in a gland of correct outside diameter, the lips are preloaded, yet the rings are easy to install. The shaft or rod diameter should be 0.002 to 0.003 inch less than the nominal inside diameter of the ring, but not more than 0.004 to 0.005 inch less. The bore of the gland should be 0.002 to 0.003 inch greater than the nominal outside diameter of the ring, but not more than 0.004 to 0.005 inch. If, as a result of wear, the rod or shaft is smaller, proper preloading of the packing will not be obtained. If a rebuilt shaft or rod is larger, the packing will be too tight, excessive friction and wear may result, and installation difficulties will be encountered which may ruin the packing. A rod or shaft worn beyond the above tolerance should be built up and refinished to the correct dimensions.

Because of the preloading, "V" packings do not require gland pressure. The gland must be deep enough to accommodate the required number of rings plus the male and female support rings and to permit the gland follower ring to be bolted or screwed metal-to-metal. Any adjustment of the gland ring made necessary because of varying stack heights, which in the case of fabricated rings may be as much as 0.010 to 0.020 inch per ring, should be made with spacers or shims.

In most applications the end of the stuffing box and the nose of the gland ring are square as shown in Fig. 14. However, where beveled stuffing boxes are used, as shown in Fig. 15, male and female adapters should match the degree of bevel.

For best results the "V" packing should be supported with metal—aluminum alloy or bronze is recommended—male and female support rings, dimensions for which are shown in Table 2.

The female metal support ring must fit both the gland and the shaft as closely as shop practice will permit. The maximum diametral clearance should not be over 0.008 inch. Larger clearances permit the rubber to extrude into the clearance space. Where the gland follower ring is used as the female support ring or where fabricated rubber support rings are used, the gland follower ring clearance should be as close as possible, commensurate with good practice for sliding fits.

Homogeneous "V" packings are furnished in the form of individual solid rings. The number of packings per set depends on the operating pressures as indicated in Table 3. These packings are not recommended for pressures over 5000 lb per sq in.

Homogeneous "V" packings can be used for both outside packed and inside packed installations and for reciprocating or rotary motion. For rotary motion application, the recommended maximum speed in

TABLE 2

DIMENSIONAL DATA FOR MACHINING METAL
SUPPORT RINGS FOR FABRICATED "V" PACKINGS

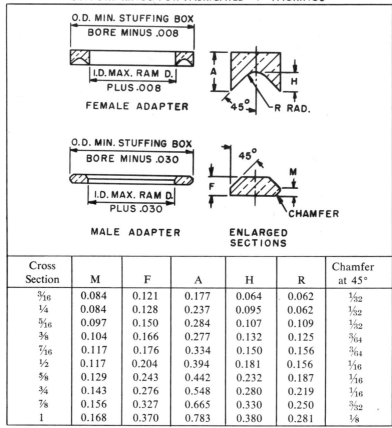

Cross Section	M	F	A	H	R	Chamfer at 45°
$\frac{3}{16}$	0.084	0.121	0.177	0.064	0.062	$\frac{1}{32}$
$\frac{1}{4}$	0.084	0.128	0.237	0.095	0.062	$\frac{1}{32}$
$\frac{5}{16}$	0.097	0.150	0.284	0.107	0.109	$\frac{1}{32}$
$\frac{3}{8}$	0.104	0.166	0.277	0.132	0.125	$\frac{3}{64}$
$\frac{7}{16}$	0.117	0.176	0.334	0.150	0.156	$\frac{3}{64}$
$\frac{1}{2}$	0.117	0.204	0.394	0.181	0.156	$\frac{1}{16}$
$\frac{5}{8}$	0.129	0.243	0.442	0.232	0.187	$\frac{1}{16}$
$\frac{3}{4}$	0.143	0.276	0.548	0.280	0.219	$\frac{1}{16}$
$\frac{7}{8}$	0.156	0.327	0.665	0.330	0.250	$\frac{3}{32}$
1	0.168	0.370	0.783	0.380	0.281	$\frac{1}{8}$

Notes. 1. Recommended material is bearing bronze.
2. Clearances shown on I.D. and O.D. of rings are maximum recommended for best packing performance.
3. Clearances on large diameters should be held as close to recommended values as shop practice will permit.
4. Female ring may be machined as part of gland ring, provided it is of the bolted type.
5. All dimensions in inches.

surface velocity is 200 feet per minute based on pressures of 750 lb per sq in. For higher pressures, the speed should be reduced proportionately depending largely upon the type of fluid, lubrication, and temperature. The maximum and minimum temperatures they can safely withstand depend upon the hardness and the compound. The manufacturer's specific recommendation should be obtained for temperatures below minus 40°F and above 200°F.

The number of "V" packings recommended per set based on solid rings is shown in Table 3.

TABLE 3

RECOMMENDED NUMBER OF "V" PACKINGS PER SET

Based on Solid Rings

Pressure, p.s.i.	Leather	Synthetic Rubber	
		Homogeneous	Fabricated
Zero to 500	3	3	3
500 to 1500	4	4	4
1500 to 3000	4	5	4
3000 to 5000	4	6	5
5000 to 10,000	5	–	6
10,000 and over	6	–	–

Typical Designs for Homogeneous "V" Packings. Three designs of interest are shown in Figs. 16, 17, and 18. Figure 16 shows an outside packed installation where the female support is an integral part of the nose on the gland ring. This is entirely satisfactory provided the gland ring is of the bolted type. If a threaded gland ring is used, regardless of whether it has an internal or external thread, the female support should be a separate unit so that the nose of the gland ring will not rotate on the packings. Figure 17 illustrates an installation of "V" packings on the end of a ram where a male ring is part of the retaining gland ring.

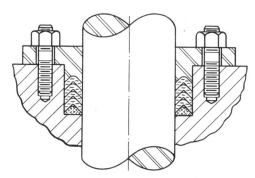

Courtesy of E. F. Houghton & Co.

Fig. 16. An outside packed "V"-ring installation in which the gland ring is bolted directly to the female support.

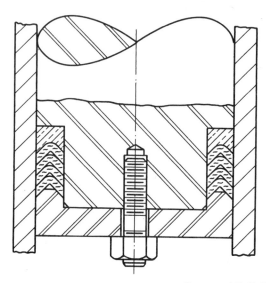

Courtesy of E. F. Houghton & Co.

Fig. 17. Installation of "V" packings on the end of a ram where a male ring
is part of the retaining gland ring.

For inside-packed, double-acting cylinders, Fig. 18 illustrates a typical design. Here the gland rings are threaded and the male and female rings are separate units.

In double-acting cylinders, opposing sets of "V" packings should always be installed so that the sealing lips face toward the pressure. The female support ring should always be adjacent to a fixed or rigid part of the piston. This will eliminate pressure loads being transmitted to the opposite set of packings. The lips of the opposing set of packings should never face each other as the result would be a locked pressure between the sets of packings.

Combinations of Leather and Synthetic Rubber "V" Packings. In some particular installations, a combination of leather and synthetic rubber "V" packings is used. With installations of this type, the leather "V" packings are always placed next to the female support ring and the synthetic rubber next to the male support ring. Alternating individual leather and rubber "V" packings is not recommended.

Two leather packings adjacent to the female support ring will back up the synthetic rubber and bridge any clearance, thereby helping to prevent extrusion of the synthetic rubber packings. This type of installation is usually used where pressures are high and friction must be held to a minimum.

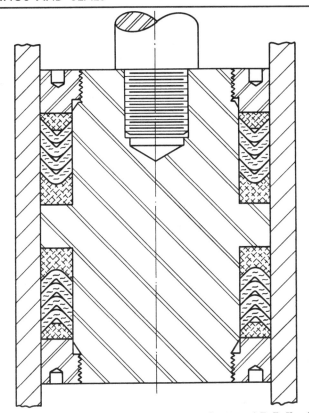

Courtesy of E. F. Houghton & Co.

Fig. 18. "V" packings are installed in the manner shown for inside-packed, double-acting cylinders. Here the gland rings are threaded and the male and female rings are separate units.

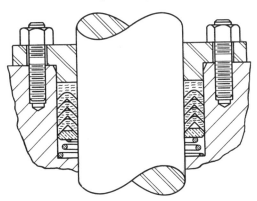

Courtesy of E. F. Houghton & Co.

Fig. 19. A set of homogeneous spring-loaded "V" packings.

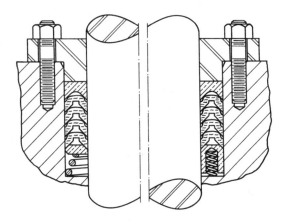

Courtesy of E. F. Houghton & Co.

Fig. 20. Two arrangements of fabricated "V" packings. The one on the left has a single (conical) spring for loading the packing while the one on the right has multiple cylindrical springs for loading the packing.

Spring-loaded "V" Packings. Shown in Fig. 19 is a set of homogeneous "V" packings which are spring loaded. Figure 20 shows two arrangements for use with fabricated "V" packings; namely, the single (conical) and the multiple spring (cylindrical). The single-spring installations are applicable for sizes up to approximately 3-inch outside diameter while the multiple-spring installations are for the larger sizes. For both single or multiple spring installations, a total spring pressure of 5 pounds per inch of mean circumference is recommended. An installation of this type automatically provides the correct adjustment and eliminates the possibility of improper adjustment of gland ring.

The Cup Packing

As piston seals, cup packings are important elements in the operation of both hydraulic and pneumatic equipment. The leather and fabricated synthetic rubber packings are the most commonly used, the homogeneous type being the least popular. Various compositions of the fabricated type are available for use with air, gas, water, and oil.

Both leather and fabricated cup packings are available in a wide range of sizes. However, a special mold is required to manufacture each size and to complicate matters, in the past, each designer has had his own ideas as to the dimensions of outside diameter, depth, hole diameter and thickness. At least one company has now established a range of standard sizes, however.

Mounting of Cup Packing on Piston

In fastening the fabricated cup packing onto the end of the piston, it is recommended that some provision be made to prevent the full pressure load from being taken on the bottom of the cup. Figure 21 shows how a cup should be supported. The shoulder cut on the bottom of the piston centers the cup and, when the proper allowance for compression of the bottom of the cup is made, the bottom of the packing is prevented from compressing under load and working loose. This shoulder also prevents the bottom of the cup from being compressed too much by excessive tightening of the inside follower plate. Too much compression of the bottom of the cup will cause the shoulder of the cup to bind against the cylinder wall. The packing life will then be shortened due to excessive friction.

As shown in Fig. 21, the piston shoulder K should be 0.005 inch less than the nominal thickness E of the packing. The tolerance on K is $+0.000$, -0.003 inch.

The packing shown in Fig. 21 is single acting, while that shown in Fig. 22 is double acting. In the latter, note the use of a back followed plate B which supports the shoulder of each cup. This plate should be a good fit in the cylinder as excessive clearance between it and the cylinder wall will result in early failure of the cups due to their extrusion into the clearance space. Allowable clearance based on cylinder diameter for pressures up to 500 lb per sq in. and 500 lb per sq in. and over are shown in Table 4.

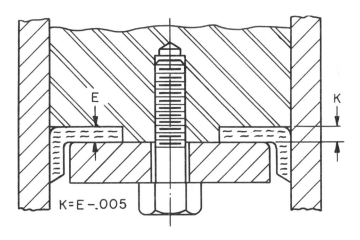

Courtesy of E. F. Houghton & Co.

Fig. 21. Correct method for supporting a cup packing.

TABLE 4

ALLOWABLE CLEARANCES FOR DOUBLE-ACTING
CUP PACKING BACK FOLLOWER PLATE

Cylinder Diameter (inches)	Pressures	
	Up to 500 lb per sq in.	Over 500 lb per sq in.
	Allowable Clearance, inch	
Under 4	0.006	0.004
4 to 8	0.008	0.006
8 to 10	0.010	0.008
10 to 12	0.012	0.010

The inside follower plate which fits inside the cup packing should *not* be a tight fit in the cup. Clearance is necessary to allow for swelling of the cup and also to allow the cup to "breathe" or "valve" under pressure.

One bolt is shown in Figs. 21 and 22 to hold the cup in place, and is sufficient for small cups. Three or more bolts are recommended for cups of medium size (6 to 12 inches in outside diameter).

Applications and Installation of the "U"-cup Packing

This type of packings is usually furnished in homogeneous rubber of 70 Shore durometer hardness for pressures of less than 1000 lb per sq in. but is also available in homogeneous rubber of 90 Shore durometer hardness for pressures above 1000 lb per sq in. However, it is not recommended for pressures above 2000 lb per sq in. Leather back-up washers are recommended for pressures over 1000 lb per sq in. These will support the shoulder of the packing and thus help

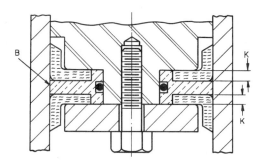

Courtesy of E. F. Houghton & Co.

Fig. 22. Correct method of supporting two cup packings for double-acting pistons.

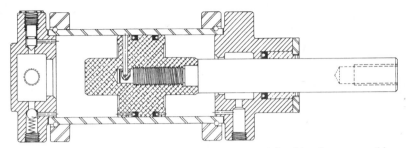

Fig. 23. The use of "U"-cup packings is illustrated in this piston assembly.

prevent extrusion. This type of packing provides very low friction and is completely automatic in action. It is molded on a flare and the nominal width is undersize at the heel but oversize at the lips so that when installed it is slightly preloaded.

It is used for air, vacuum service, on piston rods as wiper rings, and for low-pressure service. Figure 23 shows "U"-cup packings installed in a piston assembly. When used for air service, provision should be made to lubricate the packing. Note other seals in cylinder.

Figure 24 shows a typical outside packed installation. The gland ring which supports the bottom or heel of the packing must be a close fit to the rod, having a maximum clearance of 0.003 inch on a side. The surface finish of the rod should be a maximum of 16 microinches. The "U" cup should be supported by a metal ring as shown. When the gland ring is drawn up metal to metal, the support ring inside of

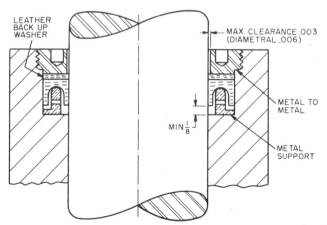

Courtesy of E. F. Houghton & Co.

Fig. 24. Typical outside-packed "U"-cup installation.

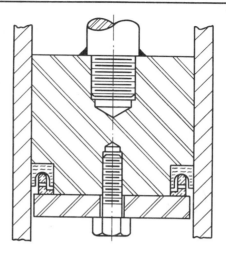

Fig. 25. An inside-packed single-acting "U"-cup installation.

the "U" cup should not exert undue pressure on the bottom of the packing as this will tend to retract the lips of the packing away from the rod and force the heel of the packing both outward and inward. It is recommended that the support ring be made as shown so that it will be properly centered.   It should be drilled crosswise so as to equalize the fluid pressure.

It will be noted in Fig. 24 that the internal threads have been offset so that the lips of the packing will not be damaged in any way when installing the packing.

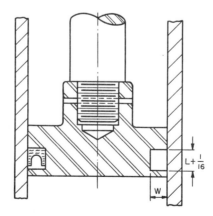

Fig. 26. Inside-packed "U"-cup installation employing a groove similar to
one used in an "O"-ring installation.

In Fig. 25 is illustrated an inside-packed installation, single acting. The same general rules apply for this type as outlined and shown in Fig. 24. Another satisfactory method of installing the "U" cup is shown in Fig. 26. Here, a groove is machined into the piston and the "U" cup stretched into the groove similar to an "O"-ring installation. The dimensions W and L refer to Table 5. When stretching is required, packing having a maximum hardness of 70 durometer should be used.

TABLE 5

DIMENSIONS OF "U"-CUP
PACKING AND SUPPORT RING

$W \& L$	D	E	F	R	N	ϕ	X	R_1
$\frac{1}{8}$	$\frac{3}{64}$	.030	.020	.020	$\frac{1}{64}$	13°	.008	.015
$\frac{3}{16}$	$\frac{1}{16}$	.038	.030	.050	$\frac{1}{32}$	13°	.008	.040
$\frac{1}{4}$	$\frac{3}{32}$	.045	.030	.070	$\frac{1}{32}$	11°	.008	.050
$\frac{5}{16}$	$\frac{1}{8}$	.050	.032	.093	$\frac{1}{32}$	10°	.008	.062
$\frac{3}{8}$	$\frac{1}{8}$	.054	.035	.125	$\frac{3}{64}$	8°	.008	.093

Note: $A = 2 \times R_1$. All dimensions in inches.

Applications and Installation of the "U" Packing

The leather or fabricated "U" packing is not nearly as popular as the "V" packing. In fact, there are few installations where a set of "V" packings would not do the job as well, if not better than a "U" packing. However, in some installations where the stuffing box or gland is shallow, which prohibits the use of "V" packings, the "U" is an ideal solution. Furthermore the "U" packing is completely automatic in its action and will produce less friction than a set of "V" packings, especially on the return stroke.

As in the case of the "V" packings, the shoulder or heel of the "U" packing is slightly undersize on the outside and inside diameters but the lips at the top are oversize. Therefore, when the packing is installed it is slightly preloaded.

When a support ring is used inside of the "U" ring, it should not exert undue pressure on the bottom of the packing when the gland follower ring is drawn up metal-to-metal.

Application and Installation of the Flange Packing

The Flange Packing. The flange packing, sometimes referred to as the "hat" packing, is the least popular of all packings. It is not used for high pressures and is recommended only where a "V" packing or "U" packing cannot be used due to lack of space. In cases where the cross-sectional width must be held to a minimum, the gland is shallow, and pressures relatively low, the flange packing can be used successfully, especially for rotary motion. The fabricated or leather class is recommended in preference to the homogeneous class.

Packing Replacement Methods

The Mobil Oil Corp. has brought together the following recommendations covering the installation, maintenance, and replacement of hydraulic packings*:

When the seal rings of a packing leak and have to be replaced, there are improper procedures that can result in an unsatisfactory installation and may even make it impossible to again make a tight seal. Correct installation of seal rings, on the other hand, can be accomplished without increased time or costs, and the packing, given good operating care, will last its expected life.

Removing Old Seals. After the gland follower ring has been removed or the piston pulled from its cylinder, the condition of the gland and sealing rings should be carefully observed, particularly if the seal being replaced has failed prematurely. Look for evidence of extrusion, lack of lubrication, dirt, undue wear, and scratches on the rod, cylinder or gland. Ordinarily, the removal of V-rings will require a tool—usually an auger with a flexible shaft or a brass or copper rod with a hook at one end. When using removal tools, utmost care should be taken not to scratch

* Mobil Oil Corp. Technical Bulletin, *Care of Packing in Hydraulic Machines.*

either the gland, shaft, rod, or piston. A scratch on the gland surface will usually cause leakage, and on the rod or shaft it will cause both leakage and seal-ring wear. A scratch on the rod, unless it is deep, can usually be stoned out and the shaft polished with fine crocus cloth. A scratch on the gland is often difficult to get at to remove. When removing sealing rings it is wise not to have the rod in its extreme outward position. If this were done and the rod were scratched while removing the rings, the scratch would most likely be in under the gland where it could not be reached, and the rod would then have to be removed from the cylinder in order to polish out the scratch.

Special tools, such as are shown in Fig. 27, can be made for removing both external and internal O-ring seals, and if carefully used permit easy removal without damage.

Selection of Replacement Packing. When ordering replacement sealing rings, it is important that the packing manufacturer be given full

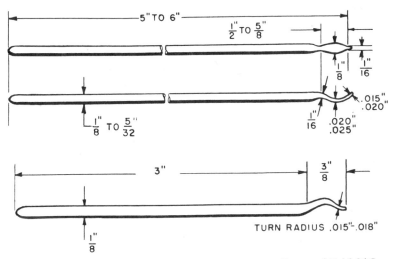

Courtesy of Mobil Oil Corp.

Fig. 27. Tools similar to these are recommended for removing "O" rings. The top tool is used to remove seals from external grooves, the bottom tool for removing seals from internal grooves.

particulars concerning rod or cylinder diameter, gland or groove diameter and gland length, operating temperatures and pressures, the fluid it is expected to seal, and the type of seal. Most packing manufacturers have a wide variety of materials from which to make sealing rings. The particular material best suited for any installation will depend on the temperatures, pressures and fluid involved. When a stock of sealing rings of various types, sizes, and material is maintained, it is obviously necessary to make sure that the replacement rings selected are correct for the job. Seal material unsuited to the conditions may soften, wear excessively, or harden, and cause premature failure.

Installing New Rings. Before installing new rings, the gland or O-ring should be scrupulously clean, and all surfaces over which the rings must slide during installation should be lubricated. It is good practice to dip the packing in hydraulic oil, particularly if it is leather, before installation.

In the case of V rod packings, the male support ring is installed first, making sure that the flat face of the ring seats squarely on the bottom of the gland. Each sealing ring is then installed separately, making sure that the sealing lips face the pressure. Wooden sticks with a V cut in their ends to fit the heel of the sealing ring have been found convenient in pushing each ring into the gland so it seats squarely on the previous ring before the next one is installed. Pointed metal tools, such as screw drivers, should never be used, because of the danger of scoring the rod or injuring the packing. The joints of split rings should be staggered alternately 180 degrees, then 90 degrees. The surfaces of the angle split should mate and overlap slightly at the joint (Fig. 28) and should not be cut. When installing split rings it is easy to get the end with the facing angle in front of instead of behind the other end as shown in Fig. 29. If this occurs, the next ring and all following rings will be cocked. If any of the rings are cocked, the seal will not give satisfactory performance. When installing V-packing in large diameter glands special care is needed to guard against twisting the packing (Fig. 30). Split rings should not be used where rotary motion is involved. Finally, the female support ring is installed, concave face toward the pressure. When solid (not split) sealing rings are used, the gland follower ring should be installed with spacers so that it can be bolted or screwed up tight, metal-to-metal, and yet exert only very slight pressure on the rings. Where the sealing rings are split, the gland ring must exert enough pressure to seal the joints. Spring-loaded glands are often recommended wherever space permits.

V-rings on double-acting pistons should not be installed with the pressure lips of the two sets of packing facing each other. This will permit pressure to be trapped between the two packings and result in increased rate of wear. The rings should be placed with the heels at the bottom of the grooves. This means that the female support ring is installed first.

When installing rings of any type, great care is needed to prevent scuffing the sealing surfaces. It is easy to do this when placing solid V-rings and O-rings over a rod or piston, because the rings are smaller in diameter and must be stretched slightly. Likewise, when pushing solid V-rings into their gland, it is easy to scuff the ring, because the outside diameter of the ring lip is slightly greater than the diameter of the gland. If the end of the rod or piston over which the ring must be started is square and without a chamfer having a taper of 30 degrees or less, damage to the ring can be prevented by using a piece of shim stock to form a cone over which the rings can be easily slid (Fig. 31). Shim stock can also be used to aid in starting a solid V-ring into its gland or any type packing into its cylinder (Fig. 32) and to protect rings against scuffing when they have to be pushed over screw threads (Fig. 33). O-rings are easily pinched (Fig. 34) if the proper chamfers (30 degrees or less) are not provided on the cylinder or rod. If the sealing surfaces of any type of sealing ring are damaged even slightly, the performance of the seal will be greatly impaired.

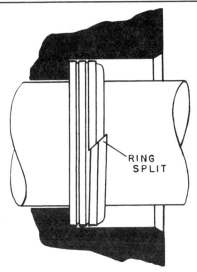

Courtesy of Mobil Oil Corp.

Fig. 28. When installing split "'V" rings on a shaft of correct size it will be found that the rings overlap slightly at the split. This slight overlap makes it possible to apply pressure to the split to stop leakage at the splits by adjusting the gland follower ring.

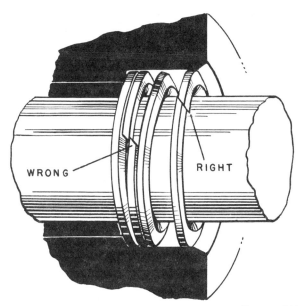

Courtesy of Mobil Oil Corp.

Fig. 29. When installing "V" rings, care must be exercised not to get the end with the facing angle in front of the other end as shown. When this happens each succeeding ring is cocked and a tight seal cannot be maintained.

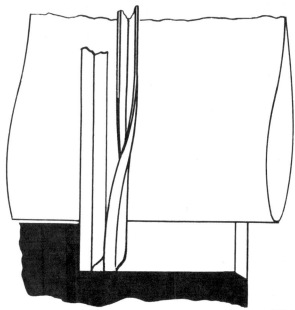

Courtesy of Mobil Oil Corp.

Fig. 30. In placing split "V" rings in large diameter glands care must be taken to prevent twisting the ring which would cause a leaky seal.

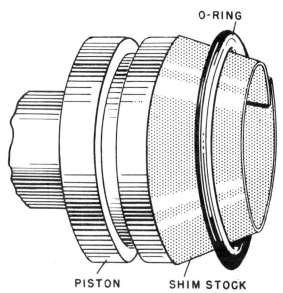

Courtesy of Mobil Oil Corp.

Fig. 31. Shim stock, shaped to form a cone, will aid in slipping an "O" ring over the edge of a piston or piston rod that has not been properly beveled.

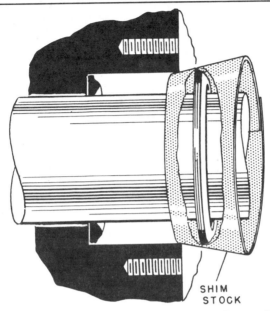

Courtesy of Mobil Oil Corp.

Fig. 32. If the edge of a gland has not been properly beveled, a piece of shim stock can be used to ease the "V" ring into its gland with minimum danger of scuffing the sealing lip.

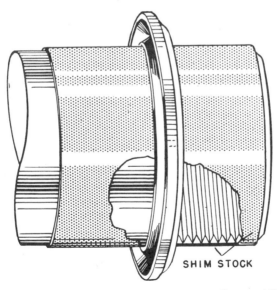

Courtesy of Mobil Oil Corp.

Fig. 33. Damage during installation to "V" rings by threads that have not been undercut can be prevented by wrapping them with shim stock.

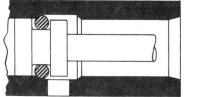 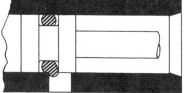

Courtesy of Mobil Oil Corp.

Fig. 34. Without proper .chamfers "O" rings are easily pinched and ruined.

Where the surface finish of a rod or shaft has been roughened, it may be desirable to change from homogeneous sealing rings to leather or fabricated rings.   When making such a change, it must be recognized that the dimensions of homogeneous and fabricated seal rings are not the same.   Homogeneous V and leather V rings are interchangeable.   In the case of V-rings, the change can be made, but it will be found that the gland will accommodate fewer fabricated rings (one less in most cases), and it may be necessary to change the spacers used in bolting up the follower ring.   Glands designed for homogeneous U-rings or flange seals may not accommodate fabricated seal rings as the dimensions of these rings have not been standardized.

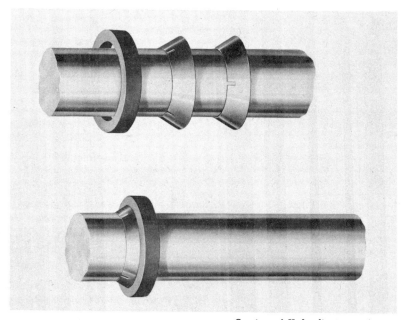

Courtesy of Hydraulic Accessories Co.

Fig. 35. Metallic rod wipers made of two flexible bronze rings mounted in synthetic rubber.   Parts before assembly are shown on top and the completely assembled unit is shown on the bottom.

Proper Operation and Maintenance. One of the principal enemies of rod-packing life is dirt, metal chips, grit from grinding wheels, etc. Any of these solids that adhere to the rod tend to be carried into packing where they interfere with the sealing action. If abrasive, the rod may be scored, or worn, which in turn will increase the rate of sealing-ring wear and leakage. Exposed rods and valve stems are sometimes damaged by tools dropped on them by operators or maintenance men. To minimize this trouble, rods should, where feasible, have protecting shields that will also keep metal chips and other solids away. Metal wiper rings or synthetic rubber wiper strips can be installed outside of the packing to keep the solids from getting into the seal itself. Figure 35 shows two bronze wiper rings with a synthetic rubber back-up ring. The use of cleaning devices is a generally recommended practice. Flange-rings and U-rings installed with their lips facing away from the pressure and leather back-up rings are also sometimes used for this purpose. To prevent damage to rods by metal and grit particles, some shops case harden the rods to 60 Rockwell C, or plate them with chromium.

All sealing rings require lubrication, particularly homogeneous rubber rings. Without lubrication, packing friction is high and ring wear rapid. Where the hydraulic fluid is oil, lubricant is always present. However, if seals are too tight, the rod may leave the packing dry and without lubricant. This results in rapid wear of the sealing rings. On rotating shafts, lack of lubrication may result in a groove being worn in the shaft as well as rapid wear of the packing. This is likely to happen if the shaft is soft (below 200 Brinell).

The importance of rod and cylinder finish, or smoothness, to packing life has been stressed repeatedly. If, for any cause, scores, scratches or roughness develops, the part should be refinished at the earliest opportunity. Packing life will be short until the condition is rectified.

Rod eccentricity in passing through the packing is detrimental to the life of any packing. In the case of O-rings and homogeneous V-rings, the maximum allowable eccentricity is 0.002 to 0.005 inch depending on rod diameter. The packing should not be part of the rod-supporting area.

When lower quality hydraulic oil is in service and temperatures are high, solid oxidation products eventually form which may collect in the packing and cause leakage, or form lacquer-like deposits on the rod or cylinder and cause leakage and increased wear.

When correctly selected sealing material is carefully and properly installed in well-designed glands of machines kept in good repair, leakage of hydraulic fluid can be kept to extremely small proportions, and packing will last its expected life. With minimum leakage, peak production can be maintained, hazards and labor costs for cleaning and oiling will be reduced. Also, advantage can be taken of the benefits of long oil life, minimized deposits, and reduced wear that result from the use of high-quality hydraulic oil.

Cartridge-type Rod Packings

Figure 36 shows a cartridge-type rod packing consisting of two sets of metal or phenolic plastic sealing rings mounted in pliable elastic

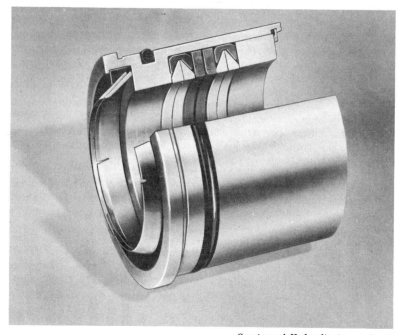

Courtesy of Hydraulic Accessories Co.

Fig. 36. Cartridge-type rod packing which is made of two sets of metal or phenolic plastic sealing rings mounted in pliable elastic cushion rings.

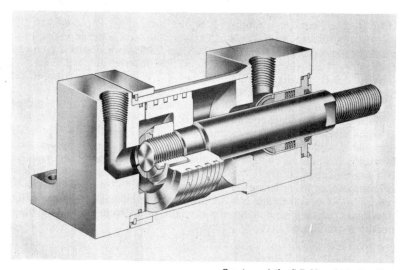

Courtesy of the S-P Manufacturing Corp.

Fig. 37. Cutaway view of a hydraulic cylinder showing the use of a cartridge-type seal, rod wiper, "O" ring, and automotive piston rings.

cushion rings. The assembly also has a metallic wiper unit at one end and uses an "O" ring as a static seal for the outside of the cartridge. Such a unit is confined in the cylinder head by a commercial snap ring that is readily accessible from the outside. No adjustment is required and the unit can be readily installed or replaced. Figure 37 shows another type of cartridge sealing unit for a piston rod. Such units are designed to meet size and pressure requirements for use in either air or hydraulic equipment.

Metallic Piston Rings

Automotive-type piston rings such as that shown in Fig. 38 are commonly used in hydraulic cylinders. An assembly of two such rings mounted on a piston is shown in the cutaway view of a hydraulic cylinder in Fig. 39. This type of piston seal is particularly useful

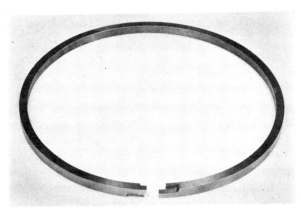

Courtesy of Grover Piston Ring Co.

Fig. 38. Automotive-type piston ring which is commonly used in hydraulic cylinders.

where, due to proximity to a furnace or other heat-producing equipment, the hydraulic cylinder is subjected to temperatures that would be harmful to nonmetallic piston seals.

In order to circumvent packing failures on hydraulic pistons, where replacement is difficult and adjustment impossible, the industry has developed certain modifications of the automotive-type piston ring similar to the cartridge-type rod packings. As shown in Fig. 40 such packings are composed of pairs of stiff metallic or synthetic sealing rings enveloped by pliable elastic cushion rings. Choice of materials is based on the various factors peculiar to the application such as pressures, temperatures, friction, etc. In operation, the action of the

Courtesy of Grover Piston Ring Co.

Fig. 39. Cutaway view of hydraulic cylinder showing an assembly of two automotive-type piston rings mounted on a piston.

fluid pressure tends to distort the cushion rings, automatically creating a tighter seal. This reaction is reduced as the pressure subsides, thus eliminating unnecessary friction.

The cushion rings are confined against extrusion by the sides of the seal rings. Their static sealing surface is convex for ease of installation while at the same time providing initial diametral compression. Designed for reciprocating motion, such packings provide effective sealing against oil, water, steam, air, gas, ammonia and other fluids, at temperatures up to 300°F. This type of seal is readily interchangeable with other types of packings. It acts as a bearing as well as a seal, tending to hold piston and rod concentric within the cylinder.

Courtesy of Hydraulic Accessories Co.

Fig. 40. Assembled, disassembled, and cross-sectional views of a modified automotive-type piston ring. It is composed of two pairs of stiff metallic or synthetic sealing rings, each pair enveloped by a pliable elastic cushion ring.

Synthetic Piston Rings

Figure 41 shows the application of filled "Teflon" piston rings used in conjunction with a rubber expander ring to effect a piston seal on hydraulic and pneumatic cylinders and valves. Some of the advantages are: friction loss is reduced, piston design is simplified, scoring of cylinder tubes or valve bodies is eliminated since there is no metal-to-metal contact, surface finish requirements are less stringent, and system damage due to fragmentation of carbon, cast iron, or phenolic rings is eliminated.

These piston rings, which are of the endless type, are available in compounds such as glass-filled, carbon-blend-filled, and bronze-molybdenum-filled.

When the piston ring is assembled over the rubber expander ring in the ring groove, this combination becomes preloaded radially, but is

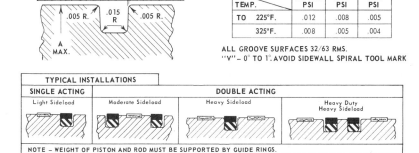

BASIC GROOVE DESIGN

"A" – AT OPERATING TEMP. AND PRESSURE

TEMP. \ PRESS.	TO 1500 PSI	3500 PSI	5000 PSI
TO 225°F.	.012	.008	.005
325°F.	.008	.005	.004

ALL GROOVE SURFACES 32/63 RMS.
"V" – 0° TO 1°. AVOID SIDEWALL SPIRAL TOOL MARK

TYPICAL INSTALLATIONS			
SINGLE ACTING	DOUBLE ACTING		
Light Sideload	Moderate Sideload	Heavy Sideload	Heavy Duty Heavy Sideload

NOTE – WEIGHT OF PISTON AND ROD MUST BE SUPPORTED BY GUIDE RINGS.

Courtesy of Halogen Insulator and Seal Corp.
Fig. 41. Synthetic piston ring data.

free of the groove axially. Because of this axial clearance, the operating fluid is permitted to enter the groove so that the fluid pressure may act on the resilient member and increase the preload in the radial direction to create a suitable seal. The piston ring has sufficient elasticity to respond to the radial force of the rubber expander ring which causes it to remain in sealing contact with the cylinder wall thus extending its wear life to the maximum. Carbon-blend-filled "Teflon" piston rings may be used in a temperature range from 225 deg. F to −40 deg. F, and for pressures up to 5000 psi (wet) and 300 psi (dry).

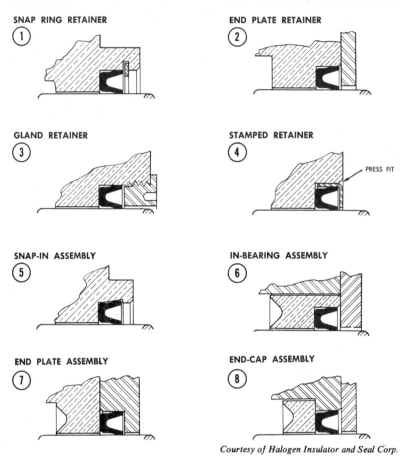

SNAP RING RETAINER

END PLATE RETAINER

GLAND RETAINER

STAMPED RETAINER

PRESS FIT

SNAP-IN ASSEMBLY

IN-BEARING ASSEMBLY

END PLATE ASSEMBLY

END-CAP ASSEMBLY

Courtesy of Halogen Insulator and Seal Corp.

Fig. 42. Assemblies for wiper rings. Chamfers shown are necessary. Where maintenance requires split-ring installation, assembly No. 1 is not recommended.

Synthetic Wiper and Scraper Rings

In Fig. 35, metal rod scrapers are shown; however, for the bulk of cylinder applications, nonmetallic wiper or scraper rings are employed to provide protection to the rod packings. These wipers or scrapers are made of various synthetic materials in various degrees of hardness, depending upon the requirements of the application. Several designs are also employed.

Figure 42 shows a number of ways in which wiper rings can be assembled into a cylinder.

Figure 43 shows the application of synthetic scraper rings. It has been found that wiper rings or scraper rings made of virgin TFE (tet-

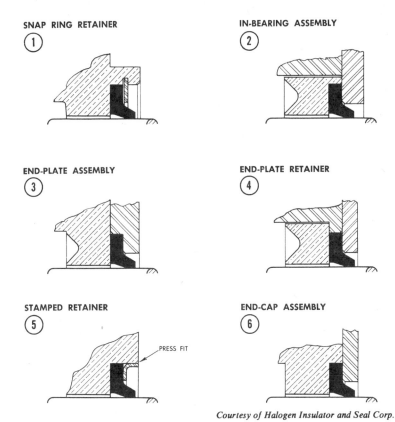

SNAP RING RETAINER ①

IN-BEARING ASSEMBLY ②

END-PLATE ASSEMBLY ③

END-PLATE RETAINER ④

STAMPED RETAINER ⑤   PRESS FIT

END-CAP ASSEMBLY ⑥

Courtesy of Halogen Insulator and Seal Corp.

Fig. 43. Assemblies for scraper rings. Chamfers shown are necessary. Where maintenance requires split-ring installation, assembly No. 1 is not recommended.

rafluoroethylene) will not adhere to the piston rod even during extended inoperative periods, and will not score piston rods made of steel or chrome-plated steel. TFE is unaffected by temperatures ranging from −100 deg. F to +450 deg. F and is not affected by grease, paint, or other coatings often used for protective packaging of cylinders and valves. Cleaning solutions used to prepare the piston rods for operation will not affect the virgin TFE material. Nearly all hydraulic and pneumatic cylinders are equipped with either a wiper ring or a scraper ring as a standard item.

CHAPTER 14

Air Filters, Lubricators, and Regulators

In selecting components for an air system, the two items which are often overlooked yet are so important to the proper functioning of an air circuit are air filters and lubricators. Probably because from an outward appearance they do not seem to perform any moving function, these items are often omitted from an air circuit. In fact, it is often difficult to convince a purchaser who has had very little experience with pneumatics that filters and lubricators are a necessary part of the air equipment. His first question is apt to be, "Why should I spend forty to sixty dollars for equipment that will increase the cost of my circuit or my machine?" Purchasers with experience in pneumatic systems won't raise this question—they know the true value of these two components.

It is almost impossible to keep the lines from the compressor to the machine or fixture on which the air circuit is being used free from moisture, pipe scale, and dirt. Remember, the compressor may be several hundred feet away from the machine and the pipe between the two may run through temperatures ranging from one extreme to the other. The pipe may be several years old and the condensation which forms inside of the pipe will cause it to rust and to form certain deposits which will eventually break loose and move toward the air circuit. It can readily be seen what would happen to the precision-made parts of valves and cylinders should pipe scale and rust be allowed to enter the system. The system would soon become inoperative.

A filter, lubricator, and pressure regulator should be placed in the inlet line to each air circuit. These may be installed as separate units but more often are in the form of a combined unit. A combined regulator, filter, and lubricator unit is shown in Fig. 1 (upper) and its location in a typical air circuit is indicated (labeled R.F.L. Unit in this diagram) in Fig. 1 (lower). In Fig. 2 such a unit (circled in

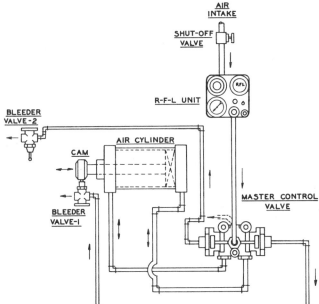

Courtesy of Logansport Machine Co., Inc.

Fig. 1. (*Upper*) Air filter unit (commonly called R.F.L. unit) for use with air-operated equipment. This unit regulates air pressure, removes dust particles and other foreign matter from air, and injects oil spray into the air intake line to lubricate valves and cylinders. (*Lower*) Air circuit· employing an R.F.L. unit.

Courtesy of LeBlond, Inc.

Fig. 2. An R.F.L. unit (shown circled in black) as it appears mounted on a machine tool.

white) is shown mounted on a modern machine tool. This combination unit supplies clean, regulated, and lubricant-carrying air to the valves and cylinders which operate the tailstock and chucking equipment.

Figure 3 shows a compact combination FRL (filter, regulator, and lubricator) unit in which the entire unit can be removed from the air line for servicing or replacing without disturbing the mounting-bracket pipe connections. This unit is available with brackets in four pipe sizes, and any of the brackets will fit on the basic model. Maximum pressure and temperature rating are 200 psig and 130 deg. F, respectively. The unit has a self-relieving regulator which provides minimum fluctuation in secondary pressures, regardless of fluctuation in primary pressure and air flow.

The case of the unit is slotted to provide visibility of liquid levels in the bowls. The two locking pins give positive visual indication as to whether the case is locked firmly in place. Locking pins cannot be removed once there is more than 3 psig air pressure on the unit.

Filter elements of the synthetic type have gained in popularity due, in part, to their compactness and their ability to remove very fine particles. Figure 7 illustrates an impregnated type filter.

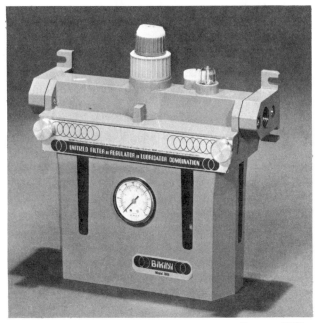

Courtesy of Wilkerson Corp.
Fig. 3. Compact combination FRL unit with metal case.

Air Filters

The ideal air line filter would be one which would remove all foreign matter and allow dry, clean air to flow without restriction to the regulator and then on to the lubricator. Filters, as used in commercially available units, range from a fine mesh wire cloth, which only strains out heavier foreign particles, to elements made of synthetic materials which are designed to remove very small particles. There are also filters that provide moving elements to remove the foreign material.

A survey made of several installations of a certain type of filter indicates that there is a vast difference in the amount of impurities found in the air lines of various industrial plants. In one plant a filter may perform satisfactorily for a year or more before it needs cleaning, while in another plant the filter may become completely inoperative after a few weeks of service. This is one of the several factors that must be considered in filter selection.

Other factors of importance are:

1. *Size of particles to be screened from the system.* Some types of valves will allow larger particles to flow through them without

causing damage to the highly precision parts. For example, a poppet-type valve usually can accommodate larger particles than a plug type or sliding piston type. By "larger particles" it should by no means be misconstrued that large chunks of scale of other foreign matter should ever be allowed to pass through any fluid power components.

2. *Capacity of the filter.* Laxity on the part of some maintenance men to clean filters at regular intervals results in the demand that the filter should be as large as possible. The capacity should also be large enough to reduce any resistance to flow to a minimum.

3. *Accessibility.* It should be possible to remove the filter from the unit in the least amount of time in order to keep down time in the system to a minimum. It is not a good idea to have a bypass so the filter can be cut out of the system during the cleaning. This would only allow scale to readily enter the system. However, on production equipment which is working around the clock, it is well to have two filters, one to be used as a stand-by which can be readily cut into service while the other is being serviced. This would eliminate any down time.

4. *Maintenance of the filter element.* Most synthetic and porous metal filters can be backwashed with a good solvent and quickly replaced in service. In plants having a large number of filtering devices, it is recommended that a few spare filter elements be carried in stock. Where the "throw away" type elements are used, spares are always important.

5. *Liquid removal.* This requires a reservoir large enough to amply handle the liquid removed, otherwise the filter element will become water logged and the filter will not perform properly. Liquid will then be forced into the air stream and the air pressure will be affected due to the water-logged filter.

Figure 4 shows a cross-sectional view of a mechanical type of filter. Compressed air moving through this device revolves four rotors at high speeds. The action is similar to that of a cream separator: foreign matter in the air stream is thrust against the side walls by centrifugal force, drops down to the base of the unit and flows by gravity to a trap.

Note that this illustration shows two rotors revolving in one direction while the other two revolve in the opposite direction. When the air stream moves through one rotor into the next, the reverse pitch of the vanes of that rotor causes violent deflection of the air stream which is a cleaning action in itself. After the air stream has passed through all four rotors, it must then pass through a group of scrubbing rings in the head of this device. The air leaves the device,

dry and clean.   The rotors in this device are made of phenolic plastic
or aluminum depending upon the size.

Figure 5 shows a cross-sectional view of a synthetic-type filter
element made of a phenolic-impregnated cellulose which is electrically
fused and polymerized for cohesion between layers so that it is impervious to liquids or gases.   This material is made up into ribbons
wound edgewise on a rotating mandrel to form the filter cylinder.
Since the air passes between these ribbons, the principle used is known
as *edge filtration*.   In operation, the air stream enters the filter at the
top and passes from the outside chamber through the filter cylinder

Fig. 4. Cross-sectional view of a mechanical-type filter.

and into the interior chamber as clean air.   Since the impurities cling
only to the outside of the element, they are easily removed by merely
reversing the air stream.   Filters of this type have been built to remove
particles as small as one-half micron (0.00002 inch).

Filter elements are also made of sintered metal shapes, porous
stones and other materials.

One of the new developments in filtering is the automatic drain
filter, shown in Fig. 6.   This provides automatic filtering and drain-

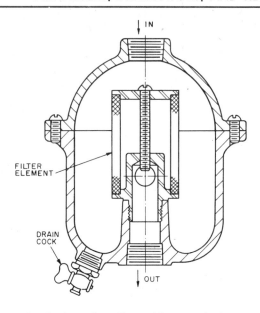

Fig. 5. Cross-sectional view of a filter with a synthetic-type filter element made of a phenolic-impregnated cellulose.

Courtesy of C. A. Norgren Co.

Fig. 6. Automatic drain filter.

ing of any condensate as long as there is air pressure on the line.   The
drain operates with constant or fluctuating air pressure, with or without air flow.   Draining the filter manually, which is necessary with
conventional-type air line filters, is eliminated with this unit.   The
possibility of having a clogged drain is remote, since the filter removes
the solids from the air stream before the air enters the drain section.
The float-operated drain mechanism discharges the collected moisture
only when the water level reaches a predetermined height.   Since
the drain does not discharge at regular intervals, but only when necessary, the wear on the drain parts is reduced.   Because of its automatic operation, the unit can be placed in locations that are not easily
accessible.

The centrifugal action which is created by passage of the air
through louvered ports wrings a high percentage of moisture from
the air stream.   The pressure drop across the filter is remarkably low.

Filter elements of the synthetic type have gained in popularity due,
in part, to their compactness and their ability to remove very fine
particles.   Figure 7 illustrates an impregnated type filter.

Where a sizable volume of compressed air must be filtered and
dried, a filter-dryer combination such as that shown in Fig. 8 can be

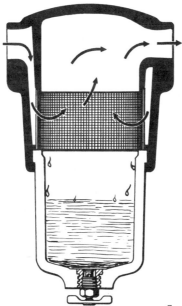

Courtesy of Wilkerson Corp.

Fig. 7. A filter using impregnated felt as the filter element.

employed.   This is a three-stage device.   Compressed air first enters a separator which spins out water drops, oil drops, and large particles such as scale and rust.   In the second stage the compressed air enters a dryer which is a desiccant type using dehydrated clay.   This draws out water and oil vapors.   Medium-size impurities are also stopped. The final stage is a filter which removes particles as small as 5 microns.

Sulfuric and hydrochloric acids, which, due to lubricant breakdown are often present in compressed air lines, will also be separated out at the filter-dryer.

In applications where the temperatures are low, a thermostatically controlled device such as that shown in Fig. 9, can be installed on the filter-dryer shown in Fig. 8.

Figure 10 shows two views of a refrigerant type dryer—at left it is shown closed, at right it is shown with cover removed.   The tempera-

Courtesy of Filter Engineering Co., Inc.
Fig. 8. A filter-dryer unit.

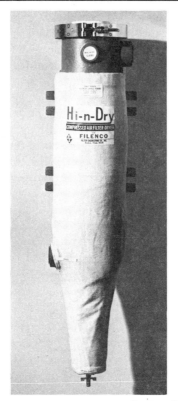

Courtesy of Filter Engineering Co., Inc.

Fig. 9. Filter-dryer unit with thermostatically controlled covering.

Courtesy of C. A. Norgren Co.

Fig. 10. Exterior and interior views of a refrigerant type air dryer.

ture of the compressed air is lowered as it passes through the dryer, water vapor is condensed and separated, and a lower dew point for the air is established. Moisture will not condense in the downstream compressed-air system as long as temperatures do not fall below the established dew point (the temperature at which water will begin to condense).

To prevent subsequent moisture condensation, the dew point should be below the coldest temperature in the air system. Since compressed-air systems operate under pressure, the pressure dew point, rather than the atmospheric dew point, is the critical consideration.

Figure 11 shows the schematic of the refrigerant air dryer. There

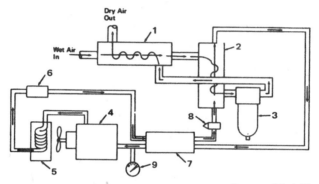

Courtesy of C. A. Norgren Co.

Fig. 11. Schematic of refrigerant type air dryer.

are two basic circuits in a refrigerant air dryer: air and refrigerant. In the air circuit, compressed air enters the dryer through the inlet port of the air-to-air heat exchanger (1) and is precooled by the cool air exiting the unit. The air then enters the air-to-refrigerant heat exchanger (2) where its temperature is further reduced by the cold refrigerant, causing water vapor to condense from the air. The separator (3) removes the liquid condensate and it is expelled by the unit's automatic drain. The dry air flows back through the air-to-air heat exchanger where it is warmed by incoming air before entering the downstream air system.

The refrigerant circuit consists of a hermetically-sealed refrigerant compressor (4), refrigerant condenser (5), a strainer-dryer (6), refrigerant-to-refrigerant heat exchanger or by-pass valve (7), expansion valve (8), and the air-to-refrigerant heat exchanger (2). The compressed-air dew point is established by the temperature of the air in the air-to-refrigerant heat exchanger.

Air Line Lubricators

Lubrication is an extremely important item in an air system if friction and wear are to be kept to a minimum. For example, lubricants, properly applied, greatly increase the life of packings used in air cylinders. It has been observed by test that two identical air cylinders, one with absolutely no lubricant and the other with the proper amount of lubricant, showed a great difference in their packing life. The packing in the one with no lubricant failed after a few hundred cycles, while the other cylinder operated for approximately six million cycles before excessive packing leakage occurred.

On large-bore air cylinders, where it is not practical from a cost standpoint to use nonferrous tubes, lubrication is a must, otherwise condensation passing through the cylinder will cause a rusting and pitting condition in the tubes. Any packing passing over these surfaces will then be quickly scored and cause leakage.

The precision parts of valves must also be properly lubricated if satisfactory service is to be expected. However, excessive lubrication is not a desirable feature as it will result in an oil mist being thrown into the atmosphere each time the air is exhausted from a valve. This condition can easily be detected and remedied by changing the setting on the lubricator.

The question that is often asked is, "What is the best type of lubricant for an air system?" Usually a good-quality, light-grade spindle oil will meet most air system requirements. Another lubricant which has been used successfully is a mixture of 50 per cent kerosene and 50 per cent S.A.E. 30 oil. Consult a competent oil dealer for his recommendations on your particular lubrication problem.

There are a number of types and designs of lubricators now available. Figure 12 shows a lubricator of the force feed type with a sight glass gauge indicating the number of drops actually entering the air stream. It has a venturi tube in which oil is atomized by air. Metering of the oil is accomplished by a needle valve which provides various sized orifices. The lubricator can be filled without manually shutting off the air pressure.

Figure 13 shows a combined lubricator, filter, and pressure regulator. The lubricator is the right-hand unit in this illustration. It provides a positive method of metering oil through a sight drop feed controlled by a needle valve. The unit is equipped with a special patented Venturi which maintains an accurate feed of oil in spite of variations in air pressure. The bowl, which is made of a transparent plastic, enables the user to know exactly how much oil is in the

Courtesy of Wilkerson Corp.

Fig. 12. Lubricator with a sight-feed glass.

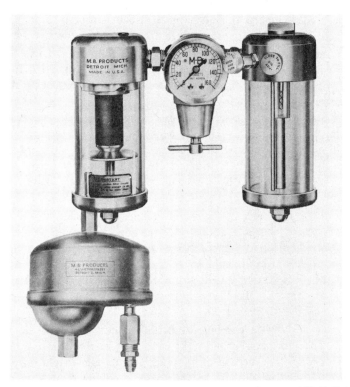

Courtesy of M. B. Products

Fig.13. A combined lubricator, filter, and pressure regulator. The lubricator is at the right. The pressure is regulated by turning the threaded member beneath the numbered gage.

lubricator at all times. This lubricator can be used for pressures ranging up to 160 lb per sq in.

However, even the finest of lubricating equipment is of little value if it is not regularly supplied with oil. It is not at all uncommon to walk through a plant and see a good percentage of the lubricators with no oil in them. Nor is it infrequent to hear the owners of such "unlubricated" air equipment wonder why it doesn't give them the service life they expect.

Air Regulating Valves

While the regulating valve, sometimes called a reducing valve, is actually an air control, it should be considered along with the filter and lubricator since most manufacturers offer these three items as a combination unit, as shown in Fig. 8, so as to properly prepare the air before it goes into the circuit.

The regulating valves provide the proper operating pressure for the circuit. While the line from the compressor may carry a pressure of 150 lb per sq in., the regulator can reduce this pressure to 0 lb per sq in. or to any point between the full line pressure and zero pressure. The regulator also acts as a pressure guard by preventing pressure surges or drops from entering the air circuit and causing equipment to malfunction at critical points in the work cycle. Regulators are of the diaphragm-type or the piston-type.

Figure 14 shows the diaphragm-type. The diaphragm, made of oil-resistant synthetic rubber with a nylon cloth reinforcement, allows the proper amount of movement for opening and closing of the valve seat yet creates no sliding friction. The lower spring-rest rides on one side of the diaphragm while the valve pin contacts the other side. With the adjusting screw in the fully retracted position, the valve is closed. As the adjusting screw is turned to compress the regulating springs, the valve is opened. The pressure at which the air leaves the valve depends upon the size of the valve opening that is maintained. This is determined by the pressure of the regulating springs on the top of the diaphragm and the pressure of the air on the under side of the diaphragm after it has passed through the valve opening. The valve acts automatically to maintain relatively constant outlet pressure. Thus, if the air pressure on the discharge side of the valve increases momentarily, it acts against the diaphragm to decrease the pressure exerted by the regulating spring on the valve. The valve moves upward toward the closed position, greater throttling action takes place, and the pressure is reduced or regulated.

The smooth contact between the rounded end of the adjusting screw

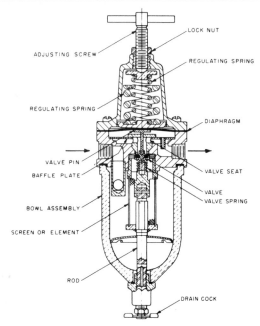

Courtesy of C. A. Norgren Co.

Fig.14. Cross-sectional view of a diaphragm-type regulator.

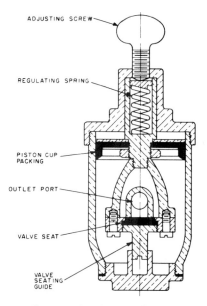

Fig. 15. Cross-sectional view of a piston-type reducing valve.

and the upper spring rest affords ease of adjustment to provide the pressure desired. After adjustment is made, the lock-nut is tightened to eliminate any chance of the setting of the adjusting screw being changed. Note that the air moves through the built-in filter element before it reaches the valve seat. This reduces the possibility of dirt or foreign matter from coming in contact with the synthetic seat.

Figure 15 illustrates a piston-type reducing valve. The piston assembly is similar to that of a single-acting air cylinder and consists of synthetic rubber cup-type packing supported by a large metal back-up disc. As in the diaphragm-type valve, the tension against the piston is created by a spring.

In this valve the air enters the chamber which is in contact with the underneath side of the piston cup packing, passes between the top of the synthetic seal and the valve seat, and on out the outlet port into the system. Note that the synthetic seal is carried by the yoke and is guided by the seating guide which works inside the bottom cap. If the pressure on the inlet side of the valve should increase momentarily, it would tend to drive the piston upward carrying the valve seal closer to the seat and thus exerting greater throttling action to regulate the pressure on the discharge side. The larger valves use two springs to overcome the load as compared with the single spring used in the small valve shown in Fig. 15. This is because of the larger piston area.

Figure 16 shows a piston-type of reducing valve in which an "O" ring is used as the piston seal. This reducing valve is built as a compact unit into a regulating, filtering and lubricating assembly. By removing the end plug, the valve seat is made readily accessible for renewal. A poppet-type seat is used as the seal. The light-weight aluminum piston and "O"-ring piston seal create little friction, making pressure adjustment very easy. The long travel of the piston assembly affords accurate pressure regulation.

Mufflers

Noise created by air exhausting from an air system may become disturbing to an operator. Where a battery of air presses or other air-operated machines are confined in a small room, the noise created by the exhausting air would be almost unbearable if it were not controlled. Exhaust air noises not only cause nervous tension and dissatisfaction among the operators, but also results in mental fatigue, lack of concentration, and inefficiency.

Exhaust noises can be greatly reduced by either of two methods. One method is to connect all of the exhaust ports in the valves to

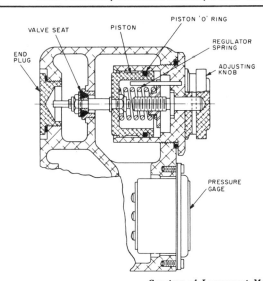

Courtesy of Logansport Machine Co., Inc.

Fig. 16. Piston-type reducing valve which employs an "O" ring as the piston seal.

a common manifold and then run this manifold through the roof of the building, letting the noise travel to the outdoors. This method may require very long runs of piping and is apt to be a pretty expensive proposition.

The preferred method is to use mufflers on each system. The efficient and economical method of silencing exhaust air is to dissipate exhaust energy. Figure 17 illustrates a cross section of a muffler which operates on a patented opposed radial-flow silencing principle. This muffler uses an aluminum deflector shield which is designed to absorb the initial impact of the pressure shock created by the high velocity air stream, diverting the air stream in a crisscross pattern with a minimum of contact with the outside section which constitutes the disseminator. The shield also acts as a condenser to precipitate the moisture present in the air system. This holds vapor spray to a minimum. The radial flow design affords a high degree of noise muffling efficiency without retarding the normal flow of exhaust air into the atmosphere. Noise elimination is obtained by reducing the velocity of the exhaust while obtaining full air flow.

The quick deceleration of the air velocity causes a lower temperature in the silencing chamber, thus causing condensation of any vapors present in the air stream. This prevents any vapors of oil and water which may be in the air stream from blowing into the atmosphere.

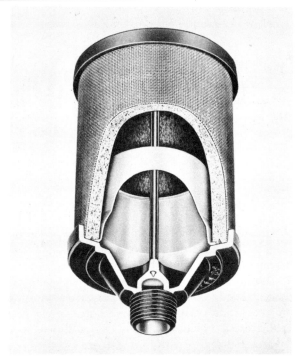

Courtesy of Allied Witan Co.

Fig. 17. Cutaway view of a muffler which operates on an opposed radial-flow silencing principle.

Any moisture condensing from the air stream collects in the form of droplets and eventually flows down and out of the disseminator.

Figure 18 illustrates a muffler in which the air stream enters one end, passes through a series of baffles, and passes out the opposite end. The end in which the air enters is tapped for a pipe connection while the opposite end has a hole large enough to allow the exhaust

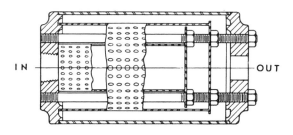

Fig. 18. Muffler which operates by passing the air stream through a series of baffles.

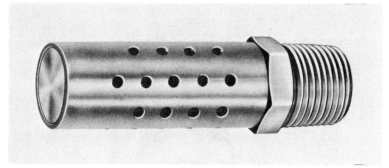

Courtesy of Allied Witan Co.

Fig. 19. Compact all-metal muffler.

air to flow without restriction to the atmosphere. The baffle tubes are perforated with a large number of small holes. The outer shell acts as a barrier and helps guide the air stream toward the exit.

Figure 19 shows another all-metal type muffler; compact and used where space is limited, it is constructed of corrosion-resistant metal to withstand shock and corrosive materials that might be in the air stream. A distribution of peripheral openings disperses exhaust air uniformly over a 360-degree pattern to provide effective reduction of objectional air-exhaust noise.

Courtesy of Allied Witan Co.

Fig. 20. Hydraulic power device which makes use of muffler in conjunction with air-operated pump.

Figure 20 shows a muffler used in conjunction with an air-operated pumping unit which produces high-pressure hydraulic fluid. This package can be used for hydrostatic testing, operating clamps in presses, actuating punches, etc.

Figure 21 shows two mufflers used to treat the exhaust air on a double-acting air-oil booster pump. Note the filter, regulator, and lubricating unit which processes the air before it reaches the control valves. Also note the heavy-duty construction of the oil booster sections, one of which contains a pressure gage with a range from 0 to 30,000 psig.

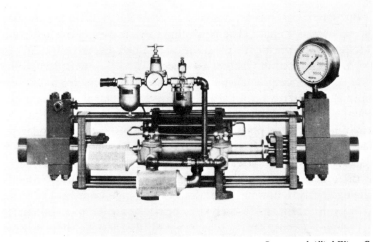

Courtesy of Allied Witan Co.

Fig. 21. Double-acting, air-oil booster pump which uses two exhaust mufflers.

CHAPTER 15

Pneumatic Controls

Pneumatic controls are designed to control and direct the air stream through an air circuit. The regulating valve is usually the first control through which the air stream passes as it enters the circuit. This control was discussed in Chapter 14 since it is so closely tied in with the filter and lubricator. Regulated air pressure flows from this valve on to the master control valve. The master control valve is a directional control of either the three-way or four-way type. The three-way type has three ports—an inlet, an exhaust, and a cylinder port. The four-way valve has four ports—an inlet, an exhaust, and two cylinder ports. Port sizes in an air control valve range from ⅛ inch to three or four inches and even larger on special valves. Standard cataloged valves, however, are usually in a range from ⅛- to 1½-inch sizes, with the ⅜- to ¾-inch sizes being by far the most popular.

Figure 1 (Top), shows schematically the porting of a three-way valve and (Bottom), two three-way limit valves, one without an actuating lever and another with a nylon roller. A three-way valve may be either of the normally-open or normally-closed type. A normally-open valve, as in top part of Fig. 1, is one in which the flow-director, when in the normal position, connects the inlet port to the cylinder port. The basic valves are the same and various types of actuators can be installed on the pin. These valves can be either normally-open or normally-closed, depending on the inlet port employed. The inlet ports shown are for normally-open operation. There is, however, an inlet port on the bottom of the valves for normally-closed operation. Figure 2 shows the porting of a four-way valve. Valves of this type are used in conjunction with pilot-operated valves of either the pressure or bleeder type. Their appearance is very much like an electric limit switch. These valves can also be employed to actuate small, single-acting cylinders; two valves can be used to operate a double-acting cylinder.

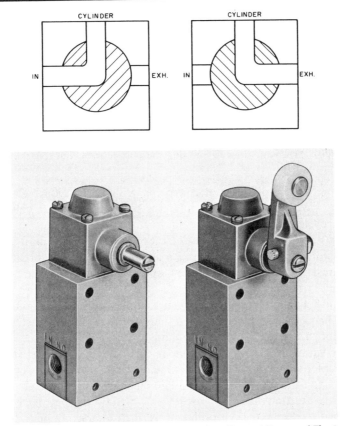

(Bottom) Courtesy of The Aro Corp.

Fig. 1. (Top) Schematic diagram showing the porting of a three-way valve. (Bottom) Three-way limit valves.

Factors Affecting Master Valve Selection

Cycling. Care must be exercised in choosing the correct master control valve for an application. For example, if the valve is required to cycle 150 to 250 times per minute, a manually-operated valve would be out of the question due to operator fatigue. An operator could not begin to cycle a valve at that rate for any length of time. A cam-operated valve is a possibility, but with cam valves there are heavy return springs which may be broken by fatigue. Unless it were quite small, a direct-operated solenoid valve would not be suitable for such high cycling, due to mechanical and electrical failures, particularly of large solenoids, when cycled at a high rate of speed. (A small solenoid-operated valve for rapid cycling is shown in Fig. 10.)

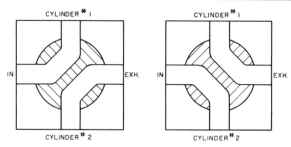

Fig. 2. Schematic drawing showing the porting of a four-way valve.

A valve of the pilot-operated type is a "natural" for high-speed cycling. With some pilot controls, it is possible to operate valves up to 600 cycles per minute.

Safety. Safety is another consideration. The designer must always keep in mind that when dealing with an air circuit, he has three things that must be considered together—speed, pressure and the human element. If a fast-moving air press ram, which will necessarily be operating under considerable pressure, makes contact with an operator's hand due to carelessness or misjudgment on the part of the operator, a serious accident is bound to result. The choice of the proper valve or valving can eliminate this hazard, as is explained in the chapter on air safety circuits. One of the jobs of plant safety men is to make certain that the air-operated machines in their plants afford ample protection to their operators. Safety devices have been designed to eliminate such problems as, for example, an operator accidently bumping into an actuator on a valve and causing it to operate. Figure 3 shows a safety shield that has been added to the three-way valve shown at left, to cover the valve actuator button.

Figure 4 shows a foot-operated four-way valve with a protective guard. The operator must put the toe of his shoe under the guard in order to contact the foot pedal.

Temperature Range. Another consideration when selecting the master control is the temperature range to which the valve will be subjected. It is often necessary to place the master control in a "hot" spot in which the temperature may be 500° to 800°F. In this case, a valve piston with packings will undoubtedly cause trouble. A sliding piston-type valve having a metal body and metal piston has been found to work out successfully in this temperature range even though the internal leakage is fairly high.

Space Limitation. Space limitation is another factor that must not be forgotten. The designer must determine whether or not the

Courtesy of The Aro Corp.

Fig. 3. Three-way valve (*Left*); and equipped with a safety shield (*Right*).

valves which he chooses will fit into the allotted space in his machine. He must be careful to make certain that the pipe connections will not interfere with some part of the machine, which may seem like a trivial item, but if not planned for, often becomes an expensive one to correct. Figure 5 shows three four-way air valves mounted on a mani-

Courtesy of The Aro Corp.

Fig. 4. Four-way foot-operated, air-control valve with protective guard.

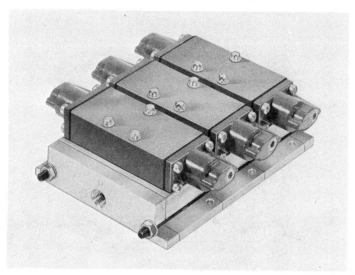

Courtesy of The Aro Corp.
Fig. 5. Three four-way air valves mounted on a manifold plate to conserve space and reduce piping.

fold which eliminates considerable piping. The pressure inlet is marked with a "*P*," and the cylinder ports are located on the bottom of the manifold plate.

Cost. Cost is always an important item which the designer must weigh carefully against other factors. In many cases he may choose a cheap control valve and experience satisfactory results; then again he may try to use a cheap valve on another application and it will quickly fail. It's the old story that the most expensive thing may be the least expensive in the end. The designer must learn to choose the most suitable valve for the application.

Types of Three- and Four-Way Valves

Three- and four-way valves may be manually, mechanically, electrically or pilot operated. Figure 6 illustrates a circuit using a solenoid-operated, three-way valve to operate a single-acting cylinder. Figure 7 shows a circuit in which a manually operated, four-way valve is used to control a double-acting cylinder.

Manually Operated Type. The term "manually operated" will be used here to cover valves that are operated by hand or by foot. The advantage of the foot-operated valve is that the operator has both hands free to handle the workpiece or perform other work while operating the foot valve. Figure 8 illustrates a four-way, foot-oper-

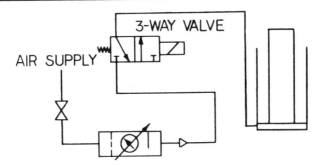

Fig. 6. Circuit using a solenoid-operated three-way valve to operate a single-acting cylinder.

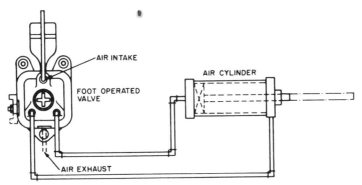

Fig. 7. Circuit using a manually-operated four-way valve to control a double-acting cylinder.

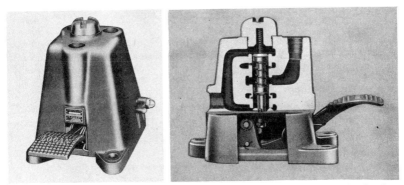

Courtesy of Logansport Machine Co., Inc.

Fig. 8. Four-way, foot-operated valve. (*Left*) Exterior view showing the latch on the right side which provides for two distinct methods of control. (*Right*) Cutaway view shows the porting and, at the bottom, the latch mechanism which provides two distinct methods of control.

ated valve. The latch on the side of the valve provides for two distinct methods of control.

When the latch lever is turned to the forward position shown, the operator depresses the foot pedal and the valve piston is raised and latched. It does not release until the pedal is pressed a second time. This allows the operator to start the cycle, then go on to another machine and start another cycle. He can then return to the first machine and complete the cycle by depressing the pedal of the valve a second time.

When the latch lever is turned to the back position, the valve piston is raised and lowered as the pedal is pressed and released. This is a desirable feature where variable control is desired, where the cycle is rapid, or where the operator is only tending one machine.

Note the large nut on top of the valve. By removing the nut, the complete piston assembly can be removed from the valve without disturbing any of the pipe lines.

Figure 9 illustrates a cross section of a hand-operated, four-way control valve. This is a five-ported valve which has an inlet port, two cylinder ports, and two exhaust ports. The valve has a spring offset configuration and is designed with a balanced spool. All of the dynamic valve spool seals operate on a Teflon-sealed, hard, anodized, non-stick aluminum surface which assures long seal life and smooth operation of the valve spool. Note the static seals between the valve liner and the valve body, and the dynamic seals on the valve spool.

Mechanically Operated Type. Three or four-way valves operated

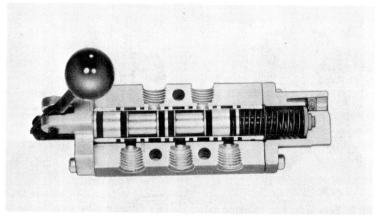

Courtesy of The Aro Corp.

Fig. 9. Four-way spring offset air control valve.

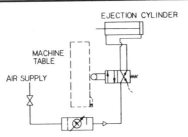

Fig. 10. Circuit in which a machine table with a swing-type cam operates the cam-operated directional control valve.

by a mechanical means are usually of the cam-roller or toggle type. The cam roller or toggle may be operated from some machine member—a moving piston rod, a rotary cam plate operated by some sort of driving mechanism or some other device. Figure 10 shows a circuit in which a machine table with a swing-type cam operates the cam-operated directional control valve. This circuit is used for ejecting the workpiece from the machine table onto a moving conveyor. As the machine table nears the end of the stroke, the swing-type cam depresses the cam roller allowing air to flow to the blind end of the cylinder and the piston moves forward very rapidly ejecting the workpiece. As the cam rides off of the roller, the cylinder piston quickly retracts. On the backstroke of the machine table, the swing-type cam does not depress the valve roller. Valves of this type work well as interlocks on complicated machine circuits where motions must be positive.

Electrically Operated Type. Electrically operated valves have solenoids as their operators. The solenoids may be either of the push type or pull type depending upon the design of the valve. In the direct-acting solenoid valves the solenoids are larger than in the pilot-operated solenoid valves, since the compressed air does most of the work in the latter.

Figure 11 shows, at left, a double-solenoid, four-way control valve that uses direct-acting solenoids of the plug-in type. The solenoids drive a lightweight aluminum spool, shown at the right. This spool has nonmetallic inserts in each end to absorb the blow from the solenoid plunger. The solenoids are of the push type. This valve is designed so that if the solenoid covers are removed, the valve will not function. Each cover has a manual override pin so that the valve can be actuated during setup periods without using the solenoids.

Figure 12 illustrates a four-way, single-solenoid, pilot-operated control valve for maintained contact actuation. The solenoid is of

Courtesy of Logansport Machine Co., Inc.

Fig. 11. (*Left*) Four-way, two-position air valve with direct-acting solenoids. (*Right*)
Spool in the four-way valve.

the plug-in type. Solenoid ratings are identified by color-keyed, molded nylon coverings that fully enclose and protect the windings. The nylon bobbin is also a bearing on which the plunger works silently without the need of lubrication. This valve combines a high response speed with the ability to handle large volumes of air over a

Courtesy of Parker-Hannifin Corp.

Fig. 12. Single-solenoid, pilot-operated, four-way control valve.

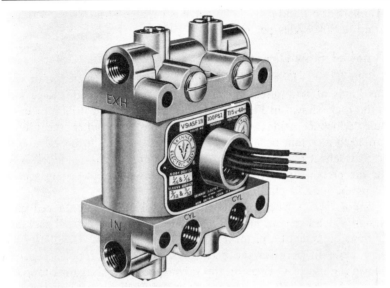

Courtesy of Skinner Electric Valve Div.

Fig. 13. Small solenoid-operated four-way valve which consists of two three-way valves so arranged as to use a common inlet and a common exhaust port.

pressure range of 15 to 150 psi while using only a small amount of electrical power.

Figure 13 shows a very small solenoid-operated, four-way valve, which is basically two three-way valves so arranged as to use a common inlet port and a common exhaust port. While this valve has small orifices, enough flow capacity is available to operate small

Courtesy of Logansport Machine Co., Inc.

Fig. 14. A plug-type directional control valve (shown disassembled) which directs an air stream.

cylinders at a rapid rate. The valve is capable of operating at a rate of up to 600 cycles per minute.

Types of Flow Directors

Plug Type. Directional control valves are constructed with various kinds of flow directors for directing the air stream. One kind is the plug, shown in Fig. 14, which has a system of drilled or cored holes which coincide with passages in the body when the plug is turned to a certain position. The plug is lapped to fit the valve body so as to effect a seal over a large surface area. The tendency is for the air pressure to try to raise the tapered plug but spring tension keeps the plug in place.

Spool Type. Another type flow director utilizes a balanced spool design which uses various sealing means: cup packing, "U" packings, O-rings, nonmetallic piston rings, and lapped spool. The latter is a very close fit between the body or liner and the spool. Figure 9 shows the use of O-rings on the valve spool to seal against the valve liner. Figure 11 (Right) shows the use of filled synthetic piston rings on the valve spool which are backed-up by O-rings.

Poppet Type. Still another type of flow director is a system of poppet valves. Poppet valves allow for quick opening yet they need travel but a little distance to allow a full pipe-volume flow. They provide a positive seal unless some foreign matter should lodge on the poppet seat thus destroying this seal. Poppet valve seats may be of a flat disc or of a washer design made of leather, rubber, synthetic material or fiber and supported by metal. A typical poppet valve design is shown in Fig. 15.

Diaphragm Type. Still another type uses a diaphragm-type seat for positive-sealing action.

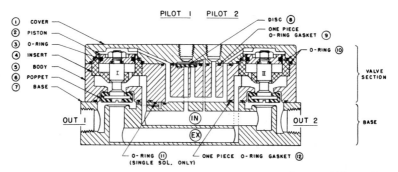

Courtesy of The Aro Corp.

Fig. 15. Diagram of a four-way poppet valve.

Thus, the designer has a wide variety from which to choose. Each type of construction has its good points and experience on the part of the designer will dictate where each can be used to the best advantage. For example, a plug-type, four-way control valve using no packing works well on high temperatures applications where valves with seals will often fail in such service.

Two-position and Three-position Control Valves

Four-way control valve flow directors are either of the two-position or the three-position type. The two-position type (shown diagrammatically in Fig. 2) is such that in one position the inlet port is connected to the cylinder port No. 1 and the cylinder port No. 2 is connected to the exhaust. In the other position, the inlet port is connected to cylinder port No. 2 and the cylinder port No. 1 is connected to exhaust. In the three-position valve, the second or intermediate position is known as the neutral position. In the extreme positions the connections are as just described for the two-position valve. In the neutral position, all ports may be blocked or the inlet port may be blocked and both cylinder ports opened to exhaust. Unlike the hydraulic valve, air valves do not make use of the open-center piston construction which would allow the pressure to flow directly to atmosphere in the neutral position. This would certainly be a waste of expensive air, especially on large port valves.

Three-position air valves with all ports blocked in the neutral position are especially useful on inching applications, or where a load is to be raised to a certain point and held there for an indefinite length of time, such as an air hoist. Valves of this type must be as nearly leakproof as possible to be of any value, since a slight leak in the interior of the valve will allow considerable drift of the piston in the cylinder which is being controlled.

Three-position air valves, with the inlet port blocked in neutral and both cylinder ports exhausting, have a somewhat limited application but play an important part when used in conjunction with machine drives where the cylinder piston must be powered by air for part of the stroke and then be free to be moved by an external means for the other portion of the stroke.

Pilot Valves and Pilot-operated Valves

Pilot-operated air control valves cover quite a wide field. They are the valves that are used in remote control systems and form the basis for many complicated safety circuits which not only provide safety to the operator but to the machine and to the workpiece as well. There should be a clear distinction made between *pilot-operated* control valves and the *pilot* valves which operate the pilot-operated control valves.

Normally we speak of the pilot-operated control valve as a four-way valve. It has the same functions as the manually-operated valves, yet the valve pistons or flow directors are shifted by air through a remote means. A two- or three-way valve could also be a pilot-operated control. Perhaps the best way to show the function of these valves is to show them in circuit diagrams.

The pilot valves may be manually, mechanically or electrically operated and may be placed adjacent to the pilot-operated control valve or at some distance away. However, the closer the pilot valve is to the pilot-operated valve, the more rapid is the response.

Use of solenoid-type pilot valves. Figure 16 illustrates a four-way, two-position, pilot-operated control valve operated by two small solenoid pilot valves mounted on top of the main valve body. Each pilot valve needs only a small porting in order to trigger the spool in the pilot-operated section. Note the manual override pin in the cover at left. There is also one in the cover at right. The main valve spool design is the same as that shown in Fig. 23.

Figure 17 shows a compact unit consisting of a stainless-steel spool valve cartridge with an integral, solenoid-operated pilot valve, a sub-plate, and a cover with two sealing members which float the symmetrical cartridge between the cover and subplate. In the main valve the cartridge and spool are made of hardened, corrosion-resistant stainless steel and are a precision fit. The pilot valve serves as a force amplifier translating a small electrical signal into a pilot-pressure force to actuate the spool. This is a five-port, two-position, four-way control valve for air service.

Figure 18 (Left) shows a double-solenoid, pilot-operated, four-way control valve with the solenoid actuators mounted in the ends of the main valve body. This valve has a sub-base mounting. Note the speed-control adjustment screws in the sub-base. This is a sliding-seal type valve. Figure 18 (Right) shows this type of valve mounted on a cylinder.

Figure 19 is a solenoid, pilot operated, in-line, four-way control valve that is designed for 150 psi maximum operation. This air-control valve is made up of the pilot and solenoid assembly, and the valve body assembly. The same pilot and solenoid assembly is employed on three, different, basic-pipe-size body assemblies. In the cross-section drawing, Fig. 19 (Right), note the manual override in the pilot section cover. Also note the poppet type pistons in the main control

Courtesy of The Aro Corp.

Fig. 16. Solenoid-actuated four-way, two-position, air-control valve.

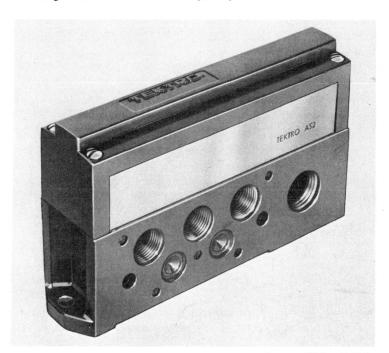

Courtesy of Tektro Fluid Power

Fig. 17. A four-way, five-port, balanced spool, solenoid-operated, lapped-spool, stainless-steel air valve.

Courtesy of Scovill Fluid Power Div.

Fig. 18. (*Left*) Double-solenoid operated, four-way air control with sub-base. (*Right*) Control valve mounted on rear cover of air-operated cylinder.

body. This valve is a four-port valve with inlet, exhaust, and two cylinder ports. The pilot section has a $\frac{1}{8}$-inch external supply port and a $\frac{1}{8}$-inch pilot exhaust port. There is also an electrical conduit port in the pilot section.

Use of Manual and Mechanical Types of Pilot Valves. Figure 20 illustrates a circuit making use of a manual-type pilot valve and a mechanical-type pilot valve used as bleeder valves and a bleeder-type pilot-operated, four-way control valve. The circuit functions as follows.

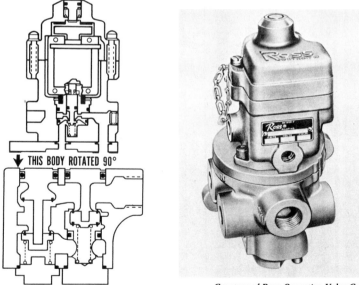

THIS BODY ROTATED 90°

Courtesy of Ross Operating Valve Co.

Fig. 19. (*Left*) Four-way, solenoid, pilot-operated, in-line, air-control valve. (*Right*) Cross section of the in-line control valve.

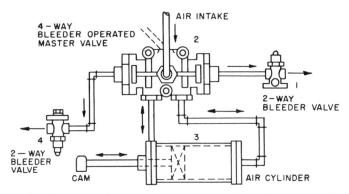

Fig. 20. Circuit which makes use of a manual-type pilot valve and a mechanical-type pilot valve as bleeder valves and a bleeder-type pilot-operated four-way control valve.

Operator places workpiece on slide and momentarily depresses button on two-way pilot valve (1), thus bleeding air from right-hand pilot chamber of pilot-operated, four-way control valve (2), shifting piston in valve (2) so that air is allowed to flow to blind end of pusher cylinder (3). As piston of pusher cylinder (3) approaches end of stroke, cam on pusher table depresses cam roller on two-way pilot valve (4). This bleeds air from left-hand pilot chamber of valve (2) and piston shifts allowing air to flow to rod end of cylinder (3) retracting piston of pusher cylinder (3). With this semi-automatic cycle, the operator is free to make preparations for loading the next workpiece while the pusher cylinder is going through its cycle.

Bleeder-type pilot-operated valves. Figure 21 shows a cross section of a small pilot-control valve, while in Chapter 19 Fig. 1 shows a

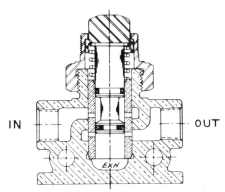

Courtesy of Logansport Machine Co., Inc.

Fig. 21. Cross-sectional view of a small pilot control valve.

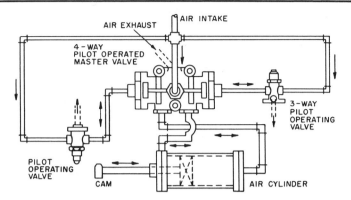

Fig. 22. Circuit which makes use of a manual type pilot valve, a mechanical type pilot valve, and a pressure type, pilot-operated, four-way control valve.

cross section of a bleeder type, pilot-operated control valve. In this Fig. 1, note the small passage (top of bottom drawing) from inlet port to valve cover (a second passage to the opposite end is not shown), which allows pressure to equalize at each end. This is called a balanced valve. When either end is bled of air the pressure becomes unbalanced and the spool assembly is shifted to the low pressure end.

Pressure-type Pilot-operated Valves. Pressure-type, pilot-operated control valves are operated by three-way valves which send pressure into the pilot chambers of the four-way valve shifting the piston. Figure 22 shows the same circuit as in Fig. 20 except that the pilot-operated valve is the pressure type instead of the bleeder type. Note the need for additional piping. The pressure-operated type gives almost instantaneous response and can be located farther away from its pilot valves than the bleeder type. Figure 23 shows a pressure type, pilot-operated, four-way control valve with five-port construc-

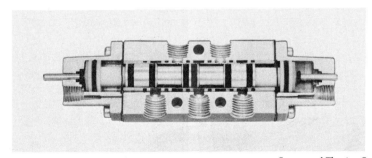

Courtesy of The Aro Corp.

Fig. 23. Pressure pilot-operated, four-way control valve.

tion. Note the manual override pins in each cover, for manual setup operation.

Use of Pilot-operated Valve with Pneumatic Timer. Figure 24 shows a circuit making use of a pilot-operated valve with a pneumatic timer. The valve is of the three-way type. This circuit is of considerable value where one operator is controlling several small presses. The operator loads one press and momentarily shifts the handle of the two-way valve (1); this allows air to flow to the pilot connection of valve (2), shifting the piston of this valve and air from the main line flows to the blind end of the single acting press cylinder (3) through valve (2) and the work is securely clamped. The air that enters the pilot connection of valve (2) is trapped in the timer chamber and can only escape through the adjustable orfice in this chamber. Meanwhile the operator can be loading and operating other similar presses. As the air bleeds off through the adjustable orifice of the timer the spring pressure exerted against the piston moves the piston up until the valve inlet port is blocked and the valve cylinder port becomes connected to the valve exhaust port and air is exhausted from blind end of cylinder. The cylinder piston, due to the load on the end of the rod, retracts to the starting position thus completing cycle.

The adjustable orifice in the timer of valve (2) may be set so as to allow the timer to quickly exhaust or to take several minutes to exhaust. A quick exhaust allows the inlet of the three-way valve to be connected to the cylinder port for only a short time while a delayed exhaust allows air to flow through the valve for a much longer period.

A good example of this type of valve is shown in Fig. 25. Note the air-control screw on top of the valve cover.

Use of Solenoid-operated Three-way Valve. Figure 26 shows a solution to a circuit problem where piston speed and compactness were important items. In studying the requirements, it was found

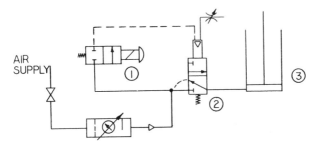

Fig. 24. Circuit which makes use of a pilot-operated valve of the three-way type together with a pneumatic timer.

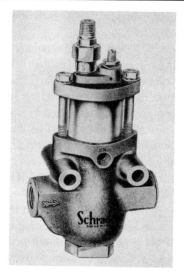

Courtesy of Scovill Fluid Power Division

Fig. 25. Pilot-operated valve of the three-way type. Note the air control screw on top of the valve cover.

that solenoid-operated, four-way control valves were either too bulky or were ported too small to allow sufficient flow of air to do a proper job. By using a very small solenoid-operated three-way valve to direct air to a cylinder-operated, spring offset-type, four-way valve, the problem was quickly and economically solved.

The circuit functions like this: The operator depresses an electric

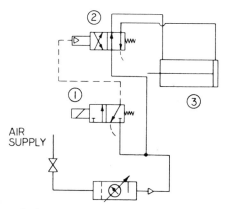

Fig. 26. Circuit employing a small solenoid-operated three-way valve to direct air to a cylinder-operated spring offset-type four-way valve which may be used when four-way control valves are either too bulky or are ported too small to allow sufficient flow of air to do a proper job.

pushbutton which energizes the solenoid of three-way pilot control valve (1); air flows to cylinder connection of valve (2) and air pressure depresses valve piston allowing air to flow to blind end of cylinder (3) and its piston rapidly moves forward. Operator releases button and solenoid of valve (1) is de-energized, closing inlet and exhausting the air pressure from cylinder of valve (2), allowing valve piston to retract to its normal position so that air pressure flows to rod end of cylinder (3) retracting its piston very rapidly.

Figure 27 shows a small, compact three-way solenoid valve.

Flow Control Valves

How Flow-control Valve Functions. Basically, a flow-control valve is a metering valve and a check valve built into one housing. In one direction the air flow is restricted as it flows through the metering valve; in the other, it flows freely through the check valve. The metering valve and check valve may be of several designs. The air flow through the metering orifice may be controlled by a pointed needle closing off the orifice, by a grooved needle in which the groove is gradually closed off, or by other gradually closing-off means. The check valve, which must be positive in its sealing action and quick

Courtesy of Skinner Electric Valve Div.

Fig. 27. Small compact three-way solenoid valve.

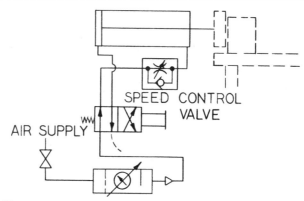

Fig. 28. Circuit showing the use of a flow-control valve to control the rate of piston movement of the outstroke.

opening when the air flow is reversed, may use a steel ball, a plunger, or a poppet for the means of sealing.

Proper Location in Circuit. The flow-control valve controls the volume of air flowing past a given point. This point is in the line that is being exhausted. This does not mean that the flow-control valve should be connected to the exhaust connection of the four-way control valve, however. Figure 28 illustrates a simple circuit showing the use of the flow-control valve to control the rate of piston movement of the outstroke. Note that it is placed between the cylinder and the master-control valve with the flow being controlled as the air leaves the cylinder.

If the flow-control valve were connected to the exhaust of the four-way control valve, the results usually would not be satisfactory for several reasons. *First,* a restriction in the exhaust of some types of master-control valves will cause them to malfunction. *Second,* even if no difficulty was experienced with the master valve, the control of the cylinder would be different on the outstroke than on the instroke. The difference is caused by the difference in volumes of air to be exhausted on the outstroke and instroke of the cylinder due to the space taken up by the piston rod. *Third,* the cylinder may be a long distance away from the master valve and, hence from the flow-control valve, thus not giving proper metering, due to the compressibility of air. The flow control should be placed as close as possible to the cylinder. *Fourth,* the use of only one flow-control valve for controlling the speed of both the in and out movement of the piston rod is not usually satisfactory as the loads are usually different in each direction.

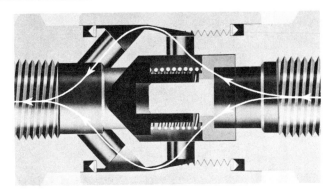

Courtesy of Rexnord, Inc.

Fig. 29. Air flow path through a speed-control valve. The check valve poppet (black V-shaped section) is sealed against the seat and the air flow is passing through the orifice.

Flow-control valves are often placed backwards in the circuit. This gives intake metering which in an air system will result in very uneven feeds.

Figures 29 and 30 illustrate the flow path through a speed-control valve. Figure 29 shows the flow path in the controlled-flow direction, while Fig. 30 shows the flow path in the free-flow direction. Note that in Fig. 29 the check valve poppet (black inverted vee-shaped section) is sealed against the seat and the air flow is passing through the orifice. In Fig. 30 the poppet has been unseated and the air flow is returning around the poppet. This allows a free flow equal to the pipe area.

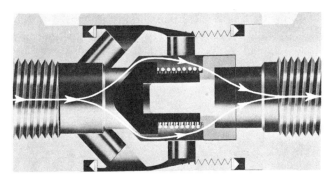

Courtesy of Rexnord, Inc.

Fig. 30. Air flow speed-control valve. The poppet has been unseated and the air flow is returning around the poppet. This allows a free flow equal to the pipe area.

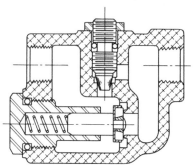

Fig. 31. Flow-control valve employing a pointed needle and a poppet for the checking action.

Figure 31 illustrates a cross section of a flow-control valve employing a pointed needle and a poppet for the checking action. The air is controlled as it flows from left to right, and flows freely as it moves from right to left. When the exhaust air leaves the cylinder and moves through the port A it meets resistance and is metered through the orifice set by the metering needle. The poppet seals off any chance for the air to escape through the check orifice. When the air flow is reversed, the poppet is quickly unseated and the air flows freely through the check orifice.

Cam-operated Flow-control Valves. Cam-operated, flow-control valves are especially useful when used as a cushioning means for a long-stroke air cylinder whose piston travels at a high rate of speed. By placing the cam-operated valve in the circuit, as shown in Fig. 32, a long effective cushion can be obtained without going to the expense of placing long cushion sections in the cylinder. The valves can be located so that by a little experimentation the desired results can be obtained. In some instances on very long stroke cylinders, traveling

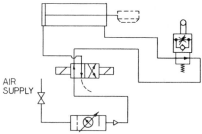

Fig. 32. Circuit utilizing a cam-operated flow-control valve which provides a long, effective cushion for a long-stroke air cylinder whose piston travels at a high rate of speed.

at high speeds and carrying a very heavy load, it may be necessary to cushion the last nine to twelve inches of stroke so as to eliminate any undue shock at the end of the stroke. The cam-operated, flow-control valve should be placed as close in the air line to the cylinder as possible (the cam being positioned at the desired point along the stroke path where the cushioning is to begin) so as to eliminate air compression in the pipe line between the valve and the cylinder.

When the cam-operated, speed-control valve is used strictly as a flow control, it is useful on skip-feed applications. The cam, which is attached to the piston rod, or machine slide contacts and depresses the roller, providing controlled flow, but as soon as the roller is released, air flows freely again through the valve. When the air stream is reversed through the valve, the air flows freely whether the cam roller is depressed or not due to an internal ball check. The controlled flow is governed by the setting of the flow-control needles.

A cam-operated shutoff valve and a flow-control valve will give the same end result.

Sequence Valves

Sequence valves provide one means of setting the sequence of operations carried out by cylinder piston movement. The principle of operation of the sequence valve is that when a predetermined pressure is reached within the valve due to the line pressure building up after the cylinder piston reaches the end of the stroke, the valve piston is unseated, allowing air to flow to a different branch of the circuit or to another air cylinder. Figure 33 shows the cross section of a typical air sequence valve. The pressure at which the piston moves away from the valve seat is determined by the spring tension against the piston. This tension is set by the adjusting screw on top of the valve. Note that the piston area which is exposed to the inlet pressure when the piston is seated is held to a minimum. This helps reduce the size of the spring. When the pressure, however, unseats the piston, the large area is exposed to the pressure and the piston is held wide open.

For reverse flow through the valve, the check is unseated and the air passes through the valve with very little resistance.

Use of Sequence Valve. Figure 34 illustrates the use of the sequence valve in a circuit. Here is how the circuit functions. Operator places workpiece on air-operated mandrel. He then depresses pedal of four-way, foot-operated valve (1); and air flows to rod end of mandrel operating cylinder (2); and piston retracts and mandrel jaws are expanded against workpiece. Pressure then builds up open-

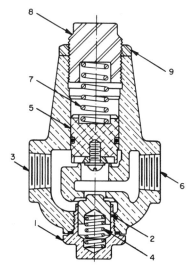

Fig. 33. Cross section of a typical air sequence valve.

ing sequence valve (3); and air flows to blind ends of drilling cylinders (4 and 5) and drills are advanced into the workpiece. When drilling is completed, operator again depresses foot pedal of valve (1) and air flows to rod ends of drill cylinders (4) and (5) and the drills are retracted. When pressure builds up, sequence valve (6)

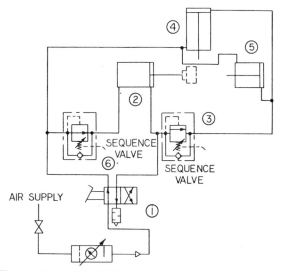

Fig. 34. Circuit which illustrates the use of a sequence valve.

opens and air flows to blind end of air cylinder (2) and pressure releases mandrel jaws; operator removes workpiece.

Other Methods of Sequencing

While the sequence valve is used very successfully in many air circuits for setting up sequential operations, there are many instances where there is a better method of setting up such a cycle. Since the air pressure in many plants is erratic due to large volumes of air being used intermittently, sequence valves may not function at the instant that they are supposed to, consequently tool breakage, machine breakage, or scrapping of the workpiece may be the result. Another instance where the sequence valve is of no value is when the line pressure to the cylinder in the second step in the sequence must be less than that in the first step. For example, a workpiece is clamped at full line pressure by using a large-bore air cylinder, then the drills for drilling small diameter holes in the workpiece must be operated by small diameter cylinders operating at low pressure in order not to break the drills. Such a sequence is easily solved by the use of other type valves; however, the circuit is more complicated than when using sequence valves. Such a circuit can be operated either by solenoid valves or by pilot-operated bleeder valves.

Use of Solenoid Valves. Figure 35 shows a circuit using solenoid valves and functions as follows. Operator places workpiece in

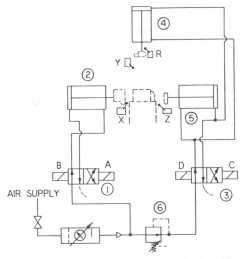

Fig. 35. Circuit which illustrates the use of solenoid valves.

fixture and momentarily depresses push button which energizes solenoid (A) of four-way valve (1) allowing air at full line pressure to flow to blind end of clamp cylinder (2). Piston of cylinder (2) moves in to clamp workpiece and just before this is accomplished, the single-acting trip dog on carriage of clamping device overrides roller on limit switch (X). Solenoid (C) of valve (3) is now energized. This allows air at a reduced pressure—since reducing valve (6) reduced the pressure, to flow to blind ends of drill cylinders (4) and (5) and their pistons move out to do the drilling at a low pressure. When the drilling is completed, limit switches (Y) and (Z) (which are hooked up in series) are contacted. Solenoid (D) of valve (3) is energized and air flows to rod ends of cylinders (4) and (5) retracting their pistons. As these pistons approach their starting position, a single-acting trip dog on one drill carriage overrides roller on limit switch (R). This energizes solenoid (B) of valve (1) and air flows to rod end of clamping cylinder (2) thus retracting piston and completing cycle. By the use of this circuit, the steps of the sequence are always in the correct order and movements are always started at the correct time. There are no limits as to the number of steps possible when using this type of sequence circuit. Line pressure fluctuation has no visible effect.

Use of Pilot Controls. The same sequence can be set up with the use of pilot controls; however, there will be considerably more piping involved, i.e., small pipe lines would be run to each of the pilot valves which would be placed in the same locations as the limit switches and push buttons, as shown in Fig. 35. Since all electrical components are eliminated, circuits using pilot controls are recommended for use in hazardous locations, such as in the presence of high explosive powders, volatile fumes and gases, and other highly inflammable substances.

Use of Timers. Another way to set up an operational sequence in an air circuit is through the use of timers. They provide a very compact and convenient method of setting up the controlling means. Timers eliminate the use of limit switches, trip dogs, cams, and other tripping devices. Figure 36 illustrates timers which are synchronous-motor-driven devices that trip their contacts after an adjustable time interval which can be set extremely accurately.

Electric timers are built for a wide variety of application. While one timer may be built to operate only one solenoid of a master valve, another may be built to operate the solenoids of a large number of valves in their required sequence. The timer may be designed to go

Courtesy of Eagle Signal Div.

Fig. 36. Two types of timers which are synchronous motor-driven devices that trip their contacts after an adjustable time interval. This time adjustment can be set very accurately.

through one cycle of operation and stop or it may be designed to keep on repeating the same cycle. A wide range of timing intervals are obtainable.

New air valves are being developed constantly to meet the ever-increasing demand for better control of air power.

CHAPTER 16

Air Cylinders and Their Design

Air cylinders are the means for converting the air pressure delivered by the air circuit into applied force and straight line motion. Their design varies from that of a small clamping cylinder which is only required to produce a few pounds of force to that of heavy-duty mill-type cylinders which produce forces of several tons. Unlike hydraulic cylinders, which can be designed for several operating pressures, air cylinders must have larger cross-sectional areas if larger forces are to be exerted. The reason for this is that in most industrial plants the maximum air pressure will not exceed 150 lb per sq in., while the average pressure in most plants is 90 lb per sq in., or less.

The question which often arises is: "Where and in what types of applications can air cylinders be used?" The dozen schematic illustrations shown in Fig. 1 present excellent examples of the type of force applications and motions obtainable. With new uses being developed almost daily, any further attempt here to enumerate all applications would obviously be both impractical and misleading. Certainly the trend toward greater and greater automation is causing a vast number of air cylinders to be applied in a multitude of ways. They are in great demand for such single-purpose applications as clamping, stamping, transferring, etc. Air cylinders are also needed for applications where high temperatures are present. Many are working where ambient temperatures range from 400° to 500°F. Speed of piston travel is another factor which favors the use of air cylinders in helping to gear up industrial plants for greater production.

Thrust and Air Consumption of Cylinders

The thrust produced by cylinders of various sizes at 60, 80 and 100 lbs per sq in. line pressure are shown in Table 1. The thrust on the return stroke of the piston is less than that shown in the table by an

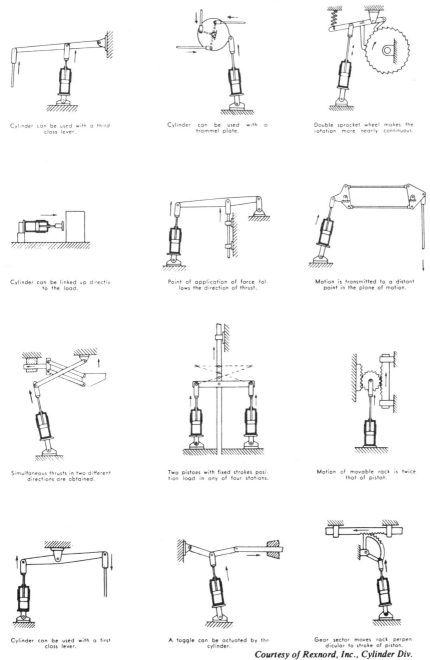

Cylinder can be used with a third class lever.

Cylinder can be used with a trammel plate.

Double sprocket wheel makes the rotation more nearly continuous.

Cylinder can be linked up directly to the load.

Point of application of force follows the direction of thrust.

Motion is transmitted to a distant point in the plane of motion.

Simultaneous thrusts in two different directions are obtained.

Two pistons with fixed strokes position load in any of four stations.

Motion of movable rack is twice that of piston.

Cylinder can be used with a first class lever.

A toggle can be actuated by the cylinder.

Gear sector moves rack perpendicular to stroke of piston.

Courtesy of Rexnord, Inc., Cylinder Div.

Fig. 1. Schematic diagrams showing typical applications of air cylinders.

amount equal to the area of the piston rod × the operating pressure. As an example, the table shows that the thrust of a 6-inch cylinder at a pressure of 100 lbs per sq in. is 2827 lb on the outstroke. If a 1⅜-inch diameter piston rod is used (area = 1.48 sq in.) the amount that should be subtracted from 2827 lb is 1.48 × 100 = 148 lb. The thrust on the return stroke is therefore 2827 − 148 = 2679 lb.

Figures given in the table do not make allowance for loss by friction or air leakage. Also, since the air pressure in a plant may vary erratically, due to intermittent use of large volumes of compressed air in all sorts of air devices, the bore size of the cylinder must be large enough to deliver the force required after allowing for any normal pressure drop. This is extremely important as insufficient force may ruin an entire operation.

The last two columns of Table 1 show the number of cubic feet of free air and compressed air consumed by cylinders of various size for each inch of piston travel.

TABLE I

THRUST AND AIR CONSUMPTION OF VARIOUS SIZE CYLINDERS *per in of travel.*

Bore Diam (in.)	Thrust at 60 psi (lb)	Thrust at 80 psi (lb)	Thrust at 100 psi (lb)	Free Air Consumed* (cu ft)	Compressed Air Consumed* (cu ft)
1½	106	142	177	0.0064	0.0010
2	188	251	314	0.012	0.0019
2½	295	393	491	0.018	0.0028
3	424	566	707	0.026	0.0041
4½	954	1,272	1,590	0.059	0.0092
6	1,696	2,262	2,827	0.106	0.0164
8	3,016	4,021	5,026	0.187	0.0291
10	4,712	6,283	7,854	0.293	0.0455
12	6,786	9,048	11,310	0.422	0.0655
14	9,236	12,315	15,394	0.574	0.0891
16	12,064	16,085	20,106	0.750	0.1164

* When compressed to 80 pounds per square inch.

Types of Air Cylinders

The applications for which air cylinders are used can be divided into three groups: light duty, medium duty, and heavy duty. The requirements of each application should be carefully scrutinized before making a decision as to the size and type of air cylinder required.

While an air cylinder of die-cast construction may prove perfectly satisfactory for use on a small clamping fixture, the same construction

Courtesy of Logansport Machine Co., Inc.

Fig. 2. A high-speed, aluminum, rotating air cylinder for rotating-spindle applications. Lettered parts are identified as follows: A. High-speed airshaft assembly; B. Airshaft end cap; C. Cylinder cover; D. Cylinder body; E. Piston rod; F. Grease fitting; G. Cylinder ports.

would not be at all suitable for the rugged service of a rollover or transfer device used in a steel mill application.

Air cylinders are built in both single- and double-acting types. The single-acting types are those in which air only moves or pushes the piston in one direction and the piston is returned by means of an external spring or by gravity. In a double-acting cylinder, air moves the piston in both directions.

Air cylinder classifications can be broken down still further; i.e., into rotating models and nonrotating models. Rotating models are used in applications requiring that the cylinder body be connected to a rotating member while the air connections to the cylinder be in a stationary housing. Such cylinders are used on lathes, grinders, spinning machines and other machines with rotating spindles. In Fig. 2 is shown a high-speed aluminum rotating type of air cylinder. An ANSI standard has been developed (ANSI B5.5-1959) to obtain interchangeability of different makes of air cylinders on the spindles of machine tools without changing the adapter or drawrod. The length of stroke of the standard cylinders, the position of the piston rod at the end of the stroke, the diameter of the piston rod, and size of the tapped hole in the piston rod have also been standardized so that air cylinder drawrods do not have to be fitted to individual air cylinders. Detailed dimensions of air cylinders and air cylinder adapters are given in the ANSI Standard for three sizes of standard adapters covering the range of standard air cylinders of from 3 inches to 18 inches inclusive. Adapter "A" fits the 3-inch and 4½-inch cylinders, adapter "B" fits the 6-inch and 8-inch cylinders, and adapter "C" fits all sizes of cylinders from 10 inches to 18 inches inclusive. A fourth size adapter "D" is also included and is provided to accommodate the 20-inch air cylinder or other power-operated devices having a drawrod pull of 26,000 to 40,000 pounds.

Nonrotating air cylinders, of course, are more widely used than rotating cylinders, since the great majority of applications do not involve a rotating member. A cross-sectional view of a nonrotating cylinder is shown in Fig. 3.

An air cylinder is composed of a cylinder tube, two covers (although one may be an integral part of the cylinder tube), a piston rod, a piston, packings, and cover screws or tie rods. If the cylinder is single-acting, a spring may be required. If it is cushioned, a cushioning arrangement and cushion collar are needed. Following are some of the construction and design features of air cylinders.

The Cylinder Tube

A number of materials may be used for air cylinder tubes. Since air, if not properly lubricated and filtered, may cause rusting of ferrous tubing, many choose drawn brass or aluminum tubing. This, however, is not possible for tubes for air cylinders of large bore. There are a number of cylinders being built today which have 28- and 30-inch bores and for these, butt welded steel tubing is being used to a large extent. Cylinder tubes are also made from castings of iron, aluminum, bronze and steel. Brass tubing is usually not available from warehouse stock in bores larger than eight inches.

Many cylinder tubes of cast construction have one cover cast integral with the tube. On nonrotating cylinders, the blind cover is usually cast integral; while on the rotating cylinder, it is the rod-end cover. Figure 4 shows a small nonrotating cylinder of cast construction.

The finish of the inside surface of the cylinder tube must be to a high degree of smoothness if satisfactory service is to be expected from the packings. Cylinder tubes with open ends are more easily bored and honed than those with one end closed.

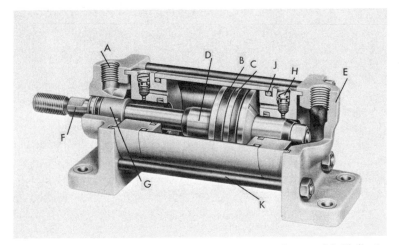

Courtesy of the Sheffer Corp.

Fig. 3. Cross-sectional view of a non-rotating air cylinder. Components are indicated by letters as follows: A. Pipe connections; B. Piston packing; C. Piston; D. Cushion collar; E. Cover; F. Piston rod; G. Rod bearing; H. Cushion needle; J. O-ring; K. Tie rod.

Tubes for heavy-duty, mill-type cylinders usually have heavy steel tube rings attached to each end of the tube. These rings are either welded to the tube or are held in place by some mechanical means. The rings contain the holes for mounting the tube to the covers.

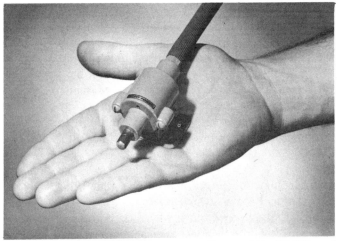

Courtesy of Mead Fluid Dynamics Div.

Fig. 4. Small non-rotating cylinder of cast construction.

Fig. 5. Typical mill type air cylinder.

Figure 5 shows a heavy-duty, mill-type air cylinder designed with interchangeable cushion adjustment valve and cushion check valve for convenience of installation. Note the through bolt construction which is employed to attach the covers to the tube rings allowing easier removal of the cylinder covers. Covers are made of alloy steel castings because of rugged service required of this type of cylinder.

Cylinder Covers

Cylinder covers are designed and built in all manner of shapes. Some are round, some are square, while others are of special shapes. There are a number of mounting (means of attaching cylinder to some stationary member) styles used for nonrotating cylinders, including rabbeted, foot, trunnion, centerline, flange on rod end, flange on blind end and clevis. Figure 6 shows some of these. A combination of these, such as flange-mounted on rod end and foot-mounted on blind end, rabbeted-mounted on rod end with flange-mounted on blind end, etc., are often used. There is no set of standards for mounting dimensions among manufacturers of nonrotating air cylinders, although there has been some discussion along this line. As previously mentioned, adapters for rotating air cylinders are now covered by an ANSI Standard, but this is not followed one hundred per cent.

The terminology for covers is somewhat confusing and perhaps should be clarified. *Rod end, stem end,* or *front cover,* all mean the same thing; i.e., the cover through which the piston rod moves. This cover contains the rod bearing and on double-acting cylinders it houses the rod packing or seal. The *blind end* or *back cover* is the one opposite the piston rod and carries no packings or bearings unless it has a cushion feature.   There are many mistakes made when ordering repair parts for cylinders because of wrong terminology.

Cylinder covers are made of die cast aluminum, cast aluminum, cast iron, cast bronze, cast steel or plate steel, depending upon the severity of service.   Front cover bearings are of bronze or cast iron. Long-stroke cylinders usually require extra long bearings.

Cylinder covers are held onto the cylinder tubes by several methods, which include tie rods, screws, threaded connection and metal inserts. Tie rods are still used by the majority of manufacturers on medium-duty cylinders.

Packing Glands

Packing glands or retainers, generally of a bronze material, are devices for holding the rod packing in place.   Their designs vary greatly.   Some are designed to act as a nonadjustable retainer for the packing.   Some are adjustable so that tension may be increased on the packing as it wears.   Some retainers are closely fitted to the cover and act as a rod bearing.   Yet others house the packing and also act as a rod bearing.   Packing glands are held in position by snap rings, cover plates, threads and screws.

Cushion Assembly

Cushioning, i.e., gradual deceleration of the piston near the end of its stroke, is a desirable feature for many applications.   It is especially helpful when the piston rod is connected to a heavy load and the piston is traveling at a high rate of speed.   It reduces the shock

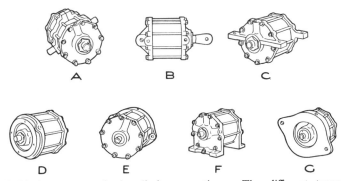

Fig. 6. Various types of air-cylinder mountings.   The different types are identified as follows: (A) Trunnion.   (B) Clevis.   (C) Centerline.   (D) Rabbeted.   (E) Blind-end flange.   (F) Foot.   (G) Rod-end flange.

that would otherwise be caused if the piston were allowed to make sharp contact with the covers without any buffer action.

A cushion is a chamber of relatively small diameter into which a cushion nose or collar enters as the piston nears the end of its stroke so that air is trapped in the cylinder tube between the piston and cylinder cover and is bled off slowly, reducing the rate of piston travel. Cushions should not be construed as speed controls, but only as shock alleviators. On some heavy-duty, mill-type cylinders, it may be necessary to install cushions having a length of six or eight inches in order to stop the load without excessive shock. Cushions on standard-type cylinders are about an inch long.

Cushions are of little value if the complete stroke of the cylinder is not used. This fact is frequently not realized, thus the user may purchase a cushioned cylinder and then install external stops which stop the piston rod when the piston is a couple of inches away from the cover.

Two parts of the cylinder are involved in the cushioning arrangement; the cover, or covers (depending on whether the cylinder is to be cushioned on one end or both ends) and the piston rod. Cushioning devices probably can best be considered in two classifications; those with metal-to-metal fit and those with metal-to-synthetic-material fit or contact. Only recently has the metal-to-synthetic-material fit or contact come into prominence. For many years the metal-to-metal fit has been the accepted method.

The cushioned cylinder cover contains either a cushion bushing or a machined cushion bore in a cast iron cover finished to close limits. Figure 7 shows the latter. As shown in Fig. 7, the cover usually contains a cushion needle which has a passage to it from the inside face of the cover that continues on to the air inlet port. The cover also contains a ball check valve which has a passage between the inside face of the cover and the air inlet port. When the cushion nose or collar (cushion nose enters blind cover and cushion collar enters rod end cover) approaches the cover and then finally enters the cushion bushing or bore, the air is trapped between the piston and the cover which is being approached. The cushion needle is set so that the air is bled off to the port connection at the proper rate to slow down the piston and reduce the shock when the piston contacts the cover. The end of the cushion nose or collar is tapered, chamfered, or rounded in order to allow it to more easily enter the cushion bushing or bore. There is a clearance of only a few thousandths of an inch between these two. When the direction of the cylinder motion is to

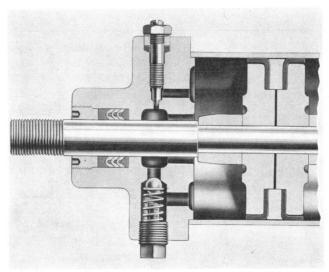

Fig. 7. Cushioned cylinder cover with a machined cushion bore in a cast-iron cover finished to close limits.

be reversed, the air enters the port opening, the ball check valve opens and air flows freely to the piston before the cushion nose or collar leaves the bushing. This allows the piston to move quickly at the start of the stroke. If it were necessary for the cushion nose or collar to move out of the bushing before a fast feed could be obtained, much time would be lost. Also, due to the fact that the air would be acting only against the area of the collar section, there might not be enough force to move the load.

Excess dirt or corrosion can be very detrimental to a metal-to-metal cushioning assembly. Pipe scale can cut the highly finished surfaces of the bushing and dirt will clog the needle.

By adjusting the position of the cushion needle shown in Fig. 7, the amount of bleed-off can be accurately controlled for various load conditions. Figure 8 shows a cylinder with a self-regulating cushion in which air is bled off between the cushion nose and the cushion bore.

Figure 9 illustrates one of the newer cushion arrangements—a metal-to-synthetic-material fit. The metal cushion collar enters the synthetic ring seal and the air is trapped between the piston and cover and released through the cushion needle as shown. Note the rounded end of the cushion collar for ease of entering into the packing ring. On the stroke reversal, the incoming air depresses the rubber seal

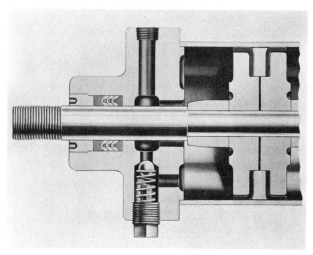

Courtesy of Galland-Henning Nopak, Inc.

Fig. 8. Cylinder with a self-regulating cushion in which air is bled off between the cushion nose and the cushion bore.

and air quickly acts on the back surface of the piston, thus eliminating the need for a ball check valve.

Figure 10 illustrates another new cushioning design which makes use of a poppet type of arrangement on each end of the piston consisting of a synthetic seal backed up by a spring. As the piston approaches the end of the cylinder, the seal contacts the flat metal surface and shuts off the free passage of air to the exhaust. A needle valve is used in both the rod and blind ends of the cylinder to bleed

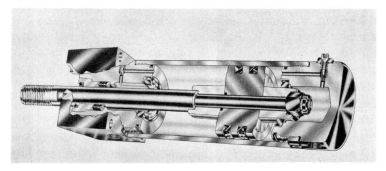

Courtesy of The Tomkins-Johnson Co.

Fig. 9. Air cylinder cushion arrangement which uses a metal-to-synthetic material fit. The metal cushion collar enters the synthetic ring seal and the air is trapped between the piston and cover and released through the cushion needle as shown.

off the trapped air. This design gives positive cushioning for a full-length cushioning stroke and provides a large sealing area. At any time the cushion packing becomes worn, it can very easily be replaced. Note the cushion needle design in which an "O" ring is used to prevent leakage past the threads.

The ideal cushioning assembly would be one which would effect a uniform deceleration from the maximum speed of the piston down to zero in the length of the cushion nose without any shock, even though the load may be greatly varied. Although some designs approach this ideal, there is usually a marked decrease in speed when the cushion collar makes contact with the recess.

Rotating cylinders do not use any cushioning device. Their port-

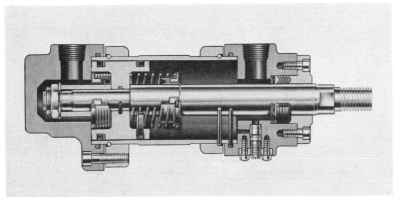

Courtesy of Rexnord, Inc., Cylinder Div.

Fig. 10. Cushion design which makes use of a poppet type of arrangement on each end of the piston consisting of a synthetic seal backed up by a spring.

ing is small and their pistons move rather slowly. In the majority of applications there is no severe shock between the piston and covers. In fact, the piston rarely contacts one cover. That cover may be either the blind end or rod end depending upon whether work is being done on the instroke or outstroke of the cylinder.

Pistons and Piston Seals

Piston design varies greatly from that used in the light-duty cylinder to that required by the heavy-duty, mill-type cylinder. There is also quite a variation within each class. Materials stay in about the same category as those in other parts of the cylinder—aluminum, brass, cast iron and steel. Where there is sliding contact with another member, materials of different hardness should be used, such as a cast iron piston and a brass tube. Pistons may be of one-piece, two-piece or three-piece construction according to the type of packings used. A double-acting piston using "U" rings such as that shown in Fig. 10, requires only a one-piece, whereas a double-acting piston using cup packings, such as that shown in Fig. 8, requires a three-piece construction.

Pistons are packed with cups, chevrons, "U" rings, formed synthetic packings, "O" rings, automotive-type piston rings and numerous other types of packings. For details of each type, see Chapter 13 on packings. Cylinders with automotive rings for piston packing are ordinarily used in "hot spots" such as on casting machines, furnaces, rolling-mill equipment where other types of packing will not stand

Fig. 11. Special type of piston rod with an eye-type rod end.

the heat. Leakage past the piston on cylinders with automotive-type rings is usually great, especially during the break-in period. When automotive-type rings are used, the piston must be closely fitted to the cylinder tube.

Piston Rods

Piston rods should be of a good ground and polished steel having a suitable tensile strength for the application intended. Standard rods are equipped with one of three types of end connections—male thread, female thread or rod head. Special rods often have rather ingenious rod ends such as the eye type shown in Fig. 11. The finish on the piston rod has much to do with the life of the rod packing. On some applications it may be necessary to have hardened or plated rods. Chromium is the plating most widely used and a thickness of 0.0005 inch is recommended. When selecting cylinders with long stroke piston rods it is important that the rod be of sufficiently large diameter to resist the buckling effect of the thrust load. Table 2 shows the proper piston rod diameters corresponding to various rod lengths and thrust loads.

TABLE 2

PISTON ROD DIAMETERS FOR VARIOUS COMBINATIONS OF THRUST AND ROD LENGTH*

Rod Diam, inches	Thrust on Rod, Pounds													
	50	100	150	250	400	700	1000	1400	1800	2400	3200	4000	5000	6000
	Maximum Length of Rod, Inches													
⅝	67	59	53	43	37	30	27	24	23	19	16	13	9	...
1		110	103	94	83	68	60	53	48	45	41	38	34	30
1⅜			146	134	118	105	92	82	75	67	63	60	56	
1¾				186	168	155	142	127	114	103	94	87	82	
2					202	190	174	160	145	130	119	110	102	
2½					275	257	244	230	213	194	175	163	152	
3						330	308	296	281	261	240	225	208	
3½							385	366	347	329	310	289	274	
4								440	415	400	378	360	342	
4½									488	461	446	426	410	
5													494	476

Rod Diam, inches	Thrust on Rod, Pounds													
	8000	10,000	12,000	16,000	20,000	30,000	40,000	50,000	60,000	80,000	100,000	120,000	140,000	160,000
	Maximum Length of Rod, Inches													
1	26	21	17											
1⅜	50	45	41	34	28									
1¾	76	70	65	57	52	39	22							
2	93	89	84	75	68	55	43	30						
2½	137	125	118	110	103	87	74	66	57	36				
3	188	172	155	142	136	120	108	96	88	71	57	45		
3½	245	222	210	188	172	156	142	130	119	104	90	77	64	47
4	310	279	269	235	218	189	177	165	154	137	120	108	98	86
4½	375	349	326	292	270	230	210	200	190	170	154	146	128	118
5	447	412	388	350	326	285	248	234	225	204	199	175	160	148
5½		482	454	420	385	330	294	269	256	240	222	207	194	182

Courtesy of Miller Fluid Power Div., Flick-Reedy Corp.

* *Example:* For a cylinder having a stroke of 48 inches and a thrust of 1800 pounds, locate the 48-inch length in the column for 1800 pounds thrust. Read horizontally left from 48 inches to the first column where the rod diameter is found to be 1 inch.

Rod Packings

While there are numerous rod packings available for use in air cylinders, the types most widely used are the chevron or "V" ring, "hat" packing, "U" ring and "sea-ring." Rod packings are discussed in Chapter 13.

Wipers or scrapers are often placed in front of the packing in order to remove foreign particles from the piston rod before it passes through the packing. Note the wipers shown at the left of the cylinder in Fig. 12 and how they are designed to fit into the end of the rod bearing. Also note that the "hat" type packing around the piston rod is secured by the end of the rod bearing to give a very compact design. This packing cannot be tampered with unless the retainer plate is removed. Heavy-duty tie rods hold the end covers.

Rotating cylinders require airshaft packing for effecting a seal between the airshaft stem and the distributor. On small diameter

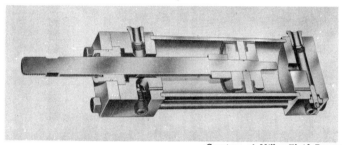

Courtesy of Miller Fluid Power Div.

Fig. 12. Cutaway view of an air cylinder in which wipers shown fitted in the end of the rod bearing at the left end of the cylinder remove foreign particles from the piston rod before it passes through the packing.

airshafts, sealing is no particular problem unless exceedingly high speeds are required, but on large diameter airshafts used in hollow center cylinders quite a difficult problem is encountered, especially if the rotating speed is very great. Some airshaft stems are as large as ten inches in diameter. This necessitates a large sealing area and the friction caused by the packing creates considerable heat. It is often necessary to water jacket the distributor housing in order to keep the packing from overheating. If the speed of the airshaft stem approaches 1000 surface feet per minute, water jackets are recommended. An airshaft stem of ten inches in diameter rotating at 400 rpm would be traveling over 1000 surface feet per minute. Rotating cylinders with hollow airshaft stems are used on machines where bar stock is being fed through the machine spindle.

Tandem-type Cylinders

Where space is a factor and a large force is required, it may be desirable to use tandem-type air cylinders. These cylinders are "stacked," using one common piston rod. They may be of two or three sections and may be either double- or single-acting. Single-acting, tandem-type cylinders are often used for die cushions on punch presses. The diameter of the recess in the press bed in which the cylinder can be located is usually limited but the depth of the recess is not limited. This is an ideal situation for multiple-section tandem cylinders.

Tandem-type rotating cylinders are used on machines where guard clearances and other obstructions will not permit a large diameter cylinder. The disadvantage of tandem-type rotating cylinders is that they have a long overhang, throwing an extra load on the spindle bearings of the machine.

Duplex Cylinders

Tandem-type cylinders are often confused with duplex cylinders. Duplex cylinders, while they are built in line the same as the tandem type, have two piston rods, one for each piston. One piston rod operates inside of the other. Applications for the nonrotating type are on press work where one piston rod moves out and holds the work and the other piston rod performs a staking, punching, pressing, or other forcing operation. The rotating type is used successfully in chucking operations where four-jaw power chucks are used. Two jaws are operated by one piston rod and the other two jaws are operated by the second piston rod. High pressure can be applied to one set of jaws while a low pressure is applied to the other set. This is ideal when it is necessary to clamp on both heavy and thin wall sections at the same time. Even though each section of the cylinder has the same bore, the pressure in one section can be reduced through a reducing valve.

Cylinders with Stroke Adjustment

A cylinder with a stroke adjustment is often required. Figure 13 shows a small air press with this feature. Note that the adjustment is provided for by a threaded section of a double-end piston rod which extends out of the top of the cylinder. The stop is provided by the check nuts. It is important that the threads are not allowed to con-

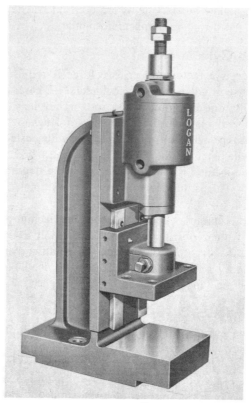

Courtesy of Logansport Machine Co., Inc.

Fig. 13. Small air press with stroke adjustment. Piston rod on this press is double-ended. The portion which extends above the top of the cylinder is threaded and adjustment is made by means of two check nuts.

tact the rod packing, otherwise excessive leakage will result. This is taken care of by a long sleeve over the rod extension.

Stroke adjustment also may be obtained by screwing a stud through the blind cover and locking it with a check nut. A wick-type packing is usually provided to act as a seal between the stud and the cover. The stud must be of sufficient strength to withstand the continual pounding of the piston. This type of stroke adjustment is shown in Fig. 14.

Cylinders with Double-end Rods

An air cylinder with a double-end rod, i.e., a rod projecting from both covers, can be used to advantage. One end of the rod may be used to transmit the force while the other end may be used for trip-

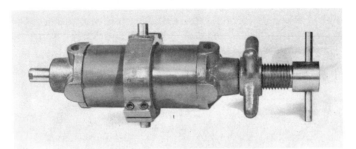

Courtesy of Anker-Holth, Power Cylinder Div.

Fig. 14. Stroke adjustment on this air cylinder is made by screwing a stud through a blind cover and locking it with a check nut. A packing is provided to act as a seal between the stud and the cover.

ping limit switches, control valves, or some other mechanism. The double-end rod also provides better rod alignment since the rod is riding between two bearings. Because of the better support provided, double-end rods are also advantageous when the cylinder is connected to a long machine bed.

Air Actuator without Moving Parts

Many force applications can be solved by the use of the "Airstroke"* actuator shown in Fig. 15. In the true sense of the definition of a pneumatic cylinder, the Airstroke actuator would not qualify as a cylinder, but it is a device that can impart a large amount of force. It is a hermetically sealed unit which has no sliding seals or moving parts, thus it provides efficient, friction-free operation. The unit is compact with axial compressed lengths down to 1.8 inches. Due to the nature of its design the Airstroke actuator has a built-in flexibility† capable of handling a considerable amount of angularity. This reduces the cost of installation. The body of the actuator is made of nylon-reinforced neoprene rubber, the end plates of corrosion-resistant plated metal. These actuators are made with one, two, or three convolutions. A two-convolution actuator of 20-inch operating diameter and a maximum usable stroke of 7.4 inches can impart 26,700 pounds force at 1-inch stroke, or 16,750 pounds force at maximum stroke, when operated at 100 psi. The minimum length of this actuator is 3.5 inches.

Some of the applications for these actuators are: press platen ac-

* "Airstroke" is a registered trademark of the Firestone Industrial Products Company.
†The Airstroke actuator expands when air pressure is applied and retracts when air pressure is released.

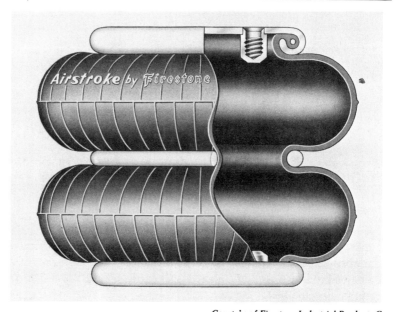

Courtesy of Firestone Industrial Products Co.

Fig. 15. Air device that provides a large amount of force without any moving parts other than end plates.

tuators, vertical lifts for conveyor sections, linear operators for rotary shafts, pivot mechanisms, clamping mechanisms, etc.

Factors Affecting Piston Speed

Keeping the master control valve close to the operating cylinder is always a desirable feature. Shown in Fig. 16 is an air cylinder with an electrically operated valve mounted into the blind cover. The plate between the electric valve and cover houses two flow-control valves. This design incorporates a complete air circuit in one package. It only requires one air line—the inlet line to the valve.

A question that often arises is, "How fast does the piston in an air cylinder move?" Due to such variables as friction, volume, and restrictions, it is almost impossible to calculate the exact answer. Under favorable operating conditions internal friction usually averages about 5 per cent. This is caused by rod packings and piston packing. However, if lubrication should fail or be omitted, the amount of friction will increase very rapidly. The amount of friction in the ways or guides to which the piston rod is connected has a marked effect upon the speed of the piston. The volume of air available is also another important factor. If there isn't plenty of air available, high speeds cannot be obtained. It may be advisable to place a surge tank adjacent to a fast cycling circuit to act as a pressure reservoir for the circuit, much like an accumulator in a hydraulic circuit. This has overcome many speed problems. Restriction is another factor that will retard the speed of a piston. There may be restrictions in the operating valves, pipe lines, or even in the cylinder covers. Often it is necessary to enlarge the pipe ports in the porting of the cylinder covers or make connections to two ports on each cover.

Installation and Maintenance Tips

There are a number of precautions that should be taken when installing or servicing air cylinders that will greatly increase their performance and operating efficiency. Some of these precautions may seem insignificant but if disregarded they can be a source of trouble.

1. Securely fasten cylinder mountings. A cylinder that is haphazardly mounted (one section of the mounting pulled down tight and another left loose or fastened to an insecure object) will throw an undue strain on ·the mounting plate, often snapping off the mountings.
2. Proper support to the piston rod. Do not leave the end of the piston rod (especially rods on long stroke cylinders) dangling out in thin air. A proper support should be placed on the end of the rod and this support should be exactly in line with the centerline of the cylinder. This will relieve strain on packing, gland, rod packing and cup packing and will increase their life.
3. Use clean piping to connect the cylinder to the master control

Courtesy of The Bellows-Valvair Div. IBEC

Fig. 16. Air cylinder with an electrically operated valve mounted into its blind cover.

valves. Any pipe scale or shavings from pipe threads will quickly damage the precision internal parts and packings of the cylinder. Many cylinders are ruined just because of this one lack of precaution.

4. Use proper lubricants in sufficient quantities. Lack of lubrication not only causes loss of power due to friction, but greatly reduces the life of the rod and piston packings. There are known instances where piston packing with proper lubrication operated satisfactorily for six million to eight million cycles; with no lubrication, it failed after a few thousand cycles.

5. If cylinders are stored for a long period of time before installation, they should be completely dismantled and the packings should be carefully checked. Cylinders should not be stored in extremely hot places or in damp places. When storing cylinders, the ends of the rods and mounting faces should be protected, and shipping plugs should be kept in cover ports in order to protect the interior.

6. When replacing covers on a cylinder having tie rods, make certain that all of the tie rods have about the same tension. If one side of the cover is pulled down tighter than the other a "cocking action" may result, causing a bind.

While air cylinders have innumerable applications, it should be mentioned that they also have their limitations. Air cylinders are not suitable on applications where extremely accurate feeds are required. They are also not practical where excessive forces are required. It would not be practical from a cost standpoint to use an air cylinder to operate a one-hundred-and-fifty-ton direct-acting press. A complete hydraulic system could be purchased for less than the price of the cylinder alone. Furthermore, the cost of the compressed air would not warrant the use of an air cylinder for such an application. Used within its limits, however, an air cylinder is a highly useful and versatile production tool.

Typical Air Cylinder Applications

The accompanying photos show some of the many applications for which air cylinders are used.

Figure 17 illustrates several air cylinders operating a multiple-station clamping device. Note air distributor in center which feeds air to air control valves and the clamping jaws around the outside of the device which hold the workpieces.

Figure 18 shows an air-operated hoist used for heavy transfer jobs. Quick action is a feature of these hoists.

Courtesy of Logansport Machine Co., Inc.

Fig. 17. A multiple station clamping device which holds the workpieces in the wedge-shaped clamping jaws around the outside. Air is introduced to the air distributor in the center which feeds air to the air-control valves of the six air cylinders making up this device.

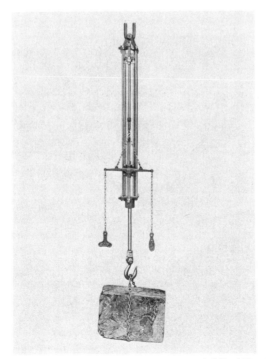

Courtesy of Galland-Henning Nopak, Inc.

Fig. 18. Air-operated hoist which is used for heavy transfer jobs.

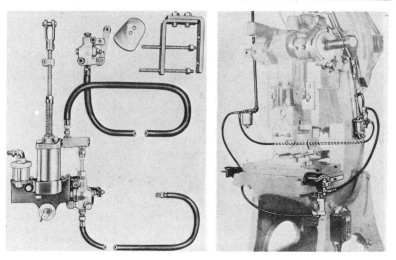

Courtesy of Scovill Fluid Power Division

Fig. 19. Air circuit as used on a power press. (*Left*) Component parts of the air system as they are shown removed from the press. (*Right*) In the installed position the two-hand control is clearly shown on either side of the press.

Courtesy of De-Sta-Co Division

Fig. 20. Toggle type clamps operated by air cylinders hold workpiece in fixture.

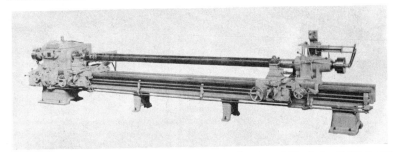

Courtesy of The Lodge & Shipley Co.

Fig. 21. Large lathe using two sets of air-operated chucking equipment. Note rotating air cylinders on each end of the spindle.

Figure 19 illustrates a packaged air circuit for use on a power press. Note the two-hand control and the compactness of the equipment.

Figure 20 shows air cylinders being used on a fixture for clamping the workpiece. A great many of the fixtures being designed for high production plants incorporate air cylinders to perform some operation. Because of their rapid operation they help to keep the loading and unloading time to a minimum.

Figure 21 shows a large lathe using two sets of air-operated chucking equipment. Note rotating air cylinder on end of each spindle.

Courtesy of The Warner & Swasey Co.

Fig. 22. Machine tool with rotating air cylinder.

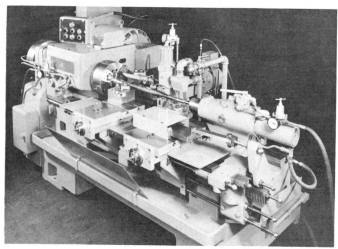

Courtesy of Sundstrand Machine Tool Co.

Fig. 23. Production machine tool which uses both the rotating and non-rotating types of air cylinders. The non-rotating type is being used as a support in the center.

Figure 22 illustrates a modern machine tool with rotating air cylinder.

Figure 23 shows air cylinders of both the rotating and nonrotating types being used on this production machine tool. In center of the illustration note nonrotating-type air cylinder being used as a support.

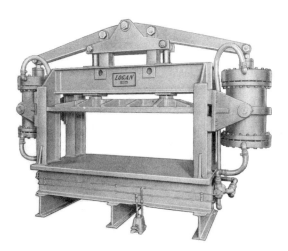

Fig. 24. Air cylinders as used to operate a large try-out press.

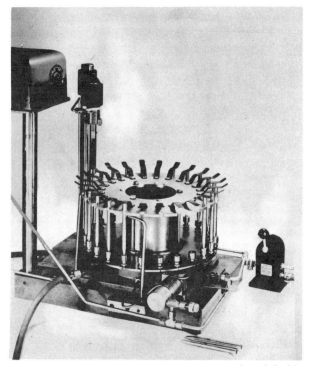

Courtesy of Mead Fluid Dynamics Div.

Fig. 25. Multi-purpose machine set-up which has been made possible by the use of air equipment.

Courtesy of Mead Fluid Dynamics Div.

Fig. 26. "One-man factory" multi-purpose machine set-up which utilizes air equipment.

Courtesy of LeBlond, Inc.

Fig. 27. Modern machine tool with air equipment that actuates two special power chucks.

Figure 24 illustrates air cylinders being used to operate a large try-out press. Note the linkage mechanism.

Figures 25 and 26 show "one-man factory" multi-purpose machine set-ups made possible by the use of air equipment.

Figure 27 illustrates the use of air equipment on a modern machine tool. One rotating air cylinder is beneath the guard on the left end of the machine; the other is attached to the right end. Figure 28 shows a machine with the guard removed. Note the two hose connections to the rotating air cylinder.

Air-to-Air Boosters

In an air-to-air booster design there is an air cylinder as the driving cylinder and a smaller bore air cylinder as the driven, or high-pressure, cylinder. They are usually connected as a unit as shown in Fig. 29. To determine the capacity of the booster in cubic feet per minute (cfm) these steps should be followed:

1. Multiply the area of the high-pressure cylinder by the stroke of the booster. This is the cubic-inch volume of air delivered per stroke.

2. Multiply the volume in (1) by the average number of strokes per minute that the booster cycles, which averages about 25. This will give the cubic-inch volume of air pumped per minute.

3. Divide the cubic-inch volume of air pumped per minute by the booster ratio which determines the volume of the high pressure air

Courtesy of LeBlond, Inc.

Fig. 28. Spindle end of machine with guard removed, showing rotating air cylinder and other mechanisms.

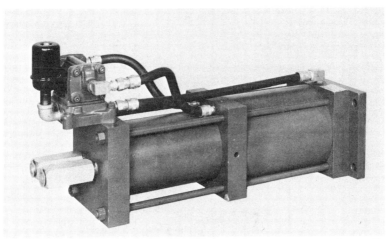

Courtesy of Lynair, Inc.

Fig. 29. Air-to-air booster with controls.

TABLE 3

AIR-TO-AIR BOOSTER

Driving Cylinder		High-Pressure Cylinder		Booster Ratio	Theoretical High-Pressure Output	
					Input	
Dia, in.	Area, sq in.	Dia, in.	Area, sq in.		80 psi	100 psi
6	28.274	3	7.069	4.00	320	400
6	28.274	3¼	8.296	3.41	272	341
6	28.274	4	12.566	2.25	180	225
6	28.274	5	19.635	1.44	115	144
8	50.265	3¼	8.296	6.05	484	605
8	50.265	4	12.566	4.00	320	400
8	50.265	5	19.635	2.56	204	256
8	50.265	6	28.274	1.77	142	177
10	78.54	4	12.566	6.25	500	625
10	78.54	5	19.635	4.00	320	400
10	78.54	6	28.274	2.78	222	278
10	78.54	7	38.485	2.04	163	204
10	78.54	8	50.265	1.56	125	156

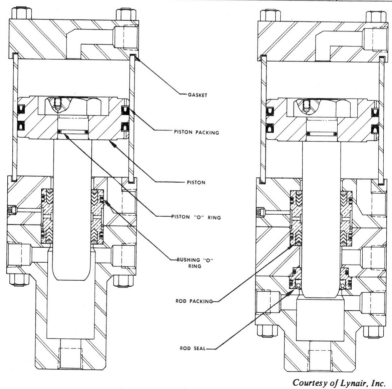

GASKET

PISTON PACKING

PISTON

PISTON "O" RING

BUSHING "O" RING

ROD PACKING

ROD SEAL

Courtesy of Lynair, Inc.

Fig. 30. Cross section of both single-pressure and dual-pressure air-to-oil boosters.

produced. To convert this answer from cubic inches to cubic feet divide by 1728 (1728 cu. in. = 1 cu. ft). This will determine the capacity of the booster in cfm.

See Table 3 for various theoretical high-pressure outputs when input pressures are 80 psi and 100 psi for driving cylinder bores of 6-inch, 8-inch, and 10-inch diameters.

Air-to-air boosters find applications in many test situations where shop air is needed at higher than normal pressure.

Air-to-Oil Boosters

In industry many applications are found for air-to-oil boosters. Some examples are shown in Chapter 17, where the boosters are employed to operate clamping devices. Some other applications are: testing of pressure vessels; expanding of metals and other materials; press operations where a hydraulic power unit is not desirable; operation of punch cylinders, etc. The air-to-oil booster can be used conveniently in explosive atmospheres as there is no need for electrical equipment.

Figure 30 shows both single-pressure and dual-pressure air-to-oil boosters. Note the heavy duty construction at the oil end of each booster.

CHAPTER 17

Power-operated Holding Devices

Nearly every workpiece must be held in some way or another before work can be performed upon it. The problem in this highly competitive manufacturing era is how to quickly grip the piece, hold it firmly during the work operation and then quickly release it after the work has been completed. This problem is being solved to a great extent with standard and specially designed fixtures operated by fluid power components. In fact, this is one of the big uses for fluid power equipment.

The holding devices may take the form of a chuck, a collet, a mandrel, a vise, a jig, or a special holding fixture. Selection of the type of fluid power is determined by the force required, space limitations, speed of actuation required, and other general conditions which may dictate as to which fluid will do the better job.

Power-operated Chucks

Power-operated chucks may be operated by two methods (1) the power-operated drawbar which is actuated by an air or hydraulic cylinder, and (2) the power chuck wrench which is an external device, actuated by an electric motor. This latter device operates the screw-locking mechanism for opening or closing the jaws of a hand-operated chuck. In this chapter we will deal only with the first type which is widely accepted for use on high-production machines.

Power chucks are used on lathes, grinders, boring machines, and other machines which are designed with a hollow spindle so that there can be a drawbar or drawtube connection between the actuating mechanism of the chuck and the rotating air or hydraulic cylinder. Figure 1 illustrates a typical spindle layout when using air-operated chucking equipment. Note how the connections are made between the chuck and the rotating cylinders. Figure 2 shows a similar arrangement using a hydraulically operated chuck.

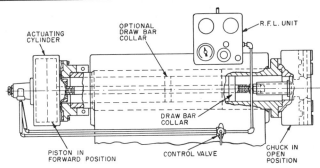

Fig. 1. Typical spindle layout employing air-operated chucking equipment.

Power chucks are manufactured in numerous styles and sizes as standard equipment. Their sizes run from 6 to 24 inches. The chucks may be 2-, 3-, or 4-jaw types with jaw styles being universal, combination, or serrated. The adaption on the back of the chuck may be of three designs; the straight recess, the American Standard Taper adaption, or the American Standard Cam Lock adaption.

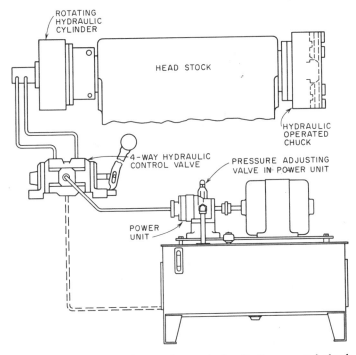

Fig. 2. Spindle layout that employs a hydraulically operated chuck.

The 3-jaw power chuck shown in Fig. 3 is frequently used when machining a forged shaft which must be held between centers. This is a compensating power chuck and is so designed that it will compensate for out-of-roundness up to ¼ inch.

Advantages of Power-operated Chucks

Three advantages of power chucks are (1) increase in production rate, (2) reduction in scrapped workpieces, and (3) lower maintenance.

A study was made to see how power chucking time compared with hand chucking time. Pieces of a round nature were chosen which were light enough to be lifted into the chucks by hand. The results are given in Table 1.

Courtesy of Sundstrand Machine Tool Co.
Fig. 3. Compensating chuck being used to machine a forged shaft which must be held between centers.

In comparing these figures it will be noted that a saving in time of as much as 80 per cent is possible and this does not take into consideration operator fatigue. In making time studies on like machines performing the same operation, it was found that operators using air- or hydraulically-operated chucking equipment showed far less operator fatigue than those using hand-operated chucking equipment. It is quite a task to wrestle a big chuck wrench all day long especially on large diameter chucks. In using fluid power equipment, a mere flick of the wrist or movement of the foot immediately gives full

TABLE I

COMPARISON OF LENGTHS OF TIME REQUIRED FOR POWER AND MANUAL
CHUCKING ON VARIOUS SIZE CHUCKS

Chuck Size, inches	Chucking Time, minutes		Time Saved, per cent
	Manual	Power	
6	0.25	0.05	80.0
8	0.35	0.10	71.5
10	0.45	0.10	77.8
12	0.50	0.10	80.0
15	0.80	0.20	75.0
18	0.80	0.20	75.0

chucking power. Where foot controls are installed, the operator has the added advantage of having both hands free for handling the workpiece.

Even on odd-shaped pieces which are to be manufactured in production quantities, there is a distinct advantage in using power-operated chucking equipment. One lathe manufacturer made a comparison between the two methods of chucking by selecting such workpieces as steering knuckles, crankshafts, drive pinions, and other odd-shaped pieces. The results are shown in Table 2. The time saved on the group averaged a little over 60 per cent which would be a considerable savings in dollars on long production runs.

TABLE 2

COMPARISON OF LENGTHS OF TIME REQUIRED FOR POWER AND MANUAL
CHUCKING OF VARIOUS WORKPIECES

Workpiece Designation	Chucking Time, minutes		Time Saved, per cent
	Manual	Power	
1	0.35	0.14	60.0
2	0.40	0.20	50.0
3	0.43	0.16	63.0
4	0.35	0.12	65.7
5	0.38	0.12	68.5

There are a number of reasons why scrap losses are reduced when using power chucking. Some of them are:

1. *Uniform Gripping Pressure.* The amount of pressure applied to the workpiece can be accurately controlled when using fluid power

to operate the chuck. This is not the case when using a hand chuck. A strong operator may apply too much pressure, while a woman operator may not be able to apply enough pressure. It is very easy to scrap a piece having a thin wall section if too much torque is applied to the chuck wrench.

With power chucking equipment, the correct gripping pressure is found when setting up the first workpiece by adjusting the relief valve which controls the fluid power pressure to the operating cylinder. The pressure then will remain uniform for the production run.

One place where power chucking has greatly reduced scrap losses is on workpieces having a very heavy wall section which, as the machining progresses, is mostly removed so that the workpiece is finished having a thin wall section. By merely shifting a valve handle, the gripping pressure can be reduced to meet the requirement without causing any irregularities in the workpiece.

2. *Reserve Movement of Jaws.* Another advantage of power chucking is that reserve movement of the jaws is present when needed. Often when chucking on a forging or a rough casting, one jaw may grip on an imperfection in the workpiece. Should the pressure break down this imperfection when the tool engages, workpiece slippage could result if it were not for the reserve movement of the jaws which keeps them constantly against the workpiece. When using a hand chuck there is no chance to tighten up on the jaws after the spindle has been put in motion. If slippage should occur, the workpiece may come out of the chuck, break the cutting tool, or ruin the workpiece.

3. *Combination of Holding Forces.* A combination of holding forces on a workpiece can be obtained by using duplex cylinders. This is often necessary when using a four-jaw chuck (the chuck consists of two sets of two jaws, each operating independently of the other) where two jaws must grip on a heavy section and two jaws must grip on a thin section. If the same force were applied by both sets of jaws, the thin section might be distorted to such an extent that the piece would have to be scrapped. The duplex cylinders (one section operating at high pressure and the other at low pressure) transfer the high pressure to the jaws gripping on the heavy section and low pressure to the jaws gripping on the thin section. This combination has solved some very complicated chucking problems.

4. *High Clamping Force.* Today's production requirements are such that the maximum amount of work must be done on a workpiece in the least amount of time. This means that on workpieces from which chips are to be removed, the rate of removal must be as high as possible. To permit a high rate of metal removal the work-

piece must be held securely and power chucks are widely used to accomplish this end. Table 3 gives the clamping forces which are obtainable when using 80 lb per sq in. operating pressure on the air cylinder.

Table 3 shows the jaw force obtained at each jaw of three-jaw and two-jaw chucks when operated at maximum recommended drawbar pull. The chucks are assumed to be in a static condition, i.e., not rotating. Of course, if higher operating pressures are used or if larger cylinders are employed, the increase in jaw force then would be correspondingly higher. This practice, however, is not recommended as it can greatly reduce the service life of the power chuck, affecting the levers; lever blocks; lever pins; and other parts of the chuck.

Since the power chuck, in nearly all power chuck applications, is connected to a revolving spindle, and since the operating rpm (revolutions per minute) of newer machine spindles are becoming ever greater, centrifugal force which acts to reduce the holding power of the chuck jaws must be taken into account. Table 4 shows the rpm at which the initial jaw force will be reduced to 75, 50, and 25 per cent, respectively. These values were calculated by using the combined weights of the master jaw and the blank false jaw with the master jaw position $\frac{1}{16}$ inch from full open. These data are for use only as a guide for average chuck operation and are not to be construed as safe operating speeds. The chuck manufacturer should be consulted for final recommendations. Factors such as type of holding surface on false jaws, irregular and unbalanced workpieces, severity of material removal, type of tooling, position of tooling in relation to the false jaws and workpiece, etc., must all be taken into consideration when determining top operating speed. The maximum height of false jaws from the face of chuck should not exceed the length of the master jaws of the chuck.

Life and Accuracy

Fluid power chucking equipment is designed for long life, even with rough usage. All parts are machined to close tolerances in order to minimize wear. It is not uncommon for a set of power chucking equipment to operate satisfactorily for 15 to 18 years with only a minimum of maintenance.

The question often arises as to how accurately a workpiece can be held in a power chuck. If false jaws, which are attached to the master jaws and are machined to grip a given workpiece, are ground in place after the chuck has been placed on the spindle, it is not uncommon to hold the concentricity to less than 0.001 of an inch.

TABLE 3

AIR AND HYDRAULIC CHUCK SIZES AND OPERATING PRESSURES FOR RECOMMENDED DRAWBAR PULLS

Chuck Size, in.	Max. Recommended Drawbar Pull, lbs.	Jaw Force at Max. Drawbar Pull, lbs. per Jaw (Static)	Cylinder Size and Operating Pressure to Obtain Max. Recommended Drawbar Pull			
			Air		Hydraulic	
			Size, in.	Press., psi	Size, in.	Press., psi
THREE-JAW CHUCKS						
6	3300	1875	8	60	4.5	185
8	6900	4180	10	90	4.5	460
10	9300	6180	12	85	6	350
12	12000	9695	14	80	6	450
15	16200	15490	16	80	8	340
18	17100	19380	16	85	8	360
21	18000	24080	16	90	10	240
24	18000	24080	18	75	10	240
TWO-JAW CHUCKS						
6	2500	1595	6	95	4.5	165
8	5625	4150	10	75	4.5	375
10	8100	6025	12	75	6	305
12	9690	7140	14	65	6	365
15	14510	12335	16	75	8	305
18	14510	12335	16	75	8	305

TABLE 4

AIR AND HYDRAULIC CHUCK RPMS FOR VARIOUS PERCENTAGES OF INITIAL JAW FORCE

Chuck Size, in.	Wt. of Top Jaws, lbs. (Std. Soft Steel Blank Jaws)	Ht. of Top Jaw from Chuck Face. in.	Jaw Force at Max. Drawbar Pull, lbs. per Jaw (Static)	Rpm for 75% of Initial Jaw Force	Rpm for 50% of Initial Jaw Force	Rpm for 25% of Initial Jaw Force
			THREE-JAW CHUCKS			
6	1.25	1.25	1875	1820	2575	3155
8	2.0	1.375	4180	1855	2575	3155
10	3.25	1.625	6180	1510	2225	2725
12	5.0	1.875	9695	1455	2060	2520
15	8.75	2.25	15490	1275	1805	2210
18	9.0	2.25	19380	1200	1700	2085
21	13.25	2.75	24080	980	1390	1700
24	13.25	2.75	24080	870	1230	1510
			TWO-JAW CHUCKS			
6	2.75	1.75	1595	1255	1775	2175
8	4.0	1.875	4150	1385	1965	2400
10	6.0	2.125	6025	1220	1720	2110
12	8.75	2.375	7140	1000	1410	1730
15	12.75	2.5	12335	965	1355	1670
18	12.75	2.5	12335	825	1165	1425

The design of the false jaws plays a very important part in power chucking. A common mistake is to allow too much gripping surface. This is especially true when using a compensating chuck. Gripping contact should be held to a minimum unless the workpiece has a very thin wall section, then it may be necessary to use a set of wrap-around type jaws to eliminate distortion.

A fine example of the use of a very small contact surface is shown in Fig. 4. The small hardened points cut through the rough skin of the casting, securely holding it for a heavy roughing operation. It has been found that this type of false jaw design is especially effective where it is necessary to break through the skin of the material to effect a good solid gripping surface.

Power-chucking Applications

The following are some typical examples of power chucking application:

Figure 5 shows a three jaw power operated chuck of the serrated jaw type for holding a large workpiece. This chuck is mounted on the spindle of a turret type machine and the turret is indexed automatically. Note the number of tools that are used.

The operator places the workpiece in a loading cradle and closes the door. An air-operated tailstock center actuated by the air cylinder shown at the right pushes the part into position on the headstock center, then the compensating power chuck jaws operated by the rotating air cylinder shown at the left close on the workpiece and the lathe is put into operation. Multiple tools quickly complete the machining operation.

Figure 6 shows a power-operated chuck of the three-jaw combina-

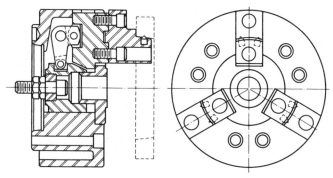

Fig. 4. Power chuck that illustrates the use of a very small contact surface.

Courtesy of The Warner & Swasey Co.

Fig. 5. Lathe with air-operated chuck of the serrated jaw type for holding large workpieces.

Courtesy of Sundstrand Machine Tool Co.

Fig. 6. Power-operated chuck of the three-jaw combination type. It is used in this instance for holding a heavy shaft during the turning operation.

Courtesy of The Monarch Machine Tool Co.
Fig. 7. Three-jaw power chuck gripping a ring section.

tion type used for holding heavy shafts during the turning operation. Note the number of cutting tools used for the heavy chip removal.

In Fig. 7 a three-jaw power chuck is shown gripping a ring section. The setup shown in Fig. 8 is for machining four-cylinder tractor engine crankshafts. This high-production lathe faces the shoulders

Courtesy of Sundstrand Machine Tool Co.
Fig. 8. Machining setup on a production lathe using an air-operated three-jaw chuck, an air-operated tailstock, and other air-operated devices.

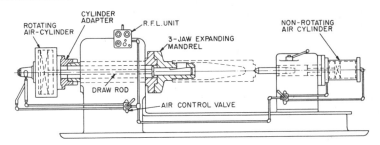

Fig. 9. Typical layout of a shell mandrel installation.

and undercuts the three main bearings on the crankshaft. The crankshaft is held between the center in the chuck and the center in the air-operated tailstock and is driven from the flanged end using a three-jaw air-operated chuck. An air-operated three-roller steady rest supports the crankshaft at the center main bearing. This machine has a straight feed-in type front carriage and a cross-feeding rear carriage. The front carriage has a single tool block which holds three facing and undercutting tools. The rear slide has a single tool block which holds two similar type tools. Both tool blocks have an air-operated tool-relief attachment in order to eliminate tool marks on the return stroke.

Power-operated Collets and Mandrels

Power-operated collets and mandrels have many uses throughout industry. Collets are used for gripping the outside diameter of the workpiece while mandrels are employed for gripping the inside diameter. Collets are used for holding shafts, tubes, shells and other items on which work is to be performed on the internal diameter or

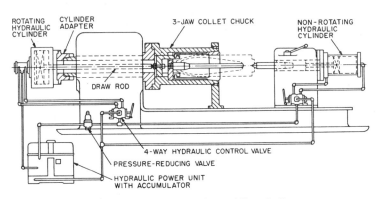

Fig. 10. Collet set-up for holding shells.

Fig. 11. Mandrel set-up for holding a shell for the first machining operation which involves the machining of a center in the end of a shell and a cutoff at the open end.

the end. Mandrels are used for gripping on the internal diameter of tubes, coils of steel, shells, gear blanks, etc., where work is to be done on the outside diameter. Mandrels are used extensively in steel processing, on slitters, coilers, and uncoilers.

Extensive use of power-operated collets and mandrels is made in the manufacture of artillery shells. Figure 9 shows a typical layout of a shell mandrel installation while Fig. 10 illustrates an installation of a collet for holding shells.

Power-operated Mandrel Applications

Mandrels are employed for heavy stock removal. The centering type of mandrel, shown in Figs. 11 and 12, holds the shell for the

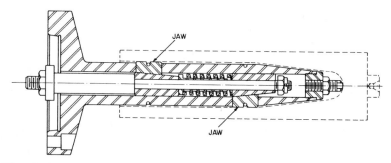

Fig. 12. Cross-sectional view of a centering type of mandrel for holding a shell.

first machining operation which is to machine a center in the end of the shell and to cut off the open end. Note that this mandrel has six jaws, three equally spaced in two locations. Centering mandrels have adjustable work locators with hardened stops for properly locating the shell with regard to the tooling. While some centering mandrels are built with work locators as part of the mandrel, others have the work locator as a separate tool which is dropped into the shell before it is placed on the mandrel. Once the locator is properly adjusted, the whole production run is usually completed without resetting.

Rough-turn mandrels, such as that shown in Fig. 13, are designed with three serrated jaws in order to produce the maximum gripping pressure for the heavy stock removal operation. The jaws must securely grip into the forging so that there will be no possibility of slippage when the tools contact the workpiece. Any slippage would quickly burn the teeth off of the serrated jaws.

Rough-turn mandrels are usually of the pull-type construction, pulling against a very shallow angle to give high gripping force. Stock removal up to ⅛ inch is made possible by the use of cemented carbide cutting tools. A 76 mm shell can be machined on the rough turn operation with multiple tools in about 45 seconds.

For support on the end of the shell a power-operated tail stock such as that shown in Fig. 10 is used. This affords ample support and can be operated very quickly.

On thin-walled shells, the rough turning operation presents more of a problem. The shell must be held tightly enough so that it will not slip, yet it must not become distorted. Jaws of the wrap-around type with small serrations have proven most successful.

Finish-turn mandrels, such as that shown in Fig. 14, have three jaws which grip in the fuze recess of the shell. This type of mandrel holds the shell while the finish taper turn, the finish turn, and the finish form turn operations are performed. On shells with small fuze recesses, it is sometimes difficult to provide enough strength at the gripping point to enable the operation to be performed. A push-type mandrel, such as is shown in Fig. 15, may be employed in these instances.

Power-operated Collet Applications

Power-operated collet chucks are usually used for the final machining operations on the shell. They are used for the nose operation and the bore operation. On large shells, collet chucks become quite bulky since the gripping must take place near the center of the shell. It is usually necessary to build a steady rest, such as that shown in Fig. 16,

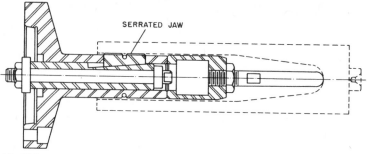

SERRATED JAW

Fig. 13. This rough-turn mandrel has been designed with three serrated jaws in order to produce the maximum gripping pressure for heavy stock removal operation.

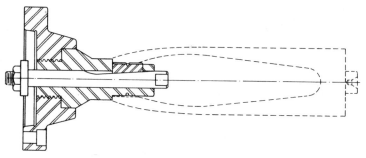

Fig. 14. Finish-turn mandrel which has three jaws for gripping in the fuze recess of the shell.

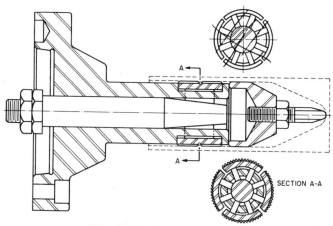

A

A

SECTION A-A

Fig. 15. Push-type mandrel.

Courtesy of The Monarch Machine Tool Co.

Fig. 16. Steady rest on the end of this collet body relieves the weight on the spindle bearings of the machine and also the side thrust created by the cutting tools.

on the end of the collet body to relieve the weight on the spindle bearings of the machine, and also to relieve the side thrust created by the cutting tools. Most of the steady rests are designed with needle or roller bearings to reduce friction and heat.

For machining operations on the nose end of the shell, the shell is located in a power-operated collet against a hardened female work locator, as shown in Fig. 17, which is attached to the collet body through a spider arrangement. This locator can be easily changed so that the same chuck can be used for the bore operation by inserting a male-type locator, as shown in Fig. 18.

Collet chucks hold the shell for the band groove operation, the face base and turn operation, and the nose threading operation. On some shell production lines the threading operation is done on tapping machines especially designed for this purpose. Figure 19 shows a large power-operated collet being used to hold a 240 mm shell for the boring and nose-tapping operation.

Special Clamping Devices

Up to this point only clamping devices which were used in conjunction with machines having rotating spindles have been described. However, it should not be inferred that these types of applications make up the majority of holding applications used in modern manufacturing methods. Air and hydraulic clamping is used in a wide variety of ways.

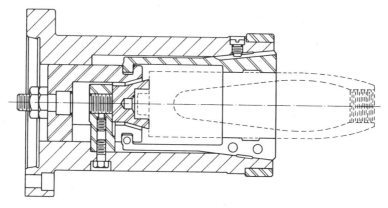

Fig. 17. Power-operated collet used for machining operations on the nose end of a shell. The shell is located against a hardened female work locator which is attached to the collet body through a spider arrangement.

The special holding fixture shown in Fig. 20 is of value on multiple-station machines. An air distributor (A) feeds the air to the master control valve (C) at each station. The master control valve, in turn, directs air to the cylinder which operates clamping device (B). Work is loaded and unloaded at one station and operations on the work-piece are performed at each of the other stations. Operations may consist of drilling, counterboring, reaming, tapping, etc.

Figure 21 shows a small air-operated holding device used to hold small parts during a milling operation. Speed of loading and unloading as well as positive gripping prove advantageous.

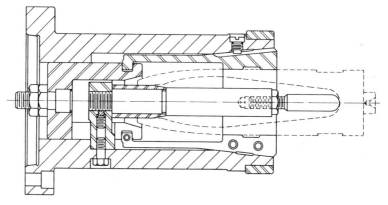

Fig. 18. Power-operated collet with male-type locator as set up for the ring-grooving and facing operation.

Courtesy of The Sidney Machine Tool Co.

Fig. 19. Large power-operated collet being used to hold a 240 mm shell for the boring and nose-tapping operation.

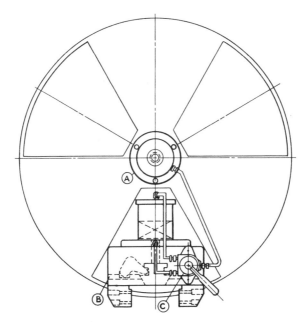

Fig. 20. Work-holding fixture as it might be set up on a multiple-station machine. Air distributor *A*, feeds air to the master control valve *C*, which in turn directs air to the cylinder which operates clamping device *B*.

Courtesy of Mead Fluid Dynamics Div.

Fig. 21. Small air-operated holding device used to hold small parts during a milling operation.

Hydraulic Clamping Devices

While we have been dwelling mainly in this chapter with air as the actuating medium, hydraulic power is used in many applications with perfectly satisfactory results. The following examples show clamping applications at the Ford Motor Co., Ltd., in Dagenham, England, where hydraulic power is used extensively for jig and fixture movements.

The clamps employed in these hydraulically operated jigs and fixtures are operated by a cam and "kicker," or a wedge, or a direct thrust from the piston rod (with locking by a wedge-actuated device). Clamping arrangements of the direct thrust type are usually applied where the movement of the clamping member is extended.

Cam- and Kicker-operated Work-holding Clamp

A cam and kicker are illustrated in Fig. 22. The piston rod (A) of a hydraulic cylinder is connected by a link (B) to the lever extension of the cam (C). This cam is mounted on the square section of the shaft (D), which carries the kicker (E) secured by a setscrew. The clamp (F) is supported by a spring on the stud (G), and is retained by plain and spherically faced lock-nuts and a spherical seating washer. A spring plunger in the bridgepiece (H), contacts the right-hand end of the clamp.

As the piston rod is retracted by hydraulic pressure, the kicker (E) engages a slot in the under side of the clamp and thrusts it forward, over the workpiece. Further movement of the piston rod causes the cam (C) to lift the right-hand end of the clamp, forcing the opposite end downward against the component. A reversal of the piston movement rotates the cam in the opposite direction, releasing the pressure on the clamp, which is then withdrawn by the kicker.

Wedge-operated Clamp

A simple wedge-operated clamp is shown in Fig. 23, the piston rod (A) being coupled to a cylindrical sliding member (B), on which is machined an inclined face. The plunger (C) is thrust downward when the piston moves to the right, and forces a spring-loaded bridge clamp (D) against the workpiece (indicated by broken lines). A self-locking action is provided by the selection of a suitable angle for the inclined face of sliding member (B).

Direct-acting Clamp with Wedge-type Lock

In Fig. 24 is illustrated a method of clamping by the direct action of a hydraulic cylinder, with a wedge for locking the ram in its final position. The piston rod (A) of the clamping cylinder passes through a slot in the wedge (B). At the upper end of the rod is mounted the ram (C). When the upward movement of the ram has clamped the workpiece against a fixed horizontal locating face (not shown), pressure is admitted to the left-hand end of the cylinder (D) so that the wedge (B) is moved to the right and the ram locked in the clamping position, independent of the hydraulic pressure supply. The

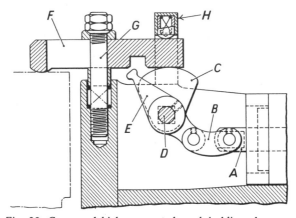

Fig. 22. Cam- and kicker-operated work-holding clamp.

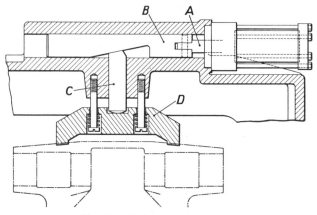

Fig. 23. Wedge-operated clamp.

application of pressure to the opposite end of the cylinder (D) releases the wedge, allowing the ram (C) to be lowered. An extended movement of this ram is permitted by the flats machined on its lower end, which enable it to pass through the larger end of the slot in wedge (B).

The timing of the oil flow to the clamping and locking cylinders is usually controlled by a sequence valve. However, it has been found

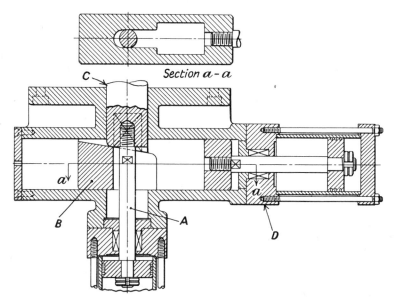

Fig. 24. Direct-acting clamp with wedge-type lock.

that, with the arrangement shown in Fig. 24, a satisfactory action is obtained if both cylinders are connected to the pressure supply simultaneously so that the wedge is held against the side of the ram until the ram assumes a position which will permit a transverse wedge movement. During the latter part of the clamping movement, the ram and the wedge move together until the workpiece is both clamped and locked.

Wedge-operated Clamp with "Butterfly" Piston

Where space for the clamping mechanism is restricted, the pistons can be made integral with the wedge, as shown in Fig. 25. This clamping device is fitted with a "butterfly" piston to rotate the clamp, when it is in the free position, so that it clears the workpiece. Also,

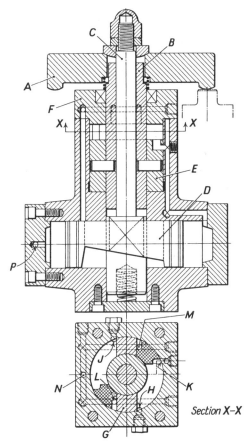

Fig. 25. Wedge-operated clamp with "butterfly" piston.

this arrangement provides for timing the sequence of the clamping operations. The spring-loaded clamp (A) rotates with the sleeve (B), on which it has a limited axial movement. When pressure is applied to the right-hand end of the piston, the clamp is pulled downward by the rod (C), due to the action of an inclined face machined on the double piston (D). Application of pressure to the left-hand end of the piston will release the clamp.

Attached to the sleeve (B) is a circular valve block (E) with its top face milled to form the butterfly profile, as indicated by cross hatching in the lower sectional view. The top plate (F) is provided with the two projections (G), and holes are drilled to form the ports (H), (J), (L), and (M) with their connecting passages. Holes (K) and (N) join with passages in the main casting that connect with the right- and left-hand ends of the cylinder, respectively, the latter end being provided with an additional inlet (P).

Initially, the clamp and piston (E) are at 90 degrees to the position shown. The ports (L) and (M) are connected through the hole (N) to the left-hand end of the cylinder and to the exhaust, and the piston (D) is at the right-hand limit of its movement. Oil under pressure is admitted by an associated four-way valve to the ports (H) and (J) on one side of the butterfly piston, the oil on the opposite side escaping through the ports (L) and (M) to the left-hand side of the clamping cylinder, which is connected to the exhaust. The piston is consequently rotated until it contacts the projections (G), as shown in the lower sectional view. Oil under pressure then passes through the port (K) to the right-hand side of the clamping cylinder, forcing the piston (D) to the left, and effecting the clamping movement.

When the setting of the four-way valve is reversed, the port (P) is connected to the pressure supply and the inlets (H) and (J) are connected to exhaust. The piston (D) then moves to the right so that the clamp is released. The oil subsequently passes through the passages in the main casting and the top plate (F) to ports (L) and (M). Rotation of the butterfly piston is thus effected in the opposite direction to the previous movement.

Rack-and-Pinion Rotated Clamp

Clamping pressure is applied in the horizontal and vertical planes by the mechanism seen in Fig. 26. The arm (A) is swung rapidly downward by means of a rack-and-pinion device, actuated by the cylinder (B). Clamping pressure is then applied by the cylinder (C) through the angular face on the piston rod (D), which engages the cam profile on the upper end of the arm (A). Both cylinders are

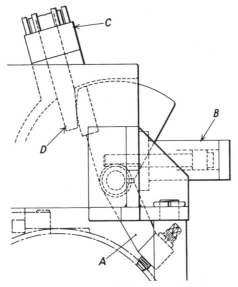

Fig. 26. Rack-and-pinion rotated clamp.

energized simultaneously, but the piston (D) cannot move downward until the clamp approaches its final position, due to the curved form of the cam.

Double-action Clamp

Thrust is applied to a component at 90 degrees to the direction of the clamping pressure by a plunger connected to the cam and kicker clamp illustrated in Fig. 27. The hydraulic cylinder (A) operates the cam (B) and the kicker (C) in a similar manner to that shown in Fig. 22. In this instance, however, the cam (B) has two lobes— one applying pressure to the clamp (D), and the other imparting motion to the spring-loaded locating plunger (E) through the ram (F). A degree of compensation is provided for the movements of the clamp and the plunger by a rectangular hole in the cam (B). This hole allows a clearance for the square shaft (G) in one direction only.

Centralizing Device

A hydraulically operated centralizing device can be seen in Fig. 28. Two spring-loaded plungers (A) expand radially inside the cored hole of a cast workpiece when oil is supplied to the lower end of the cylinder (B). The angular faces on the piston rod (C) bear directly on the inner ends of the plungers, which are fitted with adjustable

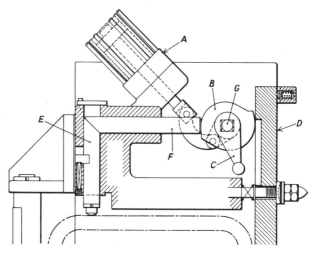

Fig. 27. Double-action clamp.

dome-headed screws. This device insures that the external machined faces are correctly positioned relative to the cored hole, and that a uniform wall thickness is maintained during a milling operation on a duplex machine.

Retractable Locating Pin Fixture

Wherever retractable locating pins are necessary in the fixtures, the standardized design seen in Fig. 29 is adopted. The design incor-

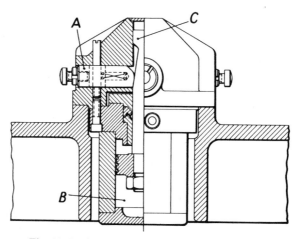

Fig. 28. Hydraulically operated centralizing device.

porates a felt sealing ring and a shroud (A) to prevent chips from entering the elevating mechanism, which comprises a rack, cut on the ram (B), meshing with a pinion (C). These devices can be inter-locked with the hydraulic circuit in various ways. In the method shown, pinion (C) engages a horizontal rack (D), connected to an eccentric (E). Keyed to the eccentric shaft is a cam which is in contact with the plunger of a two-way hydraulic valve. The cam and valve are arranged to direct the oil supply to exhaust when the locating pin is in the "down" position, and to transmit oil to the other valves in the hydraulic circuit when the workpiece is correctly located and the pin raised.

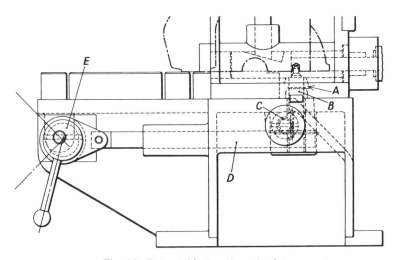

Fig. 29. Retractable locating pin fixture.

Multiple Locating Pin Fixture

Another method of interlocking is illustrated in Fig. 30. The cylindrical drum (A) can be rotated through 210 degrees by means of an attached lever, and has a helical track cut in its periphery. This track engages a fixed pin (B) so that the drum (A) moves length-wise during the rotary movement. Concentric with the drum, and moving longitudinally with it, is a shaft which is not free to rotate. Rack teeth cut on this shaft serve to rotate the pinion (C) and thus elevate multiple locating pins, one of which can be seen at (D). A Vickers rotary, four-way, manually operated valve (E) is used to control the hydraulic circuit, and is fitted with a special operating lever having an enlarged hub provided with a semicircular notch. A spring-loaded plunger (F) enters this notch when the valve is rotated through 45 degrees from the position shown, in order to cut off the hydraulic fluid supply.

The valve cannot be moved into the "live" position unless the locating pins have been raised and have entered the datum-holes in the workpiece. Under this condition, the plunger (F) is free to move to the right into a clearance hole in the drum (A) as it is depressed by the valve-lever hub during the rotary movement.

Automatic Transfer Machine for Multiple Drilling

Multiple drilling operations on transmission housings are performed by an automatic transfer machine designed by Ford engineers. The

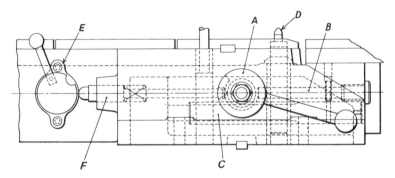

Fig. 30. Multiple locating pin fixture.

cast housings enter at one end of the machine with a flanged face resting on ways (A), Fig. 31, and is moved intermittently from one operating station to the next by the fingers (B), attached to the hydraulically operated transfer shuttles. These fingers are turned inward, move the component through one transfer stage, withdraw, and are then returned to the starting position. At certain stations, one of which is shown, the gear-box casting is rotated through either 90 or 180 degrees so that different faces are presented to the cutter-spindles.

Rams, as at (G), fitted with locating screws (H), are provided at these stations. The gear teeth cut in the lower end of each ram engage a rack (J). This rack is connected to a second rack (L) by an idler gear (K), at stations where rotation through 90 degrees is required; and by a gear train, having a 2 to 1 ratio, where a 180-degree movement is needed. The rack (L) is cut on a horizontal shaft, one end of which is coupled to the piston of a hydraulic cylinder. Cam-plates on the shaft trip limit switches for controlling the hydraulic circuit through a solenoid-operated valve, and adjustable nuts are fitted to the shaft so that its movement in both directions is limited.

In operation, a casting is positioned by the transfer mechanism over the ram (G), which is raised by the action of the lever (E)

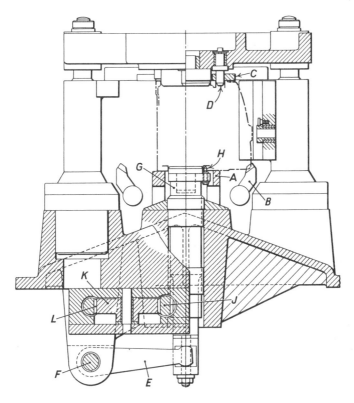

Fig. 31. Automatic transfer machine for multiple drilling.

fitted to the shaft (F), so that the screw (H) enters a hole in the casting. The workpiece is then rotated through 90 or 180 degrees. At this point, the top plate, carrying the locating pin (D), is lowered by a hydraulically operated mechanism to position the casting and clamp it securely against the fixed ways (A). A stationary stripper plate (C) is provided to ensure that the component is freed from the upper locating members as they withdraw after the drilling operation has been completed.

CHAPTER 18

Pneumatic Safety Circuits

Although the rapid expansion of compressed air makes it suitable for many applications in which speed is a factor, air must be satisfactorily harnessed to give proper results. Operators working with air devices must have ample protection, particularly when they are handling work that is brought into contact with the ram of an air press, the jaws of an air vise or other air-operated work-holding or forming device. Air circuits should have protective devices to take care of any pressure drop in the lines feeding the individual circuit, especially if the pressure drop could be dangerous to the operator or damaging to the machine in which the air equipment is being used. Pressure drop is usually caused by a greater demand for air than the compressor can produce. There have been cases where the pressure has dropped so low at the beginning of a work shift, due to all of the air devices being started at the same time, that the workpieces actually fell out of the clamping devices.

Protective devices and circuits to take care of various hazardous situations are shown in this chapter. It should be pointed out that safety devices must be so designed that when operators try to bypass them, they cannot operate the circuit. As mentioned in the chapter on hydraulic safety circuits, it is sometimes a race between the circuit designer and the operator to see who can outsmart the other. Nevertheless, both employer and employee will benefit if safety is maintained.

Lubricators and filters are important factors in the design of a safe air circuit. Their omission will cause failure in the best of circuits. Dirt will clog the orifices and passages of valves causing considerable damage. The absence of lubrication will cause otherwise safe valves to jam and wear excessively, thus tending to make the whole system unsafe. Cylinders will also be subjected to excessive wear making them unfit for safe operation.

The many basic circuits in this chapter show how different types of safety problems involving the operator, the machine, and the workpiece can be handled.

Two-Hand Control Circuit

Circuit No. 1 (Fig. 1) shows a basic air circuit for protecting the operator and for the automatic ejection of the workpiece after the operation has been completed. This is a two-hand control circuit and could be considered safe to the point that an operator can complete only the downstroke of the cylinder should he tie down the actuator of valve *C*. If the operator ties down the actuator of valve *D*, he can actuate the circuit but air would be continuously exhausting through valve *C* when it is not actuated.

The proper method of operating the circuit, however, is as follows: The operator depresses actuators of valves *C* and *D*. Air flows

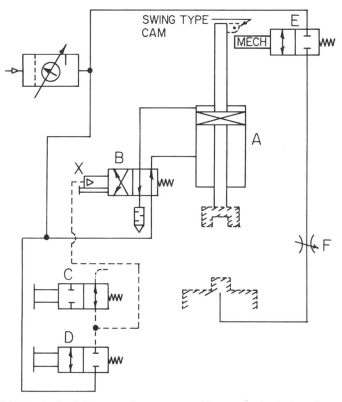

Fig. 1. Basic air circuit for protecting operator and for automatic ejection of completed workpiece.

through *D* but is then blocked at *C*. Air flows to port *X*, actuating
valve *B*, and air pressure flows to top of cylinder *A*. The piston of *A*
advances and the swing cam on the tailrod of the cylinder passes over
the cam actuator of valve *E*. When the operation has been
performed, the operator releases actuators of *C* and *D*. Air pressure
is released to pilot of *B* and spring pressure shifts spool of *B* to
normal position. Air pressure is directed to bottom of cylinder *A*
and piston of *A* retracts. The swing cam contacts cam actuator of
valve *E*, a shot of air ejects the workpiece and the cycle is completed. Needle valve *F* is used to limit the amount of air to be
directed to work ejector so that piece-part will move on to the outbound conveyor. There are a number of additions and refinements
that can be added to this basic circuit.

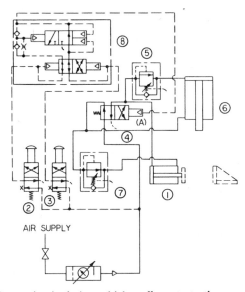

Fig. 2. Press circuit design which well protects the press operator.

Safeguarding the Operator's Hands

Circuit No. 2 (Fig. 2) shows a press circuit designed to protect the
operator even though he tries to outsmart the circuit designer. The
circuit operates as follows: operator places workpiece in air vise (1),
places hands on buttons (2) and (3) of four-way pilot valve, and depresses both buttons. These valves are of the spring-return type, requiring the operator to keep his hands on each button during the pressing stroke. Air pressure then flows to safety valve (8) and on to pilot

connection (A) of four-way, two-position valve (4). Spool in valve (4) shifts and air pressure is directed through valve (4) to blind end of air-vise cylinder (1), and piston advances, locking workpiece in vise. Air pressure builds up, and sequence valve (5) opens. Air pressure is directed to blind end of staking cylinder (6), piston of cylinder advances, and staking operation is performed.

Operator then releases both hands from buttons of valve (2) and (3). Air pressure is released from pilot chamber (A) of valve (4). Spool of valve (4) shifts, and air pressure is directed to rod end of cylinder (6). Piston of staking cylinder (6) starts to retract. When it has reached the retracted position, air pressure builds up, opening sequence valve (7), and air pressure is directed to rod end of air-vise cylinder (1). Its piston retracts and workpiece is released, thus completing the cycle.

Of course, the big feature of this circuit is its design to keep the operator from tying down either of controls (2) or (3). Also, he cannot remove either hand from the controls on the forward stroke of piston of cylinder (6), while still performing the operation. These features are controlled by safety valve (8).

This is considered one of the better pneumatic safety circuits for two-hand operation. Many variations can be made to such a circuit.

Operator and Machine Protection when Pressure Drops

Circuit No. 3 (Fig. 3) is designed to protect the machine as well as the operator when the operating pressure falls below a safe working range. This circuit is inexpensive, at least from the standpoint of making it safe, and is also simple in design. It has wide usage where pressure lines are subject to extreme pressure drop. The heart of the circuit is the air-cylinder operated four-way control valve.

The circuit functions as follows: When the line pressure is within the required operating range, air flows from the regulating, filtering, and lubricating unit to the actuating cylinder connection and inlet connection on valve (1). Normal operating pressure causes cylinder piston in valve (1) to be depressed against a spring. As it is depressed, the piston which controls the air passage inside the valve is also shifted causing the inlet port to be connected to port Y. Operator can now momentarily depress operating lever on valve (2) and allow air to flow to pilot connection A of master four-way valve (4). Piston of master valve shifts to position shown so that air flows to blind end of cylinder (5) and piston rod moves out. Operator momentarily depresses operating lever of valve (3) and air flows to pilot connection B of valve (4), shifting piston of valve back to original position so that air flows to rod end of cylinder and piston rod retracts. If the line pressure should drop below a safe operating range when the cylinder is making its outstroke, the piston of cylinder-operated valve (1) will return to its normal position and the inlet of this valve will be connected to port X and air will flow in through port Z of valve (3) and on to pilot connection B of valve (4) shifting piston of this valve, whereupon air will flow to rod end of cylinder (5) and piston will return to starting position. Circuit shown happens to be pilot-pressure operated but a pilot-bleeder circuit can be employed using the same principle.

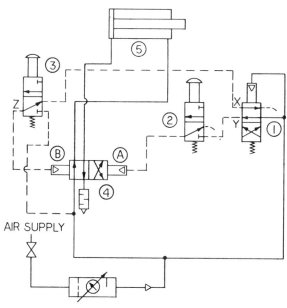

Fig. 3. Circuit which is designed to protect the machine as well as the operator when the operating pressure falls below a safe working range.

Electrically Operated Safety Controls

Circuit No. 4 (Fig. 4) makes use of electrical controls to operate the air equipment. Some safety engineers criticize the use of electrical controls in safety circuits and a great deal of the criticism is undoubtedly based upon past experiences with solenoids, limit switches, and other electrical devices. There has been a vast improvement in this type of equipment, however, and electrical controls have the advantage of being compact and of eliminating complicated piping. Usually it is less expensive to run electrical wiring than it is piping, especially if the pipe diameters must be large. The circuit shown finds application on presses used in mass production where speed is a big factor.

In this circuit the operator depresses both (PB-A) and (PB-B) control buttons closing the normally-open contacts and energizing solenoid C of three-way pilot valve (2). Air then flows to pilot connection X of four-way, three-position valve (1), shifting valve piston and directing air to blind end of press ram cylinder (4). Piston

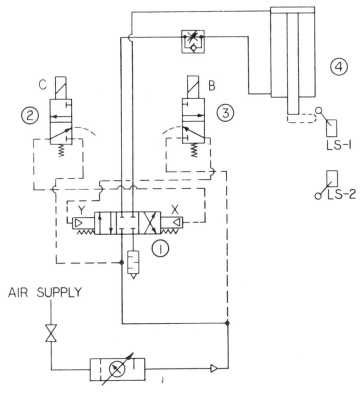

Fig. 4. Electric controls are used in operating the air equipment in this circuit.

on cylinder begins to descend. If operator removes hand from either button before end of downstroke, ram will stop. At the end of the stroke, the normally open limit-switch (LS-2) is closed, energizing electrical timer. The operator is now able to release both hands from push buttons (PB-A) and (PB-B) as contacts in timer now

hold solenoid C of valve (2) energized, maintaining constant pressure
on pilot chamber X of valve (1) which, in turn, keeps valve piston
shifted, holding air pressure on blind end of cylinder (4). When the
dwell period of the timer has elapsed, the contacts which have been
holding the solenoid C of valve (2) energized, open, releasing air
from pilot cylinder X of valve (1), and a second set of contacts
closes energizing solenoid B of pilot valve (3). The piston in valve
(3) now shifts, admitting air to pilot cylinder Y of valve (1). Valve
(1) shifts, directing air to rod end of cylinder (4) and piston starts
to rise. Limit switch (LS-2) is now restored to normally open position and the timer resets, opening second set of contacts. The piston
of ram cylinder returns to "up" position as air continues to flow to
rod end of cylinder. When piston reaches end of stroke, limit switch
(LS-1) is opened,. which breaks current to solenoid B of valve (3).
Piston of this valve shifts, blocking inlet pressure and exhausts air
from pilot chamber Y of master valve (1). All ports of master valve
are now blocked and cylinder piston remains in "up" position while
operator removes workpieces from press fixture.

Protection Against Overload

Circuit No. 5 (Fig. 5) is designed to eliminate the possibilities of
an overload which may be caused by a jam or by carelessness on the
part of the operator. The overload eliminator is automatic and relieves the operator from keeping constant watch over the mechanism.
The operator places the part on table and momentarily depresses
lever mechanism on three-way bleeder valve (2). Pressure is bled
from pilot chamber A of four-way two-position master valve (1) and
piston shifts allowing line pressure to flow to blind end of cylinder
(6). Piston of cylinder advances pushing part onto carrier. As the
outstroke is completed, cam on end of piston rod depresses lever of
three-way valve (3) bleeding pressure from pilot chamber B of valve
(1). Piston of this valve is shifted and air flows to rod end of cylinder (6) returning cylinder piston to original position. If a jam or
overload is encountered on the outstroke, the air pressure increases
and valve (5), which is a sequence valve, opens and air flows to
cylinder connection on valve (4). The spring pressure on valve (5)
is set at a point just above the air pressure required for operation
under ideal conditions. When pressure depresses piston in valve (4),
the air line to pilot connection B is opened reversing master valve
(1) and causing piston of cylinder (6) to retract. Automatic reversal
can also be accomplished by use of a pressure-operated switch in

place of valve (5) and a solenoid-operated two-way valve in place of valve (4). Whether the sequence valve or the pressure switch is used, a wide range of pressure settings is obtainable.

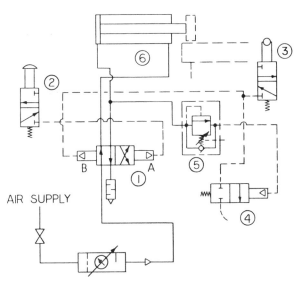

Fig. 5. This circuit is designed to eliminate the possibilities of an overload which may be caused by a jam or by carelessness on the part of the operator.

Provision of High and Low Clamping Forces in Sequence

Circuit No. 6 (Fig. 6) may be used where a workpiece is to be held with considerable force while a large amount of material is removed during an initial cutting operation, after which the holding force must be reduced to the minimum needed to hold the work for a light second operation. By reducing the holding force for the second operation, any undue distortion of the work is avoided. Another advantage afforded by this arrangement is that the workpiece need not be removed from the clamping device when the holding force is changed from high to low.

The circuit functions as follows: First, operator places workpiece with heavy-wall section in 3-jaw, air-operated chuck and shifts handle of four-way control valve (1). Air pressure at 85 lb per sq in. flows to rod end of cylinder (6), and workpiece is held at high pressure. Second, heavy roughing cut is taken on workpiece. Third, operator shifts handle of control valve (2). This directs 30 lb per sq in. air to

rod end of cylinder (6). The high-pressure air in the rod end of cylinder is reduced by being bled off through sequence valve (5). When
pressure is reduced to 30 lb per sq in., sequence valve closes and
pressure remains at 30 lb per sq in., as set by valve (3), for the finish
cut. Fourth, operator then shifts handles of valves (1) and (2) to
their original positions, air pressure flows to blind end of cylinder (6),
and moves piston forward causing chuck jaws to open.

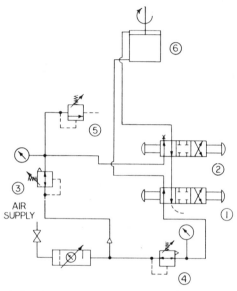

Fig. 6. Circuit designed to hold a workpiece with considerable force while
removing a great amount of material and a lesser force when removing a
small amount of material.

Protection of Two Operators

Circuit No. 7 (Fig. 7), while not foolproof, has been used on a
number of circuits where two operators must be used to tend to one
machine. This safety circuit is ideal if the operators are willing to
cooperate and do not tie down any of the levers on the three-way
valves. It is used on large presses, multiple drilling units, and other
large special machines and the components are very inexpensive compared to those required for some other types of safety arrangements.
The circuit functions as follows: Operator 1 moves large sheet of
plywood into gluing press. Both operators 1 and 2 place wood sections onto panel, one operator working from one side of the press,

the other working from the other side. When the wood sections, covered with glue on one side, are in their proper places, each operator depresses two levers. If the one operator is not ready when the other operator depresses his levers nothing will happen. When all four levers of three-way valves (2), (3), (4), and (5) are de-

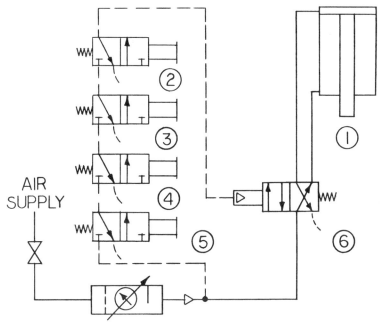

Fig. 7. Circuit for safeguarding two operators which tend one machine. This safety circuit is ideal if the operators are willing to cooperate and do not tie down any of the levers on the three-way valves.

pressed, air flows to the pilot connection of the spring-loaded four-way control valve (6), shifting its piston and allowing air to flow to blind end of press cylinder (1). Press ram starts to descend. If either operator notices that something is wrong he can release one of the levers and the press ram will return to the up position. In the normal operating procedure both operators will keep the control levers on the four valves depressed until the press ram contacts the work, then after a few seconds the operators release the operating levers, the master valve shifts to its normal position, and air flows to rod end of press cylinder, returning ram and completing cycle. The three-way valves may be manually controlled or may be of the solenoid type operated through push buttons.

Holding at Two Pressure Levels with Combination Chuck

Circuit No. 8 (Fig. 8) is a more complicated chucking circuit. When workpieces are very difficult to hold, it is often necessary to design special equipment to solve the clamping problem. This is especially true if one section of the surface to be clamped has a

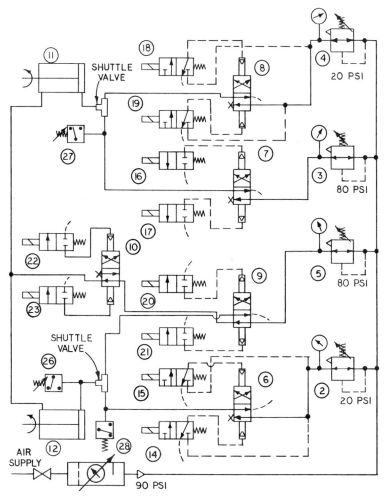

Fig. 8. Complicated chucking circuit which involves the use of special equipment to hold workpieces which have both thin and heavy sections.

heavy wall while the other sections are thin. The following circuit is a good example of what can be accomplished by choosing the right components to do the job.

The chuck used in conjunction with the cylinders in this circuit is of the four-jaw design. The skirt cylinder operates two jaws, and the boss cylinder operates two jaws. The sets of jaws are at 90° to each other.

The circuit functions as follows: At the start of the cycle the pistons in cylinders (11) and (12) are in the retracted position. Operator places workpiece in chuck and momentarily depresses electric starting button energizing solenoids of valves (15) and (22). Four-way master valves (6) and (10) shift, exhausting the "unclamp" sides of cylinders (11) and (12) and applying low pressure to "clamp" side of boss cylinder (12). Boss cylinder (12) closes chuck jaws which center work at low pressure. Pressure build-up in line closes pressure switch (28), set at 20 lb per sq in., energizing solenoids of valves (14) and (16), shifting piston of master four-way valve (6) for exhausting boss cylinder and master valve (7) for applying high pressure to skirt cylinder for locating skirt. After pressure builds up, 80 lb per sq in. pressure switch (27) closes, energizing solenoid valves (17) and (20), shifting master valves (7) and (9). High pressure of skirt cylinder exhausts to reduce any distortion of workpiece through valve (7) and high pressure is applied to boss cylinder through valve (9). Pressure build-up in boss cylinder closes 80 lb per sq in. pressure switch (26), energizing starter to lathe and solenoid of valve (18), shifting piston of four-way master valve (8), applying low pressure to skirt cylinder, completing first half of cycle. When machine work is complete, operator depresses unclamp switch energizing solenoids of valves (19), (21) and (23), shifting pistons of valves (8) and (9) for exhausting of "clamp" sides of skirt cylinder and of boss cylinder. Air at high pressure passes through valve (9) and valve (10) to "unclamp" sides of cylinders (11) and (12) for opening chuck to release workpiece and completing the cycle.

Interlock for Machine Protection

Circuit No. 9 (Fig. 9) is a safety circuit used as an interlock for machine protection. This circuit is so designed that the operator can control the motion for a portion of the cycle, then the remainder of the cycle is so controlled by valving that the operator cannot tamper with the operation. The change-over is automatic and leaves nothing to the whims of the operator.

In the starting position port A of valve (1) is open to the air supply. Line pressure is directed to pilot cylinder D of four-way pilot-operated master valve (2) and port C is connected to inlet port. Operator then depresses lever of valve (1) and port B is connected to the supply line and air flows through inlet port and port C of valve (2) to spring-loaded shuttle valve (4). The shuttle valve piston shifts and air flows to blind end of single-acting air cylinder (5). Piston of cylinder advances. A mechanism depresses roller of cam-operated three-way valve (3). Air pressure then flows to pilot cylinder E of valve (2) shifting piston of this valve since pilot con-

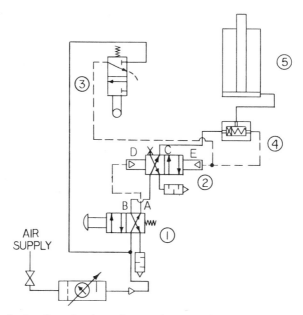

Fig. 9. A safety circuit used as an interlock for machine protection.

nection D is open to exhaust through port A of valve (1). Port C of valve (2) is then opened to exhaust and valve (4) shifts to provide air connection through valve (3) to the blind end of the cylinder, keeping it in the locked position. Mechanism then releases cam roller on valve (3) blocking pressure inlet line and cylinder is exhausted through valve (4) and exhaust of valve (3). The piston will return to its original position even if operator still has lever of valve (1) depressed. This cycle will not repeat until operator releases lever of valve (1). When operator releases lever of valve (1), valve (2) shifts back to its starting position.

Immediate Emergency Reversal

Circuit No.10 (Fig. 10) is designed so that if a jam or work defect develops the operator can depress an electrical push button and cause an immediate reversal of the traverse cylinder. A circuit of this type is used to advantage on grinding machines, milling machines, and

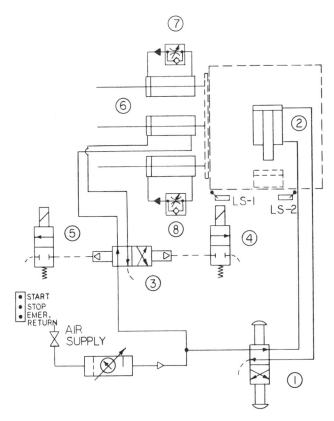

Fig. 10. This design permits immediate reversal of the transverse cylinder if a jam or work defect develops. The operator merely depresses an electrical push-button to cause the reversal.

other machines with reciprocating tables.

This circuit functions as follows: Operator places square-shaped workpieces in holding fixture and shifts lever of four-way two-position valve (1) causing air to flow to blind end of clamp cylinder (2) and piston moves jaw of clamping device forward, securely clamping workpieces in fixture. Operator then momentarily depresses push button marked "start." This energizes the control circuit and valve (3) will shift back and forth as the solenoids on valves (4) and (5) are alternately energized, allowing air to flow to one end of cylinder (6) then to the other end and causing the piston of the cylinder to reciprocate. The cylinder is designed to give smooth even feed as the piston in cylinder (6) is connected mechanically to pistons in two closed-circuit hydraulic cylinders. Limit switches on each end of the table stroke are contacted by cams which set up the electrical impulse to energize the solenoids on valves (4) and (5). When operator pushes "stop" button, cylinder (6) completes stroke in progress, then stops. If the operator should contact the emergency return button during the feed stroke, piston of cylinder (6) withdraws immediately to retracted position and then starts a new feed stroke. If the operator should contact the "stop" and "emergency return" buttons at the same time while the piston of the cylinder is in motion it returns at once to a retracted position, and comes to rest at that point. The feed on the cylinders is very smooth and accurate and can be adjusted over a wide range by means of the metering valves (7) and (8). The return stroke is rapid since the ball check of the metering valves is unseated and the oil flow is then unrestricted.

Converting Board Drop Hammer to Air Operation

Circuit No. 11 (Fig. 11) is one designed for converting board drop hammers to air operation. The circuit provides inching of the hammer on the down stroke for setup so that the operator can make certain that the two sections of the die are in direct alignment. The air is dumped from a quick-acting two-way valve on the work stroke. The circuit also provides for adjustable heights for the hammer-drop. The controls for the air circuit are all electrical. When the hammer is being adjusted, the "up" button is pressed to raise the hammer. The "up" button is connected to solenoid A of pressure-type pilot valve (1) and when this valve opens, pilot-operated valve (2) shifts and air flows to rod end of hammer cylinder (4). When cylinder piston reaches top of stroke, limit switch (LS-1) is actuated and valve (2) shifts to center position, blocking the cylinder ports. The hammer is inched down by depressing the "down" button and energizing solenoid B of pressure-type pilot valve (3) which shifts valve (2) so that air in hammer cylinder passes to exhaust. When operator's hand is removed from "down" button, air in pilot cylinder flows out to exhaust, valve (2) returns to center-blocking position and movement of hammer stops. If hammer is to be held up for a long period, shutoff valve (5) is closed.

When the hammer is ready to go into production, the "run" switch is closed and the "up" button is depressed. Solenoid A of valve (1) will be energized, valve (2) shifts and the hammer piston moves up pulling the hammer up and then limit switch (LS-1) is tripped. Relay (X) is energized through the "run" switch and limit switch (LS-1) and locks in through the "up" contact and the normally open contact of the "stop" button. Relay (X) energizes solenoid B of valve (3)

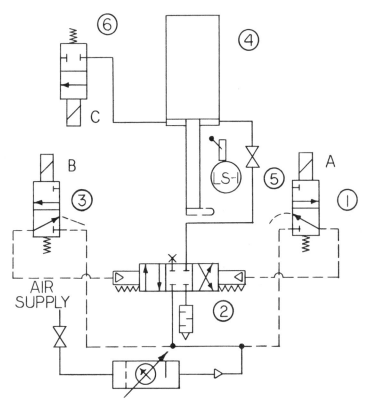

Fig. 11. Circuit for converting board drop hammers to air operation.

and solenoid C of large dump valve (6) quickly dumping air in cylinder (4) to atmosphere. The hammer drop is unrestricted. If the operator shouldn't release the "up" button at the proper moment the relay will open and the hammer will not drop. The "up" button must be pressed to start each cycle. The dump valve is the key to successful operation of an air-operated, drop hammer. It is actually a large two-way solenoid shutoff valve which must have ports that are large

Courtesy of Logansport Machine Co., Inc.

Fig. 12. Air-operated brushing machine which depends upon a rather elaborate system of interlocked air controls for proper automatic functioning.

enough to permit the air to be exhausted without back pressure as the hammer is falling. If all of the air does not quickly exhaust, the hammer will not drop properly.

Interlocked Air Controls for Complex Working Cycle

Figure 12 shows an air-operated brushing machine which depends upon a rather elaborate system of interlocked air controls for proper automatic functioning. The important part that air controls play in the operation of this machine may be seen from the work cycle:

1. Workpiece rolls into machine and contacts limit switch.
2. Feed cylinder moves workpiece over mandrel.
3. Mandrel picks up workpiece.
4. Mandrel grips workpiece.
5. Finder cylinder moves in.
6. Cylinder revolves mandrel until finder fingers locate slot.
7. Finder cylinder retracts.
8. Brushing cylinder cycles five times and retracts.
9. Mandrel releases.
10. Mandrel retracts.
11. Revolving cylinder retracts.
12. Feed cylinder retracts.
13. Unloading cylinder ejects workpiece.
14. Unloading cylinder retracts and is ready for next operation.

CHAPTER 19

Remote Control Pneumatic Systems

With the automation program gaining momentum throughout industry, the use of pilot-operated air control systems has grown by leaps and bounds. This is due largely to their great flexibility, for they can be effectively utilized in a simple basic system as well as in an extremely complicated one. They are particularly adaptable for use in safety circuits as has been described in Chapter 18, and they also lend themselves to use in circuits requiring interlocks with hydraulic systems or electrical systems.

The heart of the pilot control system is the master four-way control valve which may be either of the two-position or three-position type.

Two-position Master Control Valves

The two-position types either have two pilot connections or one pilot connection and a spring return feature. Figure 1 shows the cross section of a master control valve of the bleeder type and Fig. 2 shows a similar view of the pressure type. Note the difference in internal construction.

The *bleeder type* has two small internal passages leading from the inlet port (one is shown in the side view) to either end of the valve body so that, with the bleeder ports on both ends closed, there is equal pressure on the ends of each outer piston. When either end chamber of the valve is open to the atmosphere the piston assembly becomes unbalanced and the complete piston assembly shifts to the exhausting end. As can be seen from Fig. 1, the exhausting end is then plugged by an "O" ring seal on the piston so as to prevent further air loss if the exhausting means is allowed to remain open.

The *pressure-type* valve is designed so that pilot pressure entering one pilot chamber shifts the piston assembly. In order for the piston to shift, however, the other pilot chamber must be open to exhaust.

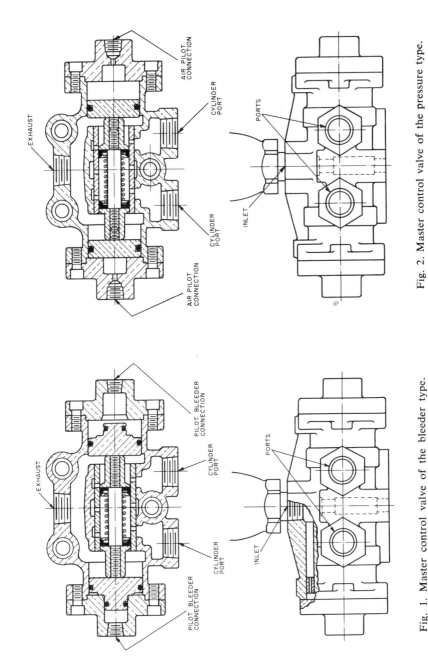

Fig. 2. Master control valve of the pressure type.

Fig. 1. Master control valve of the bleeder type.

Three-position Master Control Valves

The three-position types are spring- or pressure-centered. They may have all ports blocked in the neutral position or they may have the inlet port blocked and the two cylinder ports open to exhaust when in the neutral position. Figure 3 shows a three-position valve with all ports blocked in the neutral position and Fig. 4 shows a similar valve with cylinder ports exhausting in the neutral position. Their operation is similar to the pressure-operated two-position type except that when the pilot pressure is released, the piston centers.

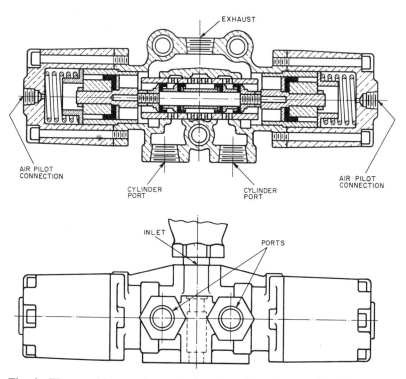

Fig. 3. Three-position master control valve with all ports blocked, in the neutral position.

Pilot Control Valves

The pilot control valves which operate the master valves are either of the two- or three-way design. The two-way valves are used to control the bleeder-operated master valves and the three-way type are used to control the pressure-operated master valves. Both the two-

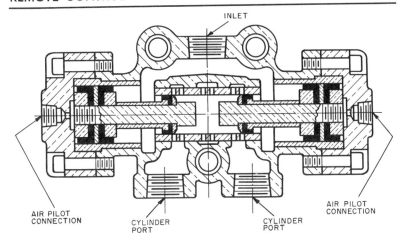

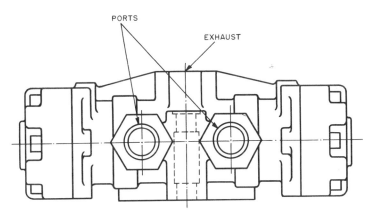

Fig. 4. Three-position master control valve with cylinder ports exhausting, in the neutral position.

and three-way pilot control valves are designed in many styles, including foot operated, push-button, toggle operated, cam-roller operated, air-cylinder operated, palm-button operated and solenoid operated. Figure 5 shows the cross section of a palm-button operated three-way pilot valve. This same valve can be used as a two-way valve by plugging the exhaust port. Note that the valve in Fig. 5 is shown in its normally closed position, that is, with the piston in its normal position the inlet port is blocked. Figure 6 shows the cross-sectional view of a two-way solenoid-operated pilot valve in the de-energized position. This valve must be mounted in a vertical position since the plunger is seated by gravity.

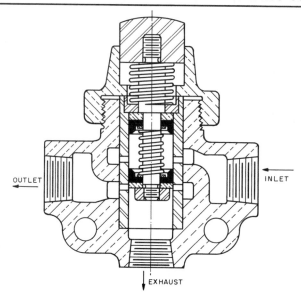

OUTLET

INLET

EXHAUST

Fig. 5. Cross-sectional view of a palm-button-operated three-way pilot valve.

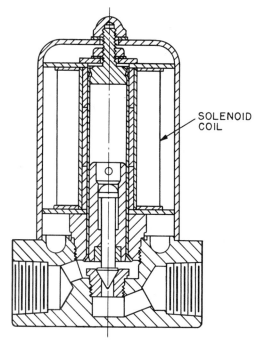

SOLENOID
COIL

Fig. 6. Cross-sectional view of a two-way solenoid-operated pilot valve in the
de-energized position.

Advantages of Pilot-operated Systems

The advantages of pilot-operated systems are:

1. *Rapid Response.* The master control valve can be located adjacent to the operating cylinder while the pilot control can be located at some distance from the master valve, yet the cylinder response is almost immediate. It is not uncommon for a pilot-controlled system to operate at a rate of more than 400 cycles per minute.

2. *Reduction in Cost of Installation.* When using direct-control systems, full-size piping is required from the master valve, which is located at the operating station, to the operating cylinder. Actually, two lengths of large piping are required, since one length is run to the one cylinder cover and the other length to the other cylinder cover. In the pilot control system, small diameter tubing is run from the pilot valve at the operating station to the master valve located next to the cylinder. Then short lengths of large piping are run to the cylinder covers. Copper or steel tubing of small diameter is much easier to assemble than long lengths of large piping. If solenoid pilot controls are used, these are usually attached to the master valve with the push-button controls located at the operator's station. The current consumption of pilot-operated solenoid controls is very small.

3. *Inexpensive Components.* The components which comprise the pilot controls are usually very inexpensive. While they are inexpensive, they are by no means gadgets. They are precision made and often withstand several million cycles of operation without any excess leakage. Mass production of these components accounts for the low cost.

4. *Safe Operation in Explosive Atmospheres.* The pressure and bleeder systems may be used in explosive atmospheres without any further precautions as there are no electrical components present. By using the combination air and hydraulic cylinder along with the components of a pilot-operated system of the bleeder or pressure type, a hydraulic system can be simulated with no electrical equipment being present. This certainly permits a substantial savings, especially around exposive atmospheres.

5. *Sequential Operations.* The use of sequence valves in a complex air system often is not satisfactory, especially if the air line pressure fluctuates to any great degree. It is usually not advisable to use two sequence valves in one portion of a circuit. By using pilot controls very accurate sequencing can be accomplished. Also the pressure in the second phase of the operation can be lower than that in the first phase. This is not true when using sequence valves.

6. *Less Operator Fatigue.* Pilot systems are very popular among women operators working on fast cycling machines since very little effort is required to shift the operating mechanism on the pilot controls. Furthermore, these controls are usually hooked up in series so that it is not possible for the operator to get her hands under a fast-acting ram on its downstroke. A pilot control in the system may also be used for an automatic work ejector.

The circuits in this chapter show the basic fundamentals of pilot control circuits. These circuits may be enlarged and they may be interlocked with other pilot control circuits.

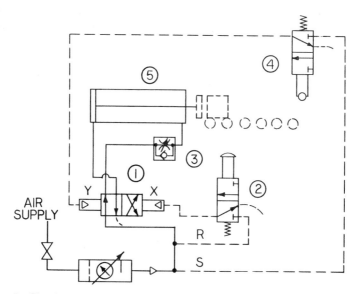

Fig. 7. Circuit which provides semi-automatic operation by using pressure-operated pilot controls.

Semi-automatic Operation with Pressure-operated Pilot Controls

In Circuit No. 1 (Fig. 7) semi-automatic operation is obtained using pressure-operated pilot controls. The same circuit can also be used with bleeder-operated pilot controls.

The circuit functions as follows. Operator moves large box onto transfer table. He then momentarily depresses operating lever of pilot-operated, three-way control valve (2) and air under pressure flows through this valve and enters pilot chamber X shifting piston of master control valve (1) so that air flows to blind end of cylinder

(5). The ram outspeed of the cylinder is controlled by flow control valve (3) through which air is being exhausted from the rod end. As the box is pushed onto the gravity conveyor, cam-operated pilot valve (4) is tripped and air flows to pilot chamber Y of valve (1) shifting piston so that air flows to rod end of cylinder (5) (passing through valve (3) without restriction) and piston retracts to original position. Operator is then ready to load next box onto transfer table. With this circuit, the operator only needs to momentarily depress the lever of one valve and the circuit automatically goes through its cycle. While the cycle is in action, the operator has time to get the next box ready to be placed on the table when the piston rod retracts.

If a bleeder-operated system is to be chosen instead of the pressure-operated system, lines R and S are eliminated. When using this system, the bleeder lines must be as short as possible due to the rapid expansion of the air within the small bleeder lines. Long lines have a tendency to cause the master valve to malfunction. The bleeder lines shouldn't exceed eight feet in length and should be much shorter if possible.

Basic Automatic Circuit with Pressure-operated Controls

Circuit No. 2 (Fig. 8) is an automatic circuit of a basic type in which pressure-operated controls are utilized. Workpieces move along power in-feed conveyor onto transfer fixture. First workpiece contacts and depresses cam roller of pilot control valve (2) and air

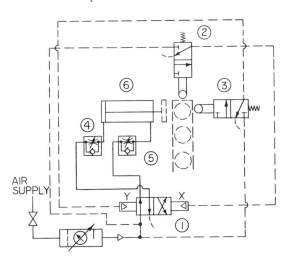

Fig. 8. An automatic circuit of a basic type in which pressure-operated controls are utilized.

flows to pilot chamber X of valve (1) shifting its piston so that air flows to blind end of cylinder (6). Flow control valve (5), through which air from rod end is exhausted, controls rate of motion. Fixture on end of piston rod moves workpiece across transfer table onto gravity conveyor. As fixture approaches end of cylinder stroke, cam roller on valve (3) is depressed and air flows to pilot chamber Y of valve (1) and valve piston shifts allowing air to flow to rod end of cylinder (6) and its piston returns to original position. Flow control valve (4) governs rate of return. Second workpiece feeds in and next cycle automatically starts. This circuit requires no operator. If, however, an operator or an observer is placed above a group of these machines so that he can watch all phases of their operations, a normally-open solenoid-operated shutoff valve may be placed in the line between the rod end of the cylinder and one cylinder port of valve (1) so that he can depress an electrical control button and immediately stop the cylinder piston on its outstroke. In this way he can easily control any jams which may occur in the handling system. Another possibility is the use of a pressure-operated switch which is set so that if a jam should occur, a solenoid valve will immediately close and stop the ram until the work can be cleared. There are so many different types of pilot controls available that the user has practically an unlimited choice.

Automatic circuits of this type have many applications in the material handling field.

Automatic Operation Controlled by Solenoid-operated Pilot Control Valves

In Circuit No. 3 (Fig. 9) automatic operation is obtained by the use of bleeder-type pilot control valves. Workpiece is conveyed onto platform. Operator momentarily depresses push button A energizing solenoid X of normally-closed pilot control valve (5) bleeding air pressure from pilot chamber C of valve (1), shifting its piston so that air flows to blind end of platform cylinder (3) and platform rises. When platform reaches top of stroke, pressure builds up in the line between the master valve and the platform cylinder opening sequence valve (2) and air flows to blind end of transfer cylinder (4). Piston rod of transfer cylinder moves forward ejecting work from platform. At end of transfer cylinder stroke, the limit switch (B) is contacted. This energizes solenoid Y of normally-closed pilot control valve (6) bleeding air from pilot chamber D of valve (1), shifting its piston so that air flows to rod ends of both cylinders and their pistons retract to starting position. If it is advisable to retract the piston of cylinder

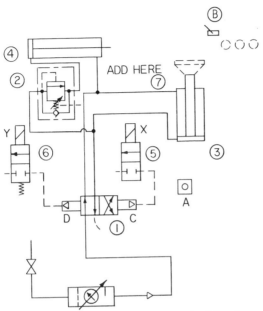

Fig. 9. Automatic operation in this circuit is obtained by the use of bleeder-type pilot-control valves.

(4) before retracting the piston of cylinder (3), then a sequence valve (7) should be added as shown.

Sequential Operation with Cam-operated Limit Switches

Where it is not advisable to use a sequence valve, the system shown in Circuit No. 4 (Fig. 10) can be used to advantage. Work is loaded from conveyor onto platform and operator depresses push button (A) energizing solenoid E of pilot valve (7) shifting piston of master control valve (1) so that air flows to blind end of platform cylinder (4). Platform rises at a speed controlled by flow control valve (3). Just before platform reaches top of rise, swing-type cam on side of platform trips limit switch (D) momentarily energizing solenoid G of bleeder-type pilot valve (8) and piston of master control valve (2) shifts. Air flows to blind end of transfer cylinder (5) and piston of transfer cylinder advances pushing workpiece off of transfer table. As the piston rod approaches end of stroke, cam on transfer device trips limit switch (C) energizing solenoid H of pilot valve (9), shifting piston of master valve (2) and piston of cylinder (5) retracts. As the piston approaches retracted position, swing-type cam on transfer device trips limit switch (B). Solenoid F of

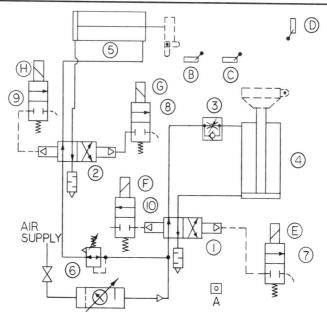

Fig. 10. Circuit which provides sequential operation with cam-operated limit switches.

pilot valve (10) is energized, shifting piston of master valve (1) to original position and piston of cylinder (4) lowers platform. By using swing-type cams, each limit switch is tripped in only one direction.

In this sequence operation the pressure required to operate each circuit is independent of the other. Thus, pressure regulator valve (6) can be used to provide a different pressure for operating cylinder (5) than that used for cylinder (4).

Circuit for Semi-automatic and Automatic Operation

By the use of bleeder-operated pilot controls, Circuit No. 5 (Fig. 11) is set up for both semi-automatic and automatic operation. The circuit functions like this: The piston of the basket cylinder (3) is normally in the retracted or loading position easily accessible to the operator for loading. Operator fills basket with parts which are to be thoroughly washed, then he momentarily trips handle of two-way bleeder valve (7), bleeding air from pressure chamber A of four-way, two-position master valve (2), shifting piston of this valve and allowing air to flow to blind end of cylinder (3). Piston of cylinder (3) advances until basket is submerged in the tank and trip arm on piston

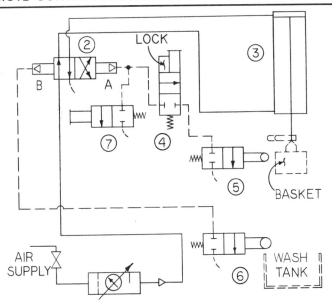

Fig. 11. The use of bleeder-operated pilot controls in this circuit makes possible both semi-automatic and automatic operation.

rod contacts cam roller and depresses piston on pilot valve (6). This opens valve (6), bleeding air from cylinder B of valve (2), causing piston of valve (2) to shift to original position and air flows to rod end of cylinder (3). Piston of basket cylinder retracts.

If it is desirable to submerge the basket several times without stopping, operator shifts lever of pilot valve (4) so that the valve is in open position and cam-operated valve (5) now functions in place of manually-operated valve (7). Piston of cylinder (3) then reciprocates automatically as the trip arm on end of piston rod alternately depresses pistons on valve (5) and (6) until lever of valve (4) is returned to original position. Operator unloads work.

Circuit for Heavy Duty Automatic Operation

Circuit No. 6 (Fig. 12) makes use of a two-way and a three-way pilot control for setting up the automatic cycle. While the circuit is rather novel in design it has many applications where the requirements are rugged, such as on tamping, vibrating, rapping, and riveting equipment. The speed of the cycle is determined by the volume of air passing the shutoff valve placed ahead of the regulating, filter, and lubricating unit. The circuit functions as follows: Operator places compound in form and depresses button on two-way pilot valve (5).

Air flows through valve (5) and on through three-way pilot valve (4) to pilot connection on four-way two-position spring-return master valve (2). Piston of this valve shifts and air pressure flows to blind end of tamping cylinder (3) and piston of this cylinder descends. Both the cylinder of valve (4) and the rod end of cylinder (3) are connected to exhaust through valve (2). Air bleeds from cylinder on pilot valve (4), through needle valve (6), causing valve (4) to shift. Pressure to cylinder of valve (2) is cut off permitting spring to restore piston of valve (2) to original position. Air then flows to rod end of cylinder (3), returning piston and at the same time bleeds through needle valve (6) to cylinder of valve (4). When pressure in cylinder is sufficiently high, valve (4) shifts and cycle is repeated. Length of piston stroke is determined by adjustment of bleeder valve (6). Cylinder will automatically reciprocate until button of valve (5) is released. The components which comprise this circuit are simple in construction, sturdy—in order to stand up under rapid cycling—and inexpensive. All of these components are very compact.

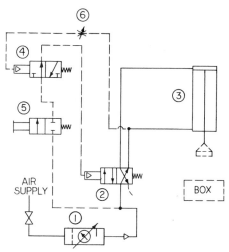

Fig. 12. This circuit makes use of a two-way and a three-way pilot control for setting up the automatic cycle.

Sequential Operation Using Pilot-operated Controls

Circuit No. 7 (Fig. 13) is a typical control circuit in which the three cylinders operate in an exact sequence, without the use of sequence valves. The cycle is started when the operator depresses

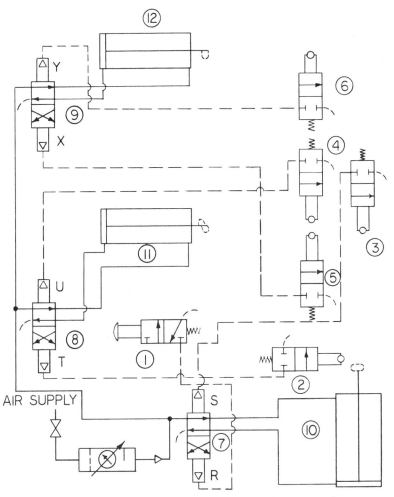

Fig. 13. A typical control circuit in which three cylinders operate in an exact sequence without the use of sequence valves.

the button on three-way bleeder valve (1). Air is bled from pilot chamber R of master valve (7) and piston shifts so that air pressure flows to blind end of pusher cylinder (10) and piston moves forward. Swing-type cam on end of piston rod contacts and depresses cam roller of bleeder valve (2), bleeding air from pilot chamber T of valve (8). Just after cam on piston rod of cylinder (10) depresses roller on valve (2), it contacts roller on bleeder valve (3), exhausting pilot chamber S of valve (7) and piston of cylinder (10) retracts. At about the same time the piston of cylinder (11) is advancing due to the shifting of valve (8). When cam on end of piston rod of cylinder (11) depresses roller of valve (5), air is bled from pilot chamber X of valve (9) shifting its piston. Cam on rod of cylinder (11) then contacts roller on bleeder valve (4) shifting valve (8) back to original position and reversing piston of cylinder (11). Piston on cylinder (12) advances when piston of valve (9) shifts, allowing air to flow to blind end of cylinder (12). When piston rod of cylinder (12) advances to the end of the stroke a cam on the end of the piston rod contacts cam roller on valve (6) and air is exhausted or bled from pilot chamber Y shifting valve (9) so that air is directed to rod end of cylinder (12) returning the piston and completing the cycle.

This cycle could be made automatic by making bleeder valve (1) cam operated and placing it along the path of the piston travel of cylinder (12). On the end of the piston rod of cylinder (12) a swing-type cam would be placed so that it would trip the lever on valve (1) as it contacted it on the "in" stroke. This cycle can be used in material handling operations to great advantage. The sequence is positive, eliminating any chance for a malfunction.

Sequential Operation with Camshaft Drive Unit Actuating Pilot Controls

The pressure-operated pilot control system shown in Circuit No. 8 (Fig. 14) has many possibilities. The speed of operation of this system depends upon the speed of the camshaft, the design of the cams, and the volume of air pressure present. This circuit makes use of two three-position, spring-centered master control valves along with four cam-operated pilot valves of the three-way type. These valves are spring-centered so that when rollers ride off the cams, the valves return to neutral position with all ports closed.

The circuit functions as follows: Operator engages camshaft drive and as camshaft rotates, cams (1) and (4) depress rollers on pilot valves (1) and (4). From pilot valve (1) air is directed to chamber C of master valve (5). This pilot pressure shifts piston of valve (5) and directs line pressure to blind ends of cylinders (7) and (8) and both cylinder pistons advance. At the same time, air flows through pilot valve (4) to pilot chamber P of valve (6) shifting the piston of this valve and air is directed to rod ends of cylinders (9) and (10) and their pistons return.

As the camshaft rotates 180 degrees, cams (1) and (4) release the rollers on valves (1) and (4) and cams (2) and (3) contact and depress rollers on valves (2) and (3). Air pressure is directed through valve (2) to pilot chamber D of master valve (5). Air pressure shifts piston and air is directed to rod ends of cylinders (7) and (8) returning their pistons to original positions. Pressure also passes through valve (3) to pilot chamber Q of valve (6) shifting valve piston and air is directed to blind ends of cylinders (9) and (10) and their pistons advance. The cylinders will continue to cycle until the operator disengages the camshaft drive.

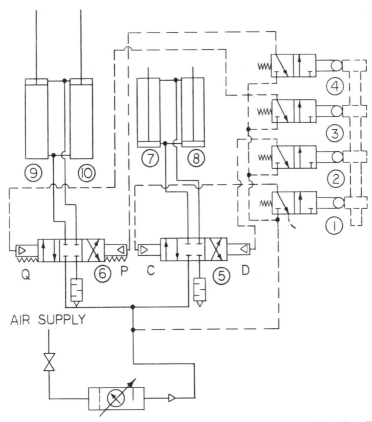

Fig. 14. Sequential operation with a camshaft drive unit actuating the pilot controls is made easy by the use of this circuit.

Pilot Control Circuit for Chucking Application

Circuit No. 9 (Fig. 15) is a chucking circuit of the basic type but shows the possibilities afforded when using pilot controls. The circuit functions as follows: Operator places casting against jaws of locking-type power chuck and depresses foot pedal of three-way bleeder valve (1) and air is bled from cylinder X of valve (2) shifting its

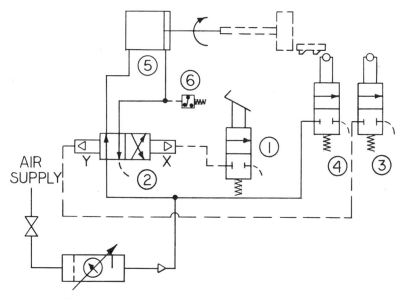

Fig. 15. A pilot-control circuit for chucking applications.

piston and directing air pressure to rod end of rotating cylinder (5). Air pressure causes piston to retract, pulling on drawbar and closing chuck jaws. Pressure builds up in line closing pressure switch (6), starting lathe motor. Machine table moves to the left mechanically and reverses in the same way. Air blast is set up as pilot valve (4) is opened by contact of cam on machine table with valve roller. When table approaches the right-hand end of the stroke, bleeder valve (3) is similarly actuated. Cylinder Y of valve (2) is exhausted and piston valve shifts so that air is directed to blind end of rotating cylinder (5). At the same time air is exhausted from the rod end and pressure switch opens, immediately stopping lathe spindle. Piston moves forward and chuck jaws open releasing workpiece and completing cycle. An electric brake may be used so that spindle will stop immediately before jaws release workpiece.

CHAPTER 20

Combination of Fluids in a Single System

Several chapters have been devoted to depicting the uses and advantages of oil as a fluid medium and several have covered the use of air as a fluid medium. This book would not be complete without going into some detail on the advantages of the two fluids used together in the same circuit. Many difficult problems have been solved by combining these fluids. By the use of the two mediums, the quick action of air and the smooth, high pressure action of oil blend ideally for the solution of many of our most difficult fluid power problems. The combination will reduce space requirements and cut maintenance costs of original equipment. It will better the performance of the equipment and allow more flexibility. It should be made clear, however, that although the two fluids are used in what amounts to the same circuit, they are not mixed together; in fact, every precaution is taken to keep them separated. Some of the many circuit possibilities are illustrated by the following examples.

Use of Combination Air-hydraulic Unit

The first circuit to be discussed is a basic one that shows the possibility of the use of air and oil in one cylinder. The air is the motivating means, and the oil is the means of control. The result is that the feeds obtained are comparable to those of a hydraulic system, yet the speeds are nearly as fast as those possible with an air cylinder. While this type of cylinder is more costly than an air cylinder, it eliminates the expense of a hydraulic power unit. A circuit using an air-hydraulic cylinder is especially suitable in explosive atmospheres because of the absence of electrical equipment. This circuit is also used to advantage on drilling machines, milling machines, induction heating equipment and other production machines.

The design of the air-hydraulic cylinder unit is simple. It may be two individual cylinders mounted side by side with a double-end rod on the hydraulic cylinder and a single rod on the air cylinder; it may be an air cylinder with a hydraulic check mounted beside it or it may be a combination cylinder as shown in the cutaway view in Fig. 1. The air cylinder should always be to the rear so that it will be the hydraulic cylinder that will have a double-end rod. The area against which the hydraulic fluid acts will then be the same on both sides of the piston.

In this type it should be noted that in order to make certain that the air and oil will not mix, a breather hole is placed between the two sets of packing in the center cover so that any leakage will be bled off to atmosphere. One thing that is very essential when using a combination cylinder is to make sure that all of the air is released from the hydraulic cylinder's closed circuit. Otherwise an uneven or "jerky" feed will be experienced.

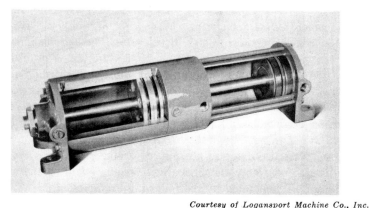

Courtesy of Logansport Machine Co., Inc.
Fig. 1. Cutaway view of a combination air-hydraulic cylinder.

Three types of feeds are available in the basic circuit: feed on the outstroke with rapid return; rapid traverse on the outstroke and feed return; or feed in both directions. Any feed from the maximum to the minimum is obtained by merely adjusting a small built-in needle valve.

Circuit No. 1 (Fig. 2) functions as follows:

1. Operator loads work into fixture and shifts handle of four-way two-position valve (1) allowing air to flow to blind end of combination cylinder (2). Piston rod connected to drilling device moves drill out at a rapid rate until cam on drilling device contacts cam-operated shut-off valve (3). This closes the high-speed oil line and speed of cylinder is then governed by the low-speed orifice as the

drill contacts the workpiece. The feed continues as the drill is fed through the workpiece.

2. When the stroke is completed, operator shifts handle of valve (1) back to original position and built-in check valves in hydraulic piston open as the air piston starts to return. This allows the piston rod to retract at a rapid rate thus completing fluid power cycle.

3. Operator then unloads workpiece.

By the use of a cam with more than one step, skip feeds can be obtained which are useful in milling and other operations of a multiple nature.

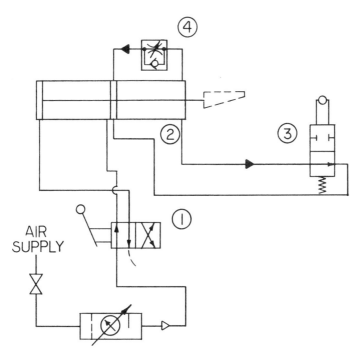

Fig. 2. Air-hydraulic circuit which provides for a rapid traverse on the outstroke, until the position for the start of the drilling operation is reached; a slow drilling feed; and a rapid traverse return.

Use of Air Cylinders and a Combination Unit

Circuit No. 2 (Fig. 3) makes use of three air cylinders and an air-hydraulic unit. This circuit functions as follows:

1. Operator places large casting into fixture.

2. He then depresses electric push button X energizing solenoid R of four-way, two-position air valve (2). This shifts piston of valve

allowing air to flow to the blind ends of the clamp cylinders (3), (4), and (5) which are located 120° apart around the fixture.

3. Pistons of cylinders (3), (4), and (5) move forward rapidly and work is securely clamped in fixture. Air pressure then builds up in the line, opening sequence valve (8) admitting air to blind end of

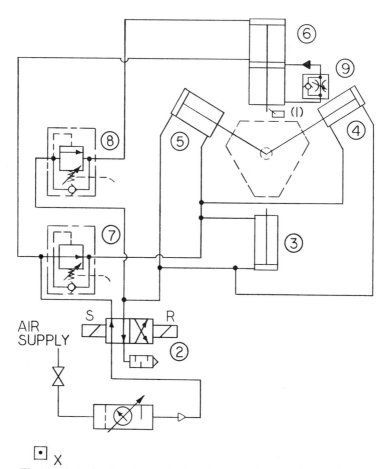

Fig. 3. Air-hydraulic circuit which makes use of three air cylinders.

air-hydraulic feed cylinder (6) and broaching fixture connected on end of piston rod of cylinder (6) moves forward and broach enters the workpiece. The flow control valve (9) mounted in the oil section of the air-hydraulic cylinder controls the speed of the piston and broaching fixture.

4. When the piston of cylinder (6) approaches the end of its stroke, limit switch (1) is contacted, energizing solenoid S of valve (2) allowing air to flow to center section of cylinder (6). Piston retracts rapidly as check valve (9) is unseated in closed hydraulic system. When piston of cylinder (6) reaches the retracted position, air pressure builds up, opening sequence valve (7) and air flows to rod ends of the three clamping cylinders (3), (4), and (5) and work is unclamped.

5. Operator removes work from fixture, reloads and is ready for next cycle.

When using air-hydraulic cylinders one must realize that they have certain feed and speed limitations. One cannot expect to have a range of from one-eighth of an inch a minute to eight hundred or nine hundred inches per minute without going into a very special setup. All of these cylinders have their limitations. Manufacturers of such cylinders will make definite recommendations and it is well to follow them.

Courtesy of Hause Machines, Inc.

Fig. 4. Combination air and hydraulic drilling unit in which air is employed as the source of motivation and oil as a means of feed control.

Combination Air and Hydraulic Drilling Unit

During the past few years there has been an increasing number of drilling units employed in which air is employed as the source of motivation and oil as a means of feed control.

In Fig. 4 is shown such a unit. It can be adjusted for rapid travel up to about 500 inches per minute on the advancing stroke and up to about 300 inches per minute on the retracting stroke. The hydraulically controlled feed rate is adjustable from 0 to 240 inches per minute on the advancing stroke and from 0 to 100 inches per minute on the retracting stroke. It has a depth accuracy within 0.005 inch with built-in controls and within 0.0005 inch with positive stop and hydraulic dwell controls.

Use of Air Cylinders and Large Hydraulic Cylinder

Circuit No. 3 (Fig. 5) makes use of air cylinders for loading and unloading the workpiece and a large hydraulic cylinder for the performance of the work on the workpiece. Workpiece moves down power-driven conveyor until limit switch A is contacted. This energizes solenoid X of four-way, two-position air valve (4). Air flows to blind end of inbound air cylinder (3). Fixture on end of piston rod rapidly pushes the workpiece into coining station. When workpiece is directly under coining ram it contacts limit switch B. This energizes solenoid Y of valve (4) and also solenoid R of valve (5). Air flows to rod end of cylinder (3) and its piston rapidly retracts. At the same time, valve (5) has shifted and oil flows to blind end of coining cylinder (6) and ram of cylinder descends and performs the operation. Pressure then builds up opening pressure switch which energizes solenoid S of valve (5). Oil flows to rod end of coining cylinder (6) and ram begins to retract. As it approaches the top of the stroke, limit switch (C) is contacted, energizing solenoid T of valve (7) and air flows to blind end of outbound cylinder (8). Piston of cylinder rapidly advances pushing workpiece onto power-driven outbound conveyor. At end of outstroke piston rod of cylinder contacts limit switch (D) and solenoid U is energized, shifting valve (7). Air then flows to rod end of cylinder (8) and piston retracts. At the same time current is allowed to flow to limit switch (A) and next cycle is set up. The cycle on this machine is completely automatic and requires no operator. One observer may watch several machines and, in case of a "jam," can immediately stop the machine motion by depressing the stop button (not shown). The circuit is designed through its electrically operated controls so that each cylinder may be operated independently during the setup period, and also in case of a "jam."

There are a number of advantages of using both air and hydraulic mediums in this circuit. The use of air cylinders for bringing the workpiece under the ram and then ejecting it speeds up the cycle time of the machine considerably. The inbound cylinder quickly deposits the work under the ram and is out of the way quickly and the same is true of the ejecting cylinder. Since only a small amount of force is required to advance the pistons, air will amply do the job. The heavy coining cylinder requires a fairly rapid approach

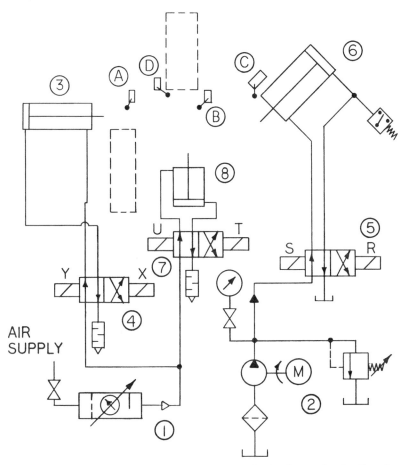

Fig. 5. Circuit which makes use of air cylinders for loading and unloading the workpiece and a large hydraulic cylinder for the performance of the work on the workpiece.

and is taken care of by the high-low oil pumping unit. By this choice of units the coining cylinder is rapidly advanced at low oil pressure until the work is met. Then high oil pressure is automatically cut in with the low volume pump and performs the work. The ram rapidly returns at low pressure. During the stand-by period for the coining cylinder, the pump is operating at low volume and high pressure, holding up the heavy coining fixture. Since only a small volume of oil is spilling through the relief valve, no excessive heat is created in the system. The whole layout offers a compact design which is self-operated and low in maintenance expense.

Use of Hydraulic Fluid as a Dampener

Circuit No. 4 (Fig. 6) makes use of hydraulic fluid as a dampener when used in conjunction with air power. Oftentimes it is not practical to use a complete hydraulic system either from a cost standpoint or for other reasons obvious in certain applications. However, hydraulic fluid as a dampener proves extremely useful. In this circuit the air-oil tank (5) is the key to the solution. The operator shifts handle of three-way air valve (2) and air rushes to the air-oil tank

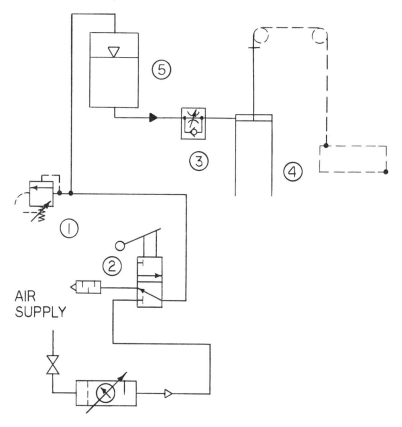

Fig. 6. The hydraulic fluid in this circuit acts as a dampener when used in conjunction with air power.

and the air pressure forces the oil through the check valve of the flow control valve (3) into the rod end of the single-acting hydraulic cylinder (4). The piston retracts quickly opening the large lid on the fixture. This lid is held in open position until operator shifts handle of valve (2). Since the lid must close very slowly and evenly the oil is slowly metered through flow control valve (3) as the weight of the lid raises the piston and forces the oil back through this valve. The piston and rod of the cylinder must be suitably packed so as to eliminate as much leakage as possible to assure satisfactory performance. By placing a hydraulic shutoff valve between the cylinder and the air-oil tank the cylinder may be stopped at any point in its travel. This valve may be remotely operated as well as valve (2) so that operator may have his station at a location some distance from the actual operation.

Use of Air to Control a Hydraulic Control Valve

Circuit No. 5 (Fig. 7) shows what can be accomplished by the use of air to control a hydraulic control valve. This arrangement is used principally on high-cycling applications where the time lag must

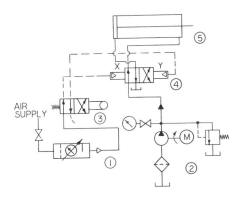

Fig. 7. Circuit which uses air to control a hydraulic control valve.

be held to the bare minimum. In this circuit, a cam operates the air pilot valve (3). Material is fed into mold and when mold is filled, operator starts cam device (not shown) which depresses and releases cam roller on valve (3). When the roller is in the "up" position, air flows to port X of valve (4) and when roller is depressed, air flows to port Y of valve (4). As the piston of four-way, two-position air-operated hydraulic control valve (4) shifts, oil is directed either to the blind end or rod end of tamping cylinder (5). The actuation of the piston rod of this cylinder is naturally controlled by the reversals caused by the cam-operating device. When the tamping time has elapsed, operator stops cam motion. Work is then unloaded.

Use of Hydraulic Equipment to Control Air Equipment

Circuit No. 6 (Fig. 8) has found a lot of applications where a number of cylinders are involved in a circuit. The air equipment performs the operation while the hydraulic equipment controls the air equipment. This is how the circuit operates. Operator depresses start button and in its normal position the piston of hydraulic feed cylinder (22) is retracted and limit switch (A) is closed. When operator depresses start button (not shown) current is fed to limit switch (A) and solenoid valve (21) is energized. Oil flows to blind end of cylinder (22), its forward piston travel being controlled by flow control valve (23). As the piston moves forward, it opens limit switch (A), then cam on piston rod first contacts two-way, cam-operated valve (11). Air bleeds from this valve allowing piston of valve (6) to shift and air flows to blind end of cylinder (1). Piston of cylinder (1) moves out. Cam on rod of cylinder (22) then contacts, in turn, cam valves (12), (13), (14), and (15), shifting valves (7), (8), (9), and (10) and pistons of cylinders (2), (3), (4), and (5) are advanced. Piston of cylinder (22) continues to advance and in turn contacts cam valves (16), (17), (18), (19), and (20). This causes valves (6), (7), (8), (9), and (10) to shift back to their original positions and air flows to rod ends of cylinders (1), (2), (3), (4), and (5) so that their pistons retract to original positions. As cylinder (22) approaches end of stroke, roller of limit switch (B) is contacted and solenoid valve (21) is energized. Oil is now directed to rod end of cylinder and cylinder begins to retract. Cam on end of piston rod is so designed that on the return stroke it will not depress rollers on valves (11) through (20). There is, however, a pin on the side of the cam which contacts limit switch (A) and the cycle is repeated. The cycle will continue to repeat until operator releases push button. Instead of using the hydraulic system to operate the cam valves, an air-hydraulic cylinder may be used with success. By using either system the operator has a wide variety of speeds available. Speeds may be controlled by the setting of the flow control needle valve (23).

Air Control of Multiple Hydraulic Cylinders

Circuit No. 7 (Fig. 9) is a variation of Circuits No. 6 and No. 7 and has wide application where multiple cylinders operating at high

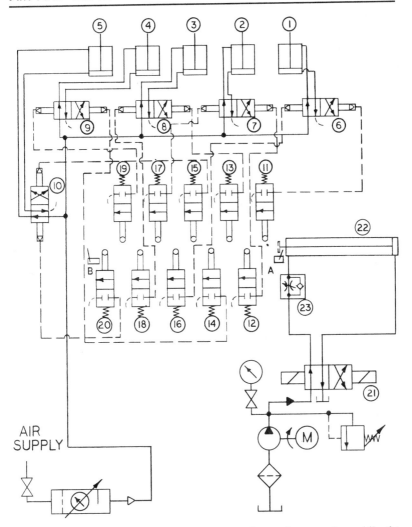

Fig. 8. Circuit in which the air equipment performs the operation while the hydraulic equipment controls it.

speeds and high pressure are involved. An electric motor with a variable-speed drive operates the camshaft which is equipped with five cam plates. These cam plates engage and disengage cam-operated, three-way, air-control pilot valves which are connected to four-way, two- and three-position oil control valves. The rapid expansion of air quickly shifts the spools in the air-operated hydraulic control valves.

The circuit operates as follows: As camshaft starts to rotate very slowly cam plate (A) contacts roller of pilot valve (F) and air flows to pilot connections of valves (G) and (H). As these valves shift, oil flows to blind end of cylinders (1) and (2) and their

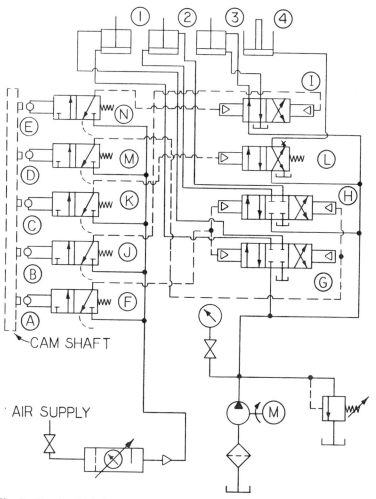

Fig. 9. Circuit which incorporates the use of air control of multiple hydraulic cylinders operating at high speeds and pressures.

pistons move forward to end of stroke. Cam plate (B) then contacts roller of pilot valve (J) and air shifts piston of valve (I) and oil flows to blind end of cylinder (3) and piston rod moves forward. Cam plate (C) then contacts roller of three-way air valve (K) and air flows to pilot connection of four-way spring-return valve (L) and piston shifts supplying oil to blind end of single-acting cylinder (4). Cam plate (D) contacts roller of three-way valve (M) and air flows to pilot connection of valve (G) and oil is then directed to rod ends of cylinders (1) and (2) and, as their pistons retract, cam plate (E) contacts roller of valve (N) and air flows to pilot con-

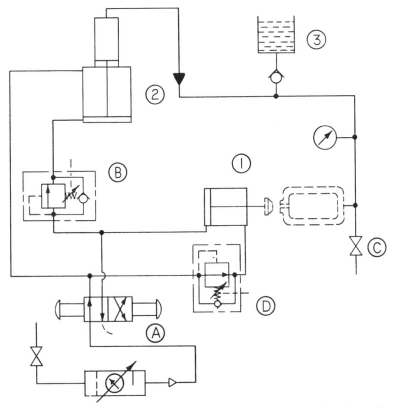

Fig. 10. This circuit illustrates the use of air employed in conjunction with oil for pressure testing applications.

nection of valve (I) and piston shifts allowing air to flow to rod end of cylinder (3) and piston retracts. As piston of cylinder (3) retracts cam plate rides off of roller of valve (K) and spring return valve (L) shifts to original position and cylinder (4) returns to original position, thus completing the cycle. The speed of the cycle can be varied through the variable drive mechanism. The speed of the cylinders can be altered by inserting flow control valves in the cylinder lines.

Converting Air Pressure to High Hydraulic Pressure

Circuit No. 8 (Fig. 10) shows the use of air used in conjunction with oil for pressure testing applications. This circuit is of special benefit where the elimination of the hydraulic power unit is desirable yet very high pressure is required. Operator loads piece to be tested into fixture, shifts handle of four-way, two-position air control valve (A) and air flows to blind end of plug-locking cylinder (1). Piston moves out and plug is locked in the end of workpiece. Pressure in cylinder (1) builds up and sequence valve (B) opens and air flows to blind end of intensifier cylinder (2) which has a ratio of 50 to 1. The piston starts forward and the intensified oil pressure is forced from the nose of the intensifier and flows into the test piece. When pressure test is completed, operator opens exhaust valve (C) and high pressure exhausts from test piece. Operator then shifts valve (A) back to original position and air flows to rod end of intensifier cylinder (2) and piston retracts to original position. Pressure builds up opening sequence valve (D) and air flows to rod end of clamp cylinder (1) and piston retracts to original position. A prefill chamber (3) is necessary to replenish the oil in the line between the nose end of the intensifier and the workpiece. The intensified pressure is dependent upon the line pressure and the ratio of the area of the air cylinder to the intensifier cylinder. Thus, in this case, with an intensifier ratio of 50 to 1, if the air pressure were 80 lb per sq in., then the intensified fluid pressure would be 4,000 lb per sq in.

Air Used with Hydraulic System to Apply Even Pressure and Absorb Shock

Circuit No. 9 (Fig. 11) shows one of the ways in which air is used in conjunction with a hydraulic system to smoothly apply an even pressure or to absorb shock. Many applications on hydraulic presses use an air die-cushion to absorb the shock of impact. The amount of cushioning is dependent upon the operating pressure of the air system.

This circuit operates as follows:

Operator places workpiece on lower platen which is connected to piston rod of air cylinder (1). In the normal position its piston rod is always extended. Operator then shifts handle of four-way two-position valve (2) and oil is directed to blind end of cylinders (3) and (4) and the pistons of these cylinders move forward carrying rollers which are to make contact with the workpiece. Operator

then shifts handle of four-way, two-position valve (5) and oil flows
to blind end of forming-die cylinder (6).   Piston starts to descend
and forming-die contacts workpiece lying on top of platen on air
cylinder.   Since air is compressible, the ram of cylinder (6) forces
ram of cylinder (1) to retract.   During the downstroke the rollers
on ends of cylinders (3) and (4) keep tension on the workpiece as
it is formed and as the rollers move down the pistons are slightly

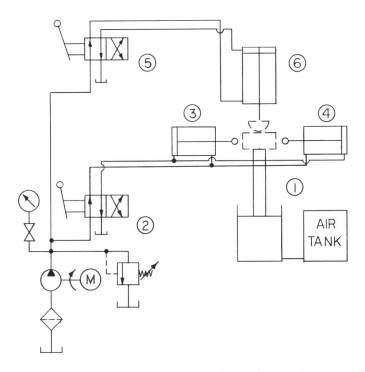

Fig. 11. One of the ways in which air can be used in conjunction with a
hydraulic system to smoothly apply an even pressure or to absorb shocks is
illustrated in this circuit.

retracted.   At the end of ram stroke of cylinder (6), operator shifts
handle of valve (2) to original position and oil flows to rod ends of
cylinders (3) and (4) and rollers retract.   Operator then shifts
handle of valve (5) to original position and oil flows to rod end of
cylinder (6) and ram retracts.   Operator then unloads finished workpiece and is ready for next cycle.

A circuit using such cylinders as (1) and (6) has application
where smooth even pressure must be applied to both sides of the

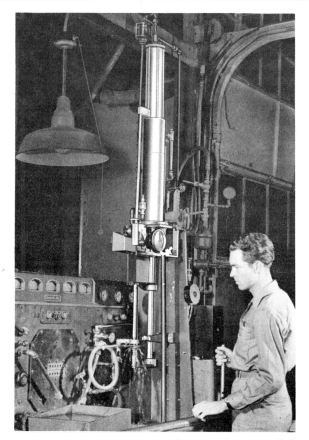

Courtesy of Logansport Machine Co., Inc.
Fig. 12. Air-hydraulic cylinder as used on a heat-treating machine.

workpiece while it is being lowered into a solution, as, for example, a flat gear blank being lowered into quenching oil. This eliminates undue distortion during the cooling period.

As automation is increased, there will be more and more applications for the two mediums in a single circuit.

Figure 12 shows an air-hydraulic cylinder being used on a heat treating machine application.

CHAPTER 21

High-Pressure Hydraulic Systems

In Chapter 3 we briefly discussed the trend toward higher pressures for hydraulic systems. Pressures are being increased as the result of better component designs, better seals, and the ability to produce closer tolerances and finer machined finishes. Internal leakage, which has always been somewhat of a problem in hydraulic systems, must be held to an absolute minimum when working with high pressures. Even a few drops per minute of high-pressure fluid cannot be tolerated in some critical applications.

What do we mean by high pressure? A couple of decades ago 2000 psi was uncommon in industrial hydraulics. Today, this is considered quite common. Three-thousand-pound systems are being widely used and would be listed at the low end of the high-pressure range. Many systems are now being developed for 5000, 6000, 8000 and 10,000 psi. As fluid power component designs progress we can expect even higher pressures. The major advantages of high-pressure systems are compactness of components, portability, and use of small quantities of oil.

High-Pressure Pumps

High-pressure pumps are usually of the piston type. Piston pumps are able to withstand high shock loads. They have a relatively low noise level and high efficiency rating. Their service life is good. Pumps with check-valve piston units have a great deal of reliability because each piston is a pump in itself. A breakdown of one piston will not put the pump out of service. In most instances the larger volume hydraulic pumps have a lower pressure rating. For example a 20-gallon-per-minute pump might be rated for 3000 psi, whereas a one-gallon-per-minute pump would be rated for 10,000 psi.

There are two distinct categories of high-pressure pumps—those known as hand-operated pumps and those actuated by some rotary

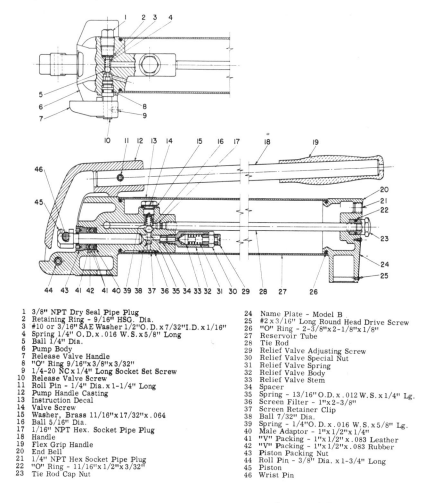

1	3/8" NPT Dry Seal Pipe Plug
2	Retaining Ring - 9/16" HSG. Dia.
3	#10 or 3/16" SAE Washer 1/2"O.D. x 7/32"I.D. x 1/16"
4	Spring 1/4" O.D. x .016 W.S. x 5/8" Long
5	Ball 1/4" Dia.
6	Pump Body
7	Release Valve Handle
8	"O" Ring 9/16"x 3/8"x 3/32"
9	1/4-20 NC x 1/4" Long Socket Set Screw
10	Release Valve Screw
11	Roll Pin - 1/4" Dia. x 1-1/4" Long
12	Pump Handle Casting
13	Instruction Decal
14	Valve Screw
15	Washer, Brass 11/16"x 17/32"x .064
16	Ball 5/16" Dia.
17	1/16" NPT Hex. Socket Pipe Plug
18	Handle
19	Flex Grip Handle
20	End Bell
21	1/4" NPT Hex Socket Pipe Plug
22	"O" Ring - 11/16"x 1/2"x 3/32"
23	Tie Rod Cap Nut

24	Name Plate - Model B
25	#2 x 3/16" Long Round Head Drive Screw
26	"O" Ring - 2-3/8"x 2-1/8"x 1/8"
27	Reservoir Tube
28	Tie Rod
29	Relief Valve Adjusting Screw
30	Relief Valve Special Nut
31	Relief Valve Spring
32	Relief Valve Body
33	Relief Valve Stem
34	Spacer
35	Spring - 13/16"O.D. x .012 W.S. x 1/4" Lg.
36	Screen Filter - 1"x 2-3/8"
37	Screen Retainer Clip
38	Ball 7/32" Dia.
39	Spring - 1/4"O.D. x .016 W.S. x 5/8" Lg.
40	Male Adaptor - 1"x 1/2"x 1/4"
41	"V" Packing - 1"x 1/2"x .083 Leather
42	"V" Packing - 1"x 1/2"x .083 Rubber
43	Piston Packing Nut
44	Roll Pin - 3/8" Dia. x 1-3/4" Long
45	Piston
46	Wrist Pin

Courtesy of Owatonna Tool Co.

Fig. 1. Hand-operated hydraulic pump.

means such as an electric motor, an internal combustion engine or a steam engine. There are many uses around an industrial plant for hand-operated pumps such as that shown in Fig. 1. While only a small amount of hydraulic fluid is moved each time the pump handle is actuated, it often requires only a small amount of fluid under high pressure to perform certain tasks that may be found in experimental laboratories and in maintenance departments. For example, in an experimental laboratory such a pump might be used in load testing and in maintenance departments such a pump could be used to actuate a wheel puller.

A hand pump usually has a built-in overload device since it is possible, depending upon the strength of the operator, to produce pressures even higher than those for which it was designed. In this type of pump the sealing surfaces and sealing devices are most important. The fluid is moved in small quantities, 0.1 cubic inch per actuation of the pump handle, for example, and leakage cannot be tolerated. A rotary power-operated, atmospheric inlet, axial piston pump that delivers up to 70 gpm and which has a low noise level of operation is shown in Fig. 2. This pump is designed for 3500 psi continuous operation and 5000 psi intermittent operation, and can be driven at speeds up to 2600 rpm.

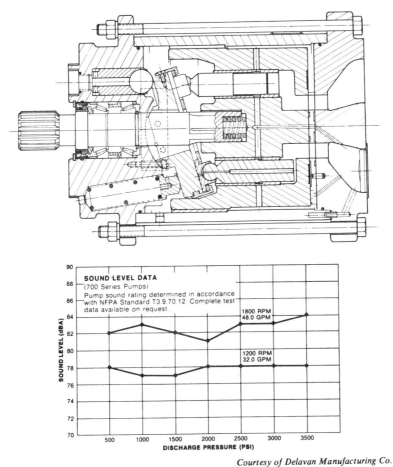

Courtesy of Delavan Manufacturing Co.

Fig. 2. Atmospheric inlet axial piston pump and sound-level data table.

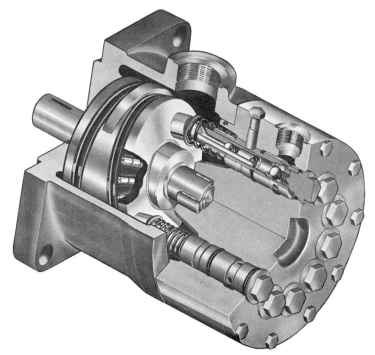

Courtesy of Dynex Div., Allied Power, Inc.
Fig. 3. Check-valve type of high-pressure pump rated at 50 hp.

A unique porting arrangement in the rotor provides excellent flow to fill the piston cavities. Complete filling is critical to the efficient operation of a large displacement atmospheric inlet pump. The pistons are hydraulically returned to eliminate impact loading on shoes and other related sources of noise generation. The table shows the pump sound-level data at various discharge pressures.

Figure 3 shows a check-valve type of high-pressure pump that is

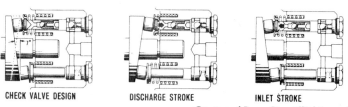

CHECK VALVE DESIGN   DISCHARGE STROKE   INLET STROKE

Courtesy of Dynex Div., Allied Power, Inc.
Fig. 4. Schematic diagrams showing check-valve design and two stages in the operation of rotating axial pump.

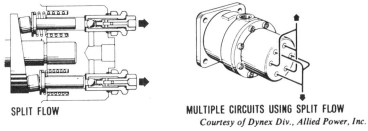

SPLIT FLOW   MULTIPLE CIRCUITS USING SPLIT FLOW
Courtesy of Dynex Div., Allied Power, Inc.

Fig. 5. Schematic diagrams showing how flow from pump in Fig. 4 can be split for use in multiple circuits.

rated for 8000 psi intermittent pressure or 6000 psi continuous pressure. It can be operated at speeds up to 2000 rpm continuously. This pump converts the rotating input motion to a reciprocating axial piston motion through a cam plate, one face of which is inclined to a plane at right angles with the center line of the shaft. This cam plate is keyed to and revolves with the input shaft and actuates ten pistons in succession. Figure 4 shows the check valve design and the action of the pump as the shaft revolves. In the left-hand view, the face of

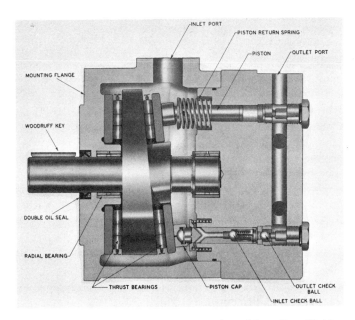

Courtesy of Dynex Div., Allied Power, Inc.

Fig. 6. Cross-section of fixed delivery pump with a capacity of 22 gpm at 6000 psi and 1800 rpm. It can be rotated in either direction.

the cam plate is shown as having a 7-degree angle. In the center view, as the cam plate assembly is rotated the thick section slides behind the upper piston cap which, in turn, pushes the piston toward the discharge port. Pressurized fluid operates the check valve arrangement and fluid is delivered to the system. The right-hand view shows that during the next half-cycle the thin section of the cam slides behind the upper piston cap and the return piston spring forces the piston toward the inlet side. This causes a vacuum in the tightly sealed piston cavity, and the inlet check ball operates to allow fluid to enter the piston chamber.

Figure 5 shows how the flow from this pump can be split. This allows the designer considerable flexibility. He can obtain a pump that will provide two different operating pressures, two different flows, two equal flows, or can obtain other variations that will help him in his circuit designs.

Figure 6 names the parts that make up a fixed-delivery pump while Fig. 7 shows the parts in a variable-volume pump. This design incorporates integral valving that permits incremental changes in volume. An eccentric cam rotating in response to the pump control operates poppet valves that bypass fluid from the discharge side back to the inlet side.

A 10,000 psi pump is shown in Fig. 8. This is a two-stage pump in which the first stage raises the pressure to about 600 psi and then the second stage takes over and boosts the pressure to 10,000 psi. The pump is driven by a high-speed electric motor rotating at 3450, 3600 or 12,000 rpm. The motor drives a small gear (A) of the gear pump in the direction shown by the arrow. Gear (A) drives the larger gear (B). The wobble plate (C) of the piston pump is driven by a coupling connected to the shaft of gear (B). (In this schematic diagram the gear pump has been rotated around to a vertical plane to show the flow of fluid more clearly. Actually, the axis of the large gear coincides with the axis of the wobble plate.) The gear ratio is approximately 2:1 or 6:1 depending upon the drive, developing a speed of about 2000 rpm to the wobble plate (C). As the wobble plate rotates it imparts reciprocating motion to the pistons (D) and (E). As shown, piston (E) is starting the intake stroke and piston (D) the pressure stroke.

Oil is taken into the gear pump at (F) and is then trapped between the gear teeth and the housing. It is carried around the periphery and discharged at the outlet (G). Pressure up to 600 psi can be developed by the gear pump.

The low-pressure oil is discharged to the system through spool (H)

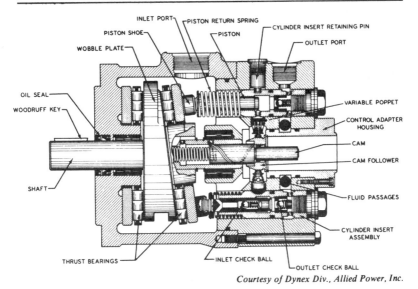

Courtesy of Dynex Div., Allied Power, Inc.

Fig. 7. Cross-section of variable displacement type of piston pump.

of the unloading valve and check valve (J). Back pressure developed by forcing the oil through the check valve provides pressure to supercharge the piston pump. Two functions are served by supercharging the intake (K). During the intake stroke oil is forced through the intake valve (L) to fill the pressure chamber (R). This

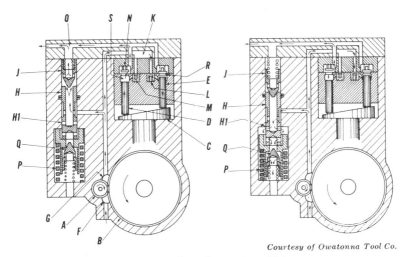

Courtesy of Owatonna Tool Co.

Fig. 8. Schematic cross-section of two-stage high-pressure pump.

Courtesy of Owatonna Tool Co.

Fig. 9. Gasoline engine driven hydraulic pump operates power unit with two cylinders for pulling utility poles.

allows the use of smaller spring-loaded intake valves and smaller oil passages. At the same time, pressure forces the piston out as the wobble plate recedes and assures contact at all times between the pistons and the surfaces of the wobble plate. This eliminates the need for springs or some other mechanical means to retract the pistons.

Piston (D) is starting the pressure stroke as the piston is forced into the cylinder barrel so that pressure is developed in the pressure chamber (S). Intake valve (M) closes and the discharge check valve (N) is forced open delivering oil from the piston pump to the system which joins the oil from the gear pump at (O).

When pressure builds up it closes valve (J), as shown in the right-hand view, and forces spool (H) downward to compress spring (P). Oil from gear pump (A) is now directed to the super-charge relief valve (Q) which is set at about 130 psi or just enough to insure that oil from the gear pump (A) is fed to the high-pressure pump without causing cavitation and with sufficient pressure to keep the pistons in contact with the cam plate. The excess oil from the gear pump not required to feed the piston pump is by-passed through relief valve (Q) to the reservoir at low pressure thus reducing the amount of heat initially produced and releasing additional power for the development of high pressure.

High pressure continues to be developed by the piston pump and is held at check valve (J). Such a pump will deliver about 50 cubic inches per minute at 10,000 psi.

Pumps are generally furnished as part of a power unit which includes oil reservoir, filter, relief valve and other components as discussed in Chapter 3. However we find that in the high-pressure units a portable type unit is often used such as those shown in Figs. 9 and 10. This unit is light enough to be transported manually to

Courtesy of Owatonna Tool Co.

Fig. 10. Power unit with pump driven by electric motor used in post-tensioning operation on huge I-beams.

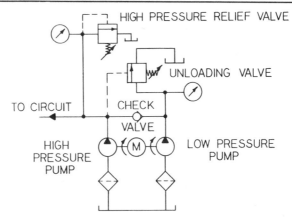

Fig. 11. Schematic diagram for a high-low pressure circuit.

the job site. In Fig. 9 the pump is driven by a gasoline engine while in Fig. 10 it is driven by an electric motor.

With a high-pressure pump available it is very easy to work up a good high-low circuit. A basic schematic for such a circuit is shown

Courtesy of Circle Seal Products Co.

Fig. 12. High-pressure shut-off valve.

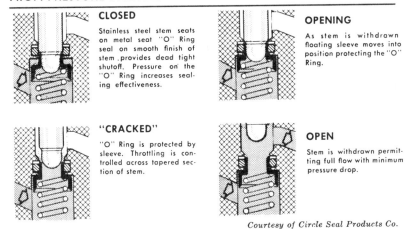

CLOSED

Stainless steel stem seats on metal seat "O" Ring seal on smooth finish of stem .provides dead tight shutoff. Pressure on the "O" Ring increases sealing effectiveness.

OPENING

As stem is withdrawn floating sleeve moves into position protecting the "O" Ring.

"CRACKED"

"O" Ring is protected by sleeve. Throttling is controlled across tapered section of stem.

OPEN

Stem is withdrawn permitting full flow with minimum pressure drop.

Courtesy of Circle Seal Products Co.

Fig. 13. Three steps in operation of high-pressure valve.

in Fig. 11. Here we make use of a large gear pump to deliver the fluid for the low-pressure portion of the operation and a very small piston pump for the final portion of the operation. Both are coupled to an electric motor of the double-end-shaft variety.

High-Pressure Controls

Control valves that operate in conjunction with high-pressure pumps must not only withstand the pressures developed but also have a minimum of internal leakage. In high-pressure systems we find valves that perform similar functions to those in the low-pressure systems (see Chapter 6) but are generally much more compact.

Some of the features in high-pressure valves are shown in the following illustrations. In the shutoff valve shown in Fig. 12 an "O" ring is used as the high-pressure seal. One of the problems in high-pressure valves is the probability of "wire drawing." This design overcomes that problem with a dead tight shutoff between the "O" ring and the stem. Figure 13 shows various steps in the operation of the valve from closed to wide open. Note how the floating sleeve protects the "O" ring when the stem is withdrawn.

Check valves used in high-pressure systems range from those with lapped metal-to-metal seats and poppets to those with a synthetic seal. Here again the leakage factor is very important. Check valves must close quickly and be able to withstand high shock loads. Another desirable feature of a check valve is that it should be compact yet pass sufficient fluid. Most high-pressure checks are of the "inline" type that fits easily into various piping arrangements. Figure 14

OPEN

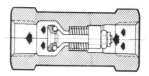

Flow passes smoothly over poppet
head with minimum turbulence.

CLOSING　　　　**CLOSED**

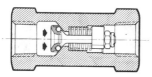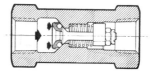

O-Ring automatically establishes line
of contact with spherical seat to cushion closing and insure perfect sealing.

O-Ring only seals. Full pressure is
carried by metal to metal seat. Increasing pressure increases sealing
efficiency.

Courtesy of Circle Seal Products Co.

Fig. 14. Three positions of a check-valve assembly that makes use of an "O"
ring type of seal.

shows three different positions of a check assembly that makes use
of an "O" ring for the seal.

Check valve bodies may be made of steel, aluminum or stainless
steel. Buna-N, Teflon and Neoprene are commonly used as "O"
ring compounds for seals, depending upon the service requirements.
Various pipe connections are provided in the valve bodies.

Relief valves used to protect the pump or some portion of the

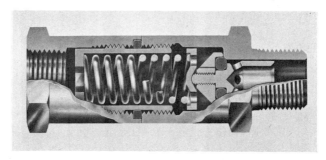

Courtesy of Circle Seal Products Co.

Fig. 15. Direct-acting relief valve.

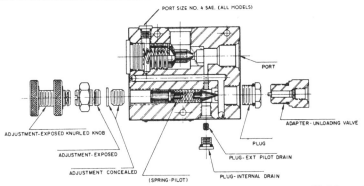

Courtesy of Dynex Div., Allied Power, Inc.

Fig. 16. Pilot-operated relief valve.

system are of numerous designs.   It is desirable that the spool or
plunger quickly reseats and that chatter or squeal is eliminated.   Two
types of valves are available—direct-operated and pilot-operated.
Figure 15 shows one of the direct-acting type and Fig. 16 shows one
of the pilot-operated type.   The basic configuration of Fig. 16 can
also be adapted for a sequence or unloading operation.

Four-way controls must be able to withstand high pressures without
distortion as tolerances must be very close in order to hold leakage
to an absolute minimum.   It is important on a spool type or slide

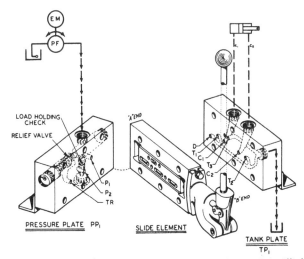

Courtesy of Dynex Div., Allied Power, Inc.

Fig. 17. High-pressure slide valve.

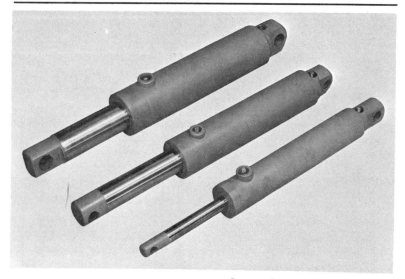

Courtesy of Dynex Div., Allied Power, Inc.

Fig. 18. High-pressure double-acting hydraulic cylinders.

type valve that the pressure be balanced.   Generally the high-pressure four-way valves are manually or mechanically actuated.

Figure 17 shows a slide type valve which is made in sections that are manufactured to very close tolerances and have a high degree of finish.   The relief valve and load-holding check is built into the pressure plate.   This valve may also be built up into a bank of six four-way segments.

Four-way valves for high pressure are also manufactured with poppets, sliding discs, or shear seals.

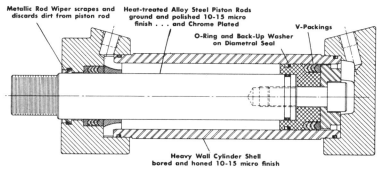

Courtesy of Owatonna Tool Co.

Fig. 19. Tie-rod type high-pressure cylinder.

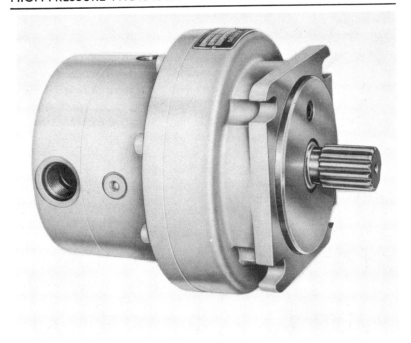

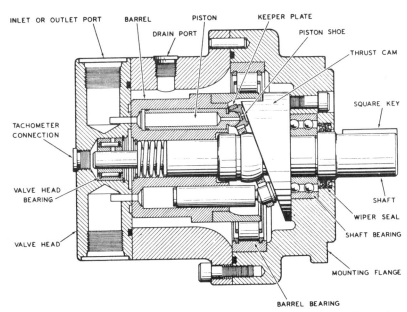

INLET OR OUTLET PORT — BARREL — PISTON — KEEPER PLATE

DRAIN PORT — PISTON SHOE

THRUST CAM

SQUARE KEY

TACHOMETER
CONNECTION

VALVE HEAD
BEARING

SHAFT

WIPER SEAL

SHAFT BEARING

VALVE HEAD

MOUNTING FLANGE

BARREL BEARING

Courtesy of Dynex Div., Allied Power, Inc.

Fig. 20. A 50 hp hydraulic motor.

High-Pressure Cylinders and Motors

To transform the power created by the high-pressure hydraulic pumps into linear or rotary motion, use is made of cylinders and motors. High-pressure cylinders are usually of a small bore in order to conserve space and weight. In the application shown in Fig. 9 the cylinders are quite compact yet they impart a very high force. Thus, although such cylinders are compact, they must be of rugged design.

Figure 18 shows three high-pressure hydraulic double-acting cylinders. Note the relatively large rod diameters in comparison to the respective cylinder diameters. Also note the size of the clevis holes in the cylinders.

Figure 19 shows a cross section of a high-pressure cylinder that uses tie rod construction. Note the heavy wall cylinder tube. The internal surface of this tube has a 10 to 15 micro-inch finish. The piston rod has a seal at the piston. V-packings are used to effect a

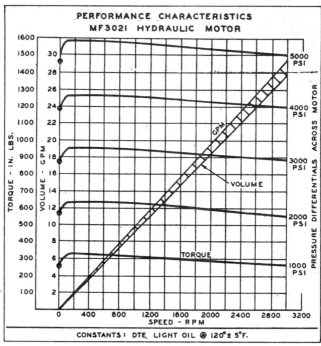

Courtesy of Dynex Div., Allied Power, Inc.

Fig. 21. Performance curves for hydraulic motor shown in Fig. 20.

seal between the piston and cylinder tube and also between the cover and piston rod. "O" rings with back-up washers are employed to make a seal between the tube and covers.

Hydraulic motors for high pressures are of the piston type and operate somewhat in the reverse manner of a piston type pump. Oil under high pressure causes a number of pistons axially located around the main shaft to impinge on a thrust cam. This results in rotary motion being imparted to the output shaft. Figure 20 shows a motor that will produce 50 hp yet it is only 7⅜ inches in diameter and 7⅝ inches long, less the shaft extension. Compare this with the size of an electric motor of 50 hp. A large radial bearing absorbs the piston shoe reaction forces, eliminating any force coupling and provides balanced loading. Performance curves are shown in Fig. 21.

With the advent of high-pressure hydraulic components, many are being found in such applications as on earth movers, hydraulic presses, metal forming machines, mining machines, food processing equipment, marine equipment, construction equipment, test equipment, power line maintenance equipment, snow plows, portable cranes, lift trucks and many others. The compactness of the high pressure components is especially desirable when applied to mobile or portable equipment.

CHAPTER 22

Rotary Actuators

A rotary actuator is a device that produces output torque over a limited range of rotation. In fluid power systems this torque varies directly with the fluid pressure applied at the inlet port. The fluid which operates the actuator may be in the form of air, oil, or water. The actuator is often part of a system in which other motions are performed.

Generally the rotation of the shaft of an actuator is less than 360 degrees. Some have a rotation of 90 degrees, others, 180 degrees, and still others, 270 degrees. There are, however, some special types of actuators that may have a shaft rotation of several revolutions. These are somewhat rare.

The actuator is a compact device which can be used in place of exposed mechanisms such as a rack and pinion operated by a cylinder or a long lever arm operated by a cylinder. Actuators are suitable for use in adverse atmospheric conditions. They can be used for lifting, tilting, clamping, opening, closing, metering, mixing, turning, swinging, counterbalancing, bending and many other operations. Specific applications are on roll-over devices, conveyors, valve operators, printing presses, backhoes, rock drills, die clamps, etc.

There are three basic types of actuators—the vane type, the cylinder type and the type with helices machined in the piston and drive shaft which mesh with each other.

Vane Type Actuators

Figure 1 shows a cross-section of a foot-mounted single-vane, roller-bearing type actuator. This design is suitable for 3000 psi operating pressure and provides a total angular travel of 280 degrees. It has a free travel of 270 degrees and a snubbed travel of 5 degrees on either end.

There are also double-vane type actuators but the angle of shaft rotation is limited to 100 degrees. The torque output is approximately twice that of a comparable single-vane actuator since the output torque of a vane type actuator is proportional to the size and number of vanes multiplied by the input pressure.

The "breakout pressures" required to overcome friction caused by seals and other parts are quite low and range somewhere around 25 psi. Teflon is often used for dynamic seals. Actuators are designed to operate satisfactorily over quite a wide temperature range. Some

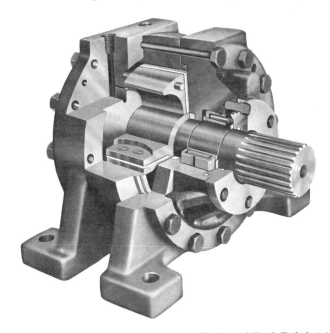

Courtesy of Houdaille Industries, Inc.

Fig. 1. Foot-mounted, single-vane, roller-bearing type actuator for 3000 psi pressure.

are used in applications where temperatures could conceivably range from −40 degrees F to 275 degrees F.

In planning for the installation of vane-type actuators, end loading as well as side loading of the output shaft of the actuator should be eliminated. In instances where side loading cannot be eliminated, outboard supports should be incorporated into the design.

High-Pressure Cylinder Type Actuators

There are many designs of cylinder type actuators available. In these the degree of rotation of the output shaft is dependent upon

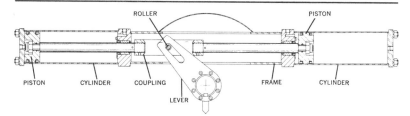

Courtesy of Ledeen Inc.

Fig. 2. Tandem actuator.

the stroke of the cylinder and the device which the cylinder operates. The device may be in the form of a rack and pinion, a chain drive, a lever arm or some other mechanism.

Figure 2 shows the construction of a tandem actuator which consists of two fluid-powered cylinders, one mounted on each end of a common frame. The piston rods of the cylinders are coupled together so that the pistons function together. A hardened roller in the coupling is free to articulate in the lever slot and linear motion is converted into rotary motion. The lever is bolted to the lever hub. Such actuators are used to position the plug in large plug valves.

Figure 3 shows the construction of a quad actuator which consists of two tandem units mounted symmetrically on a common bracket and positioned on each side of the stem. The lever with a double slot extends through both couplings so that the actuators when piped together deliver their combined and balanced torque output to the stem.

Tandem actuators and quad actuators may be constructed for either one- or two-fluid systems. In the case of the tandem actuator for a

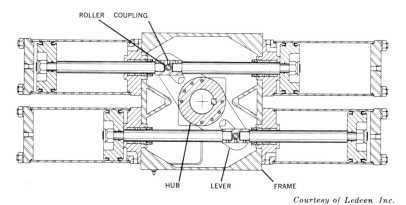

Courtesy of Ledeen Inc.

Fig. 3. Quad actuators.

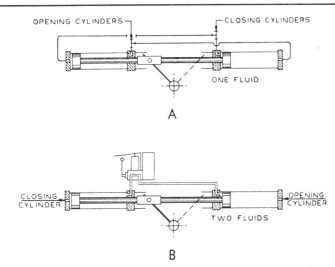

A

B

Fig. 4. A. Tandem actuator using one fluid.   B. Tandem actuator using two fluids.

one-fluid system the power medium or fluid is applied in two active areas as shown in Fig. 4A.   In the two-fluid system for tandem actuators the power medium or fluid is applied to one active area as shown in Fig. 4B.   The other area is used for a separate hydraulic emergency manual operation and speed control.   Figure 5 shows a two-fluid tandem actuator complete with controls and hand pump.

In one-fluid systems quad actuators have the fluid applied to four active areas in double-acting units as shown in Fig. 6A.   Figure 6B

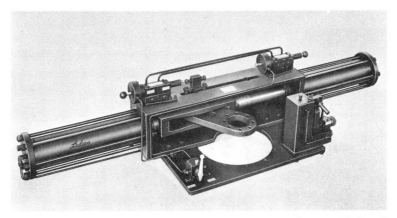

Fig. 5. Tandem actuator with hand pump used for emergency shut-down service.

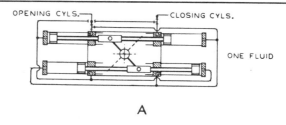

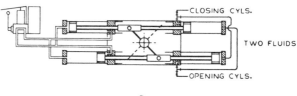

Courtesy of Ledeen Inc.

Fig. 6. A. Quad actuator using one fluid. B. Quad actuator using two fluids.

shows the two-fluid system where the power medium is connected to two active areas. The other areas are used for hydraulic manual operation and speed control.

The use of an independent hydraulic system (two-fluid system) permits effective hydraulic speed control, permits "skip" type circuits, and actuation in case of power failure. Figure 7 shows a quad actuator for a two-fluid system. Note the dual-pressure hand pump which provides for high pressure and low volume at breakaway and lower pressure with higher volume during the balance of the rotary motion.

Another cylinder type actuator is shown in a cutaway view in

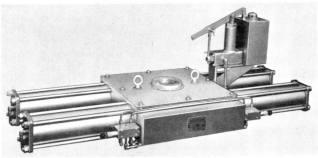

Courtesy of Ledeen Inc.

Fig. 7. Quad actuator with hand pump for two-fluid system.

Fig. 8. This is a Rota-Cyl® rotary actuator and is suitable for air or hydraulic oil service. This actuator uses two cylinders in parallel. The larger bore cylinder is the power cylinder and the smaller cylinder is the chain return or seal cylinder. These cylinders, instead of being fastened in tandem on a rigid piston rod, are connected in series by an endless chain which passes over the sprocket as shown at the left of Fig. 8 and over an automatically tensioned idler at the right. Pressure is applied to either end with the opposite end vented to atmosphere or drain. The large piston moves away from the pressure due to the differential area of the two pistons. Movement of the large piston pulls the chain causing the sprocket and output shaft to rotate. Thus the linear motion is transferred into rotary motion.

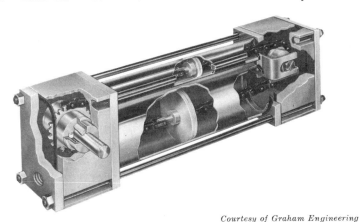

Courtesy of Graham Engineering

Fig. 8. Cutaway view showing internal workings of a rotary actuator.

The Rota-Cyl operation is virtually free of backlash. An automatic tensioning device on the idler causes the idler to move away from the output shaft so that constant chain tension is maintained. The slightest motion of the actuation fluid is detected and transmitted to the output shaft. In this design no precision parts are exposed to the atmosphere.

Figures 9A and 9B show a Rota-Cyl actuator which is designed for operating pressures up to 3000 psi. Hydraulic pressure is applied at inlet port (A). Through a pressure equalizing tube (B) and the operation of shuttle valve (C), pressure acts equally on both sides of the floating piston (D), and on only one side of the power piston D'. To reverse the action, pressure is applied at the alternate port A', the shuttle valve reverses and piston (D) becomes the power piston. Two hundred degrees of rotation can be obtained.

Cylinder-operated actuators are also adaptable for use with gear-and-rack mechanisms and cam mechanisms.

Cylinder type actuators can be cushioned by employing a cushion mechanism similar to that used on non-rotating cylinders. Stroke adjustment mechanisms can also be applied.

Helix Type Actuator

Figure 10 shows a cutaway view of an actuator that probably

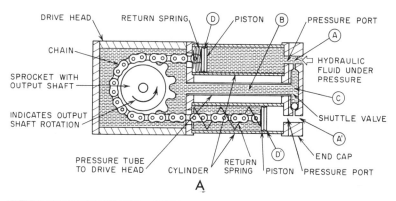

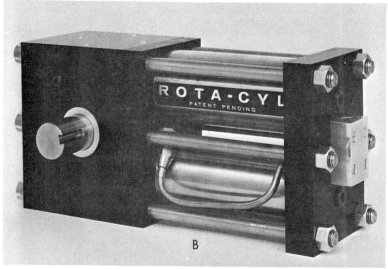

Fig. 9. A. Rotary actuator with two pistons connected by chain to actuate output shaft. B. Rotary actuator capable of 10,000 pound-inch torque and 200-degree rotation.

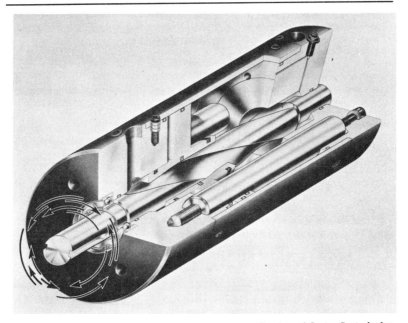

Courtesy of Carter Controls, Inc.

Fig. 10. Compact actuator in which helically grooved piston rod is rotated by longitudinal motion of piston.

could be classed as a cylinder type actuator since it has a piston, a piston rod and a tube. However, it has a unique feature in that the piston rod does not move in and out. The piston rod is designed with an O.D. helix and the piston has an I.D. helix. Two guide rods which go through the piston and run the length of the actuator keep the piston from rotating. As the force of the fluid moves the piston, the piston rod is caused to rotate.

Cushioning is obtained at both ends of the piston stroke by the cushion device on both sides of the piston. This device enters the cushion recess and is sealed by a synthetic cushion packing. The speed of the cushioning is controlled by a cushion needle which meters the trapped fluid. A ball check allows for a fast start even though the cushion device is engaged. The self-locking helix angle of the piston and the rod eliminates the possibility of external torque causing any rotary movement of the piston rod. This is true even if there is a complete power failure. This feature is beneficial on roll-over devices where there is a great amount of external torque applied when the load passes the center position.

CHAPTER 23

Air Motors

The characteristics of air motors are such as to make them more suitable for certain industrial applications than electric motors and, in some instances, than hydraulic motors as well. Some of the advantages offered by air motors are:

1. *Explosion-proof.* The nature of the medium which operates the motor and the construction of the motor, eliminates sparks. This also makes the air motor ideal in adverse atmospheres. Workmen may use them in wet locations without any possibility of electric shock.

2. *No overheating.* They can be overloaded or stalled without burning out. There is no heat build-up when they are stalled.

3. *Reversible.* Many industrial applications require a rotary shaft that can be revolved in either direction.

4. *Variable speed.* By the use of simple valving the shaft of the air motor can be rotated at speeds ranging from a few hundred rpm to several thousand rpm.

5. *Light and compact.* Normally, air motors are considerably lighter and smaller than electric motors of the same horsepower rating. They develop a high torque per pound of weight.

6. *Cool running.* Due to the cooling action of the expanding exhaust air they are suitable for use in locations where the ambient temperature is relatively high.

7. *Low price.* This means of rotary motion is less expensive than other explosion-proof types of motors.

8. *Relatively clean.* When using hydraulic motors a shaft leak becomes quite messy and can result in damage to material which is being processed. This is not apt to happen with air motors unless excessive lubrication is used.

9. *Low operating pressures.* The pressures are lower than are used in hydraulic systems and hence the piping is less expensive.

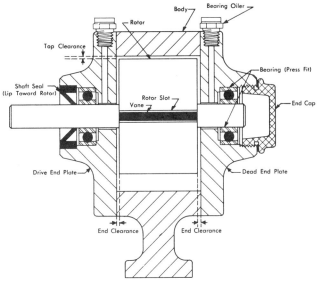

Fig. 1. Cross-sectional view of rotary vane type air motor.

Air motors are of various designs but the most common are the vane, axial piston, and radial piston types. Other types are turbine, lobed rotor, and v-motor design. Many speed ranges are also offered, extending from as low as 300 rpm up to as high as 25,000 rpm.

Vane Type Motors

Figure 1 shows a cross-section of a vane type air motor. Such motors may be designed to give shaft rotation in one direction only or in either direction. In this illustration are shown the shaft seal, bearing arrangement, rotor with its clearance in the motor body and a vane. Figure 2 shows an end view of a reversible motor with the vane arrangement and the porting. Note the use of push-pins and leaf spring in the vane slots. Figure 3 shows a cross-section of a non-reversing vane type air motor. The air enters the left-hand port and is released through the right-hand port. As each rotor slot passes across the kidney-shaped port the air follows the path shown in the drawing.

Vane type air motors as shown in Figs. 1, 2, and 3 have longitudinal vanes and these vanes fit into the radial slots in the rotor. Depending upon the design of the vane type motor, there may be anywhere from three to ten vanes. Generally the number of vanes

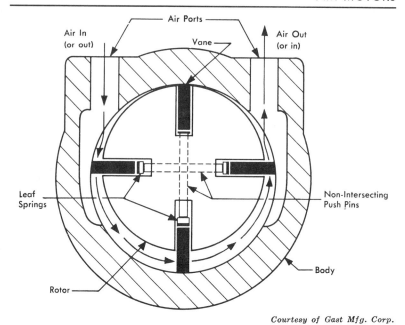

Fig. 2. Reversible rotary vane type air motor.

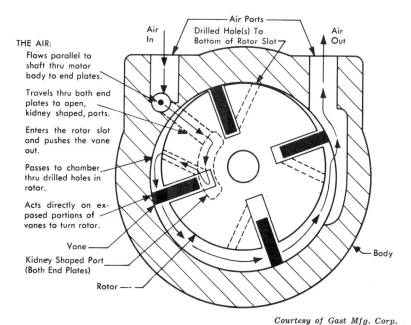

Fig. 3. Non-reversing rotary vane type air motor.

in an air motor design is not critical if uniformity of torque at low speed is not critical. The starting torque is increased as the number of vanes is increased. Also the starting torque may be increased by increasing the line pressure to the motor, by increasing the exposed vane area, and by increasing the diameter of the bore in which the vanes ride. Vane motors are available in sizes up to 25 hp.

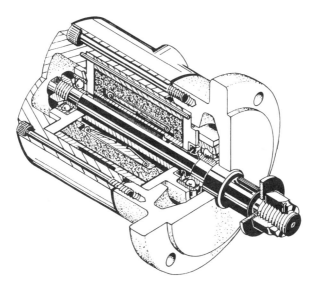

Courtesy of Gardner-Denver Co.

Fig. 4. Rotary vane type of air motor with flange mounting.

Figures 4 and 5 show sectional views of rotary vane type motors. Note the difference in shaft ends and in the internal bearing construction.

One of the graphs in Fig. 6 shows characteristic curves for horsepower related to speed and torque related to speed; the other shows curves relating horsepower to air consumption.

Piston Type Air Motors

Piston type motors may be of the axial or radial piston design. Figure 7 shows a cross-section of an axial piston type motor. The operation of this type of air motor is somewhat similar to the axial piston type of hydraulic motor. As the pistons reciprocate in sequence they actuate a wobble plate as shown in Fig. 7 and this in turn imparts a rotary motion to the output shaft through a gear train. Piston type motors may have four, five or six cylinders. The power

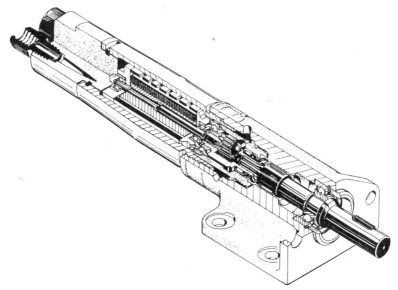

Courtesy of Gardner-Denver Co.

Fig. 5. Rotary vane type of air motor showing gearing and bearing
arrangements.

developed by these motors is dependent upon the inlet line pressure,
the number of pistons, the area of the pistons, the stroke of the pistons, and the speed. Axial piston motors are generally manufactured
in sizes up to 3½ hp while the radial piston motors are made in sizes

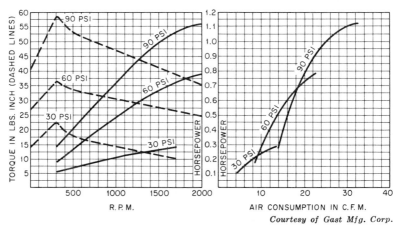

Courtesy of Gast Mfg. Corp.

Fig. 6. Typical horsepower-speed, horsepower-torque and horsepower-air
consumption curves for rotary vane air motor.

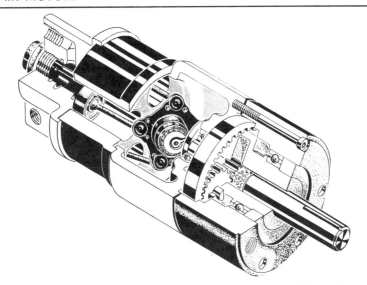

Courtesy of Gardner-Denver Co.

Fig. 7. Cross-section of axial piston type air motor.

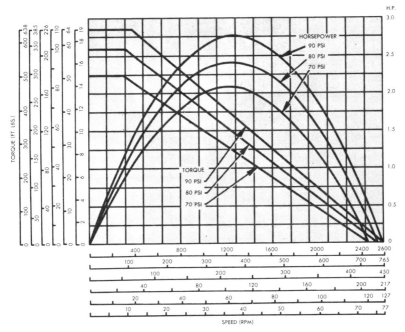

Courtesy of Gardner-Denver Co.

Fig. 8. Typical torque-speed and horsepower-speed curves for axial piston
type air motor.

up to 25 hp.   Radial-piston motors are low speed mechanisms which have a top speed somewhere around 3000 rpm.   They are ideal for low speed operation where high starting torques are desired.

Piston type air motors will reach full speed in milli-seconds.   Vane type motors are generally not quite as fast but will reach full speed in one-half revolution.

Figure 8 shows typical axial piston type air motor torque-speed and horsepower-speed curves for three different air pressures.   The torque and speed figures closest to the graphs are used together and each successive speed line is used in conjunction with the successive torque line.

Figure 9 illustrates the sequence of power impulses in the five-cylinder piston design air motor.   The five-cylinder design provides

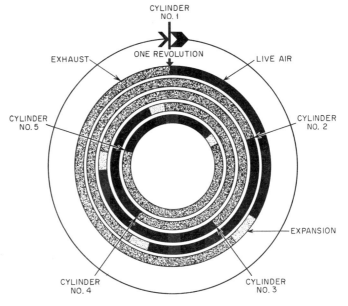

Courtesy of Gardner-Denver Co.

Fig. 9. Diagrammatic representation of power impulses in five-cylinder piston design air motor.

even torque at any given operating speed due to the overlap of the five power impulses occurring in each revolution of the motor.   At least two pistons are on the power stroke at all times.   The smooth overlapping power flow and accurate balancing make these motors vibrationless at all speeds.   The smooth operation is especially noticeable at low speeds when the flywheel action is negligible.

The direction of rotation and the speed of air motors may be controlled by various valves such as hand, foot, pilot-operated, solenoid or pendant. Whatever the type, the operating valve should have full-flow air passages to utilize the full power of the motor.

The speed of the air motor shaft is controlled by the volume and pressure of air admitted into the motor and may be regulated from

Courtesy of Gast Mfg. Corp.

Fig. 10. Foot mounted type of rotary vane motor. A muffler is shown on outlet port.

minimum to maximum without harming the motor. Reversible type motors may be rapidly reversed without damage.

In most applications air line pressures should not exceed 100 psig. Generally shop pressures are not in excess of 90 psig.

Air motors are manufactured with several types of mountings; among them are flange, foot, and NEMA. Figure 10 shows one type

Courtesy of Gast Mfg. Corp.

Fig. 11. Rotary vane type air motor with **NEMA** type flange mounting.

of foot mounting and Fig. 11 shows a NEMA type flange mounting.

Applications for air motors are many. They may be used in conjunction with hydraulic power units, bench grinders, conveyor belts, agitators and mixers, feeding devices, hoists, liquid pumps, machine feeding devices, pipe threaders, remote control operations, tool drives, vibrators, and many others.

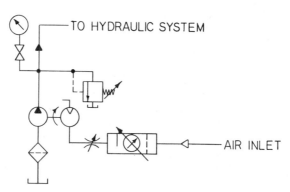

Fig. 12. Schematic diagram for hydraulic power unit driven by air motor.

Figure 12 shows an air drive arrangement for a hydraulic power unit. Here the air motor can be regulated to give various shaft speeds as when driving a hydraulic pump of a fixed displacement type to simulate the action of a variable displacement type. This setup is also advantageous in explosive atmospheres as no electrical equipment would be required. The unit is quite compact and can be made portable. Only one air connection is required. By using a length of hose and a quick disconnect coupling the unit can be moved around and "plugged in" to many of the outlets in a shop.

Courtesy of Gast Mfg. Corp.

Fig. 13. Air motor used for rotating a lapping and grinding table.

Figure 13 shows an application of a rotary vane type air motor used for driving a rotary lapping and grinding table. The workpieces are clamped to the table and are rotated under the grinding wheel. A lapping plate may be clamped to the table for lapping parts requiring a fine micro-finish. By using an air motor, speeds may be varied between 40 and 100 rpm.

Figure 14 shows the use of an air motor on an automated screw-

Courtesy of Gast Mfg. Corp.

Fig. 14. Air motor used for powering screwdriver.

Courtesy of Gast Mfg. Corp.

Fig. 15. Air motor used as drive unit on conveyor belt.

driver. Air is also used to drive the hopper agitator and the rotary brush which aligns screws in the feed tube for delivery to driver. Figure 15 illustrates the use of an air motor as a drive for a conveyor belt. The conveyor belt carries fragile material with a varying drag, depending upon load. As the drag increases, the speed of the air motor driving the conveyor belt is automatically reduced. This action maintains constant tension on material in process. Should a jam occur, the motor would stall but no damage would occur to the material in process. Also there would be no damage to the air motor.

Courtesy of Gast Mfg. Corp.

Fig. 16. Air motors used as power devices on printing press.

Figure 16 illustrates the use of air motors in the printing industry. This is a silk screen printer for steel drums and it uses two air motors. A single rotation motor is used to drive the drum rotating mechanisms and a reversible motor drives the printing frame to coincide with drum rotation thus maintaining close register. Even though these motors are used near volatile paint and printing ink fumes, they are safe since they are explosion-proof.

CHAPTER 24

Pneumatic Logic Controls

Webster defines logic as the science that deals with the canons and criteria of validity in thought and demonstration; the science of the formal principles of reasoning. Logic dates back to Aristotle, the Greek philosopher, who was the first philosopher to examine in detail the form of reasoning called syllogism or deductive logic. This is a logical analysis of a formal argument which consists of the major premise, the minor premise, and the conclusion.

Although Aristotle lived from 384–322 B.C., very little was done with logics in regard to science and mathematics until George Boole, an Irish logician and mathematician who lived from 1815 to 1864, worked out a mathematical system in 1847 to solve problems in logic, probability, and engineering. It is known as Boolean Algebra, an "algebra of logic." This system formulated logical statements symbolically so that they could be written and proven in a similar manner as used in ordinary algebra.

Relationships between sets which may be of objects or groups of ideas are the basis of Boolean Algebra. This is the algebra of propositions where only two possibilities are allowed: true or false.

Boolean Algebra had been used very little until recent years when it was first applied to electrical circuits and now to pneumatic circuits.

To a point, logic has been employed since the conception of fluid power, but it was not called such. Since the art of fluid power circuits is becoming more and more sophisticated, the term "logic" is becoming more prevalent. All logic is performed in binary systems. Binary is a number system of radix of base two which means two values: ON-OFF, Passing-Non-Passing, etc.

The only two numbers used in Boolean Algebra are 0 and 1. A · (dot) between letters means AND; a + (plus sign) between the letters means OR. The following are theorems of Boolean Algebra:

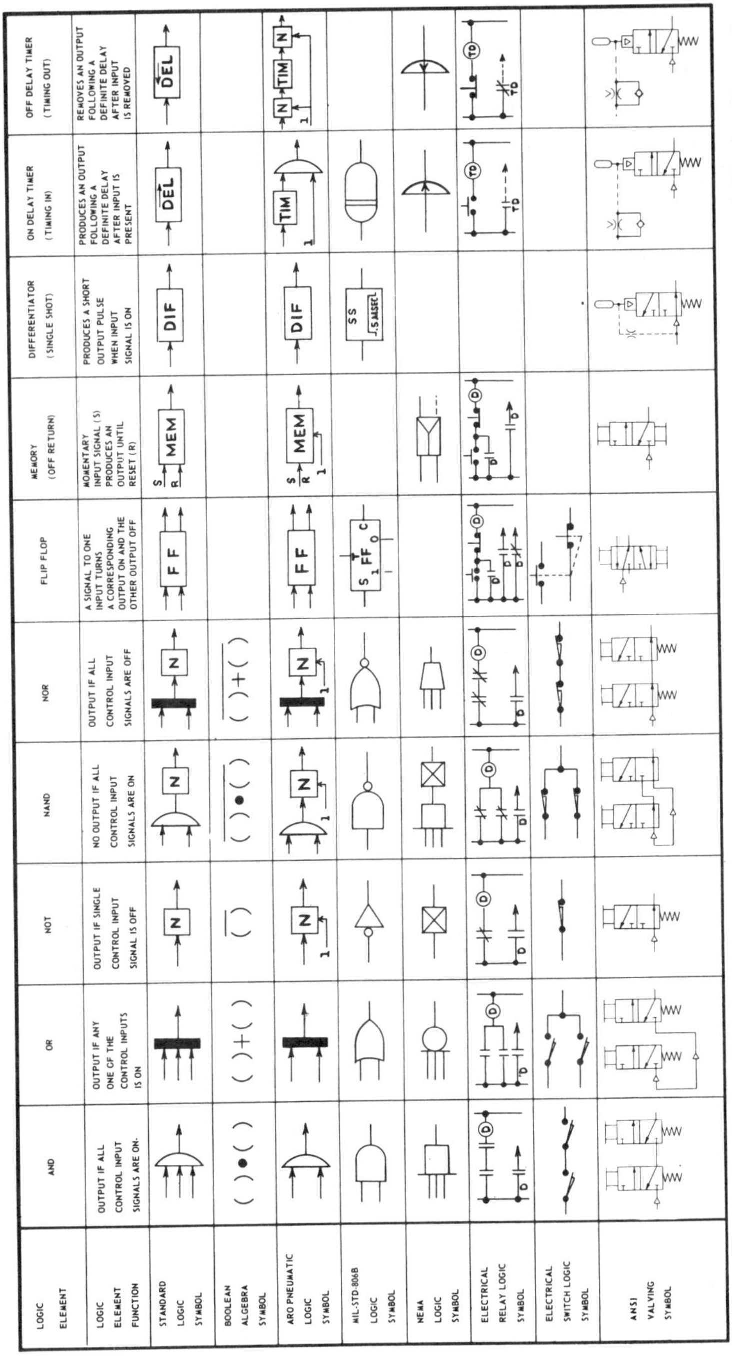

LOGIC ELEMENT	AND	OR	NOT	NAND	NOR	FLIP FLOP	MEMORY (OFF RETURN)	DIFFERENTIATOR (SINGLE SHOT)	ON DELAY TIMER (TIMING IN)	OFF DELAY TIMER (TIMING OUT)
LOGIC ELEMENT FUNCTION	OUTPUT IF ALL CONTROL INPUT SIGNALS ARE ON	OUTPUT IF ANY ONE OF THE CONTROL INPUTS IS ON	OUTPUT IF SINGLE CONTROL INPUT SIGNAL IS OFF	NO OUTPUT IF ALL CONTROL INPUT SIGNALS ARE ON	OUTPUT IF ALL CONTROL INPUT SIGNALS ARE OFF	A SIGNAL TO ONE INPUT TURNS A CORRESPONDING OUTPUT ON AND THE OTHER OUTPUT OFF	MOMENTARY INPUT SIGNAL (S) PRODUCES AN OUTPUT UNTIL RESET (R)	PRODUCES A SHORT OUTPUT PULSE WHEN INPUT SIGNAL IS ON	PRODUCES AN OUTPUT FOLLOWING A DEFINITE DELAY AFTER INPUT IS PRESENT	REMOVES AN OUTPUT FOLLOWING A DEFINITE DELAY AFTER INPUT IS REMOVED
STANDARD LOGIC SYMBOL			N	N	N	F F	S R MEM	DIF	DEL	DEL
BOOLEAN ALGEBRA SYMBOL	( )•( )	( )+( )	(‾)	( )•( )‾	( )+( )‾					
ARO PNEUMATIC LOGIC SYMBOL			N	N	N	F F	S R MEM	DIF	TIM	N TIM N
MIL-STD-806B LOGIC SYMBOL						S FF C		S S ASSET		
NEMA LOGIC SYMBOL										
ELECTRICAL RELAY LOGIC SYMBOL										
ELECTRICAL SWITCH LOGIC SYMBOL										
ANSI VALVING SYMBOL										

Fig. 1. Logic symbol cross reference chart.

Courtesy of The Aro Corporation

Sum Form (OR)	Product Form (AND)
$A + A = A$	$A \cdot A = A$
$A + 0 = A$	$A \cdot 0 = 0$
$A + 1 = 1$	$A \cdot 1 = 1$
$A + \overline{A} = 1$	$A \cdot \overline{A} = 0$
$A + B = B + A$	$A \cdot B = B \cdot A$
$A + A \cdot B = A$	$A \cdot (A + B) = A$
$A \cdot \overline{B} + B = A + B$	$(A + \overline{B}) \cdot B = A \cdot B$
$A + B + C = (A + B) + C$	$A \cdot B \cdot C = (A \cdot B) \cdot C$
$A \cdot B + A \cdot B = A$	$(A + B) \cdot (A + \overline{B}) = A$
$A \cdot B + A \cdot C = A \cdot (B + C)$	$(A + B) \cdot (A + C) = A + (B \cdot C)$
$(\overline{A + B + C}) = \overline{A} \cdot \overline{B} \cdot \overline{C}*$	$\overline{A \cdot B \cdot C} = \overline{A} + \overline{B} + \overline{C}*$

    * Form of DeMorgan's theorems.

To illustrate the relationship between some of the terms used in pneumatic logical systems, see Fig. 1.

Pneumatic logic is a moving parts system, whereas fluidics have no moving parts. Fluidics will be discussed in Chapter 25.

Actually, pneumatic logic controls in the strictest sense are small valves.

The logic element OR has two or. more Inputs and one Output. The Output is ON if any one or more of the Inputs are ON. The Output is OFF only if all Inputs are OFF. The Inputs may be actuated by any of the normal means of actuation such as mechanical, manual, pilot, etc. While Fig. 1 shows two three-way, two position, normally closed valves to perform the function, Fig. 2 employs two or more (three shown) two-way, two position normally closed valves. Note that these valves are hooked up in parallel with the supply line connected to each inlet, and the output line is connected to each outlet. Figure 3 shows a cross section of an OR element.

Logic element AND has two or more Inputs and one Output. The Output is ON only if all Inputs are ON, and the Output is OFF if one

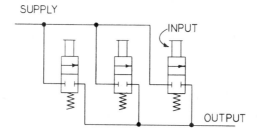

Fig. 2. Logic element OR.

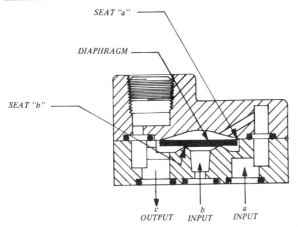

Courtesy of The Aro Corporation

Fig. 3. Cross section of OR element.

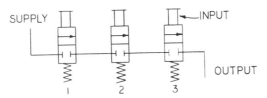

Fig. 4. Logic element AND.

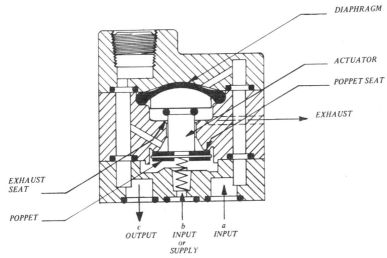

Courtesy of The Aro Corporation

Fig. 5. Cross section of AND element.

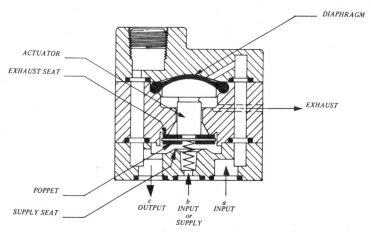

Courtesy of The Aro Corporation

Fig. 6. Cross section of NOT element.

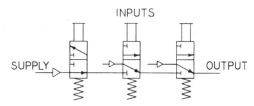

Fig. 7. Logic element NAND.

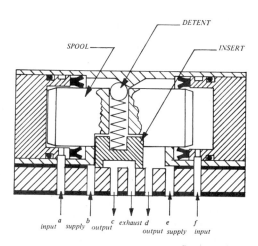

Courtesy of The Aro Corporation

Fig. 8. Cross section of Flip-Flop element.

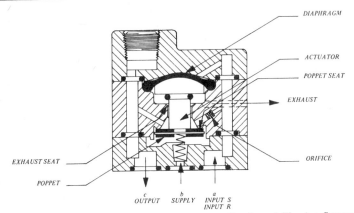

Courtesy of The Aro Corporation

Fig. 9. Cross section of a MEMORY element.

or more Inputs are OFF. Note that in Fig. 1 two three-way, two position normally closed valves connected in series perform this function.
This function can also be performed by two or more two-way, two
position normally closed valves as shown in Fig. 4. These control
valves are also connected in series with the supply connected to the
inlet of valve 1, the outlet of valve 1 connected to inlet of valve 2 and
the outlet of valve 2 connected to inlet of valve 3. The outlet of valve
3 will be connected to the device to be actuated. Figure 5 shows a
cross section of an AND element.

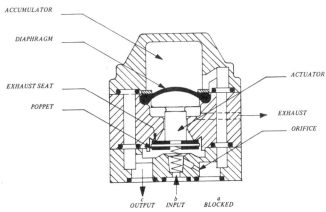

Courtesy of The Aro Corporation

Fig. 10. Cross section of a Differentiator.

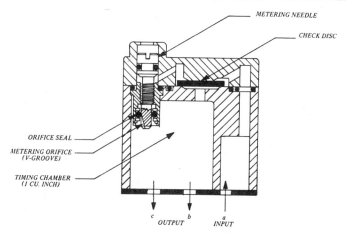

METERING NEEDLE

CHECK DISC

ORIFICE SEAL

METERING ORIFICE
(V-GROOVE)

TIMING CHAMBER
(1 CU. INCH)

c   b   a
OUTPUT   INPUT

Courtesy of The Aro Corporation

Fig. 11. Cross section of a timing element.

The logic element NOT has one Input and one Output. The Output is ON if the Input is OFF, and the Output is OFF if the Input is ON. This is accomplished by the use of a three-way, two position normally open (NO) control. Input *a* can be pneumatic or non-

Courtesy of The Aro Corporation

Fig. 12. Circuit board assembly.

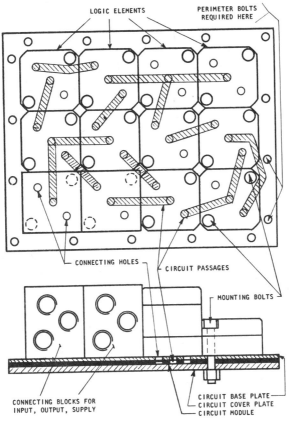

LOGIC ELEMENTS   PERIMETER BOLTS
                        REQUIRED HERE

CONNECTING HOLES   CIRCUIT PASSAGES

MOUNTING BOLTS

CONNECTING BLOCKS FOR   CIRCUIT BASE PLATE
INPUT, OUTPUT, SUPPLY   CIRCUIT COVER PLATE
                               CIRCUIT MODULE

Courtesy of The Aro Corporation

Fig. 13. Pneumatic logic circuit board with logic elements.

pneumatic depending upon the requirements. A cross section of a NOT element is shown in Fig. 6.

The logic element NAND which is made up of NOT-AND circuits has two or more Inputs and one Output. The Output is OFF only if all Inputs are ON. The Output is ON if one or more Inputs are OFF. To accomplish this, two three-way valves may be used as shown in Fig. 1, or three three-way valves as shown in Fig. 7. Figure 7 employs two normally closed or non-passing three-way valves and one normally open or passing three-way control valve.

The logic element NOR consists of NOT-OR circuits. NOR has two or more Inputs and one Output. The Output is OFF if any one or more Inputs are ON. The Output is ON only if all Inputs are OFF.

Courtesy of The Aro Corporation

Fig. 14. Cutaway of a circuit board assembly.

As shown in Fig. 1, this can be accomplished by two three-way normally open or passing control valves connected in series.

The logic element Flip-Flop has two Inputs and two Outputs. One Output is always ON, and the other is always OFF. When a signal is applied to one of the Inputs, it turns the corresponding Output ON and turns OFF the other Output. When the corresponding Output is already ON, the Flip-Flop will remain in that state. The Input signal must not be applied to both Inputs at the same time. The four-way valve to accomplish this function is diagrammed in Fig. 1. An element that is employed to accomplish the Flip-Flop function is shown in cross section in Fig. 8. The Flip-Flop is a memory device and is often used in digital control. For ordinary sequencing circuits a single Output memory device is generally employed.

The logic element MEMORY may be of various configurations. The one shown in Fig. 1 has two Inputs and one Output. The Output switches to and remains ON if a signal is applied to Input S(Set). The Output switches to and remains OFF, if the signal is applied to R(Rest). Nothing will happen if the Output already corresponds to the Input signal. A cross section of a MEMORY element is shown in Fig. 9.

The Differentiator, ON Delay Timer and OFF Delay Timer elements are quite similar in operation as shown by the ANSI valve dia-

Courtesy of The Aro Corporation

Fig. 15. APLC control panel replacement for the electrical panel shown in Fig. 16.

grams in Fig. 1. The cross section of a Differentiator is shown in Fig. 10, whereas a timing element is shown in Fig. 11.

To obtain compact logic systems, the elements are mounted on a logic circuit board assembly, see Fig. 12. These circuit board assemblies are similar in principle to the sandwich-type manifolds that are so popular in machine tool hydraulic circuits. In logics, however, one is dealing with much smaller flows and lower pressures. Pneumatic logic circuit boards eliminate piping as all of the flow passages are cut in the center section or circuit module which is sandwiched in between the circuit base plate and circuit cover plate as shown in Figs. 13 and 14. The passages in the circuit module are sealed by the mounting bolts which hold the elements to the circuit board.

To compare the physical size of logic elements and a circuit board assembly to an electrical panel that performs the same function, see

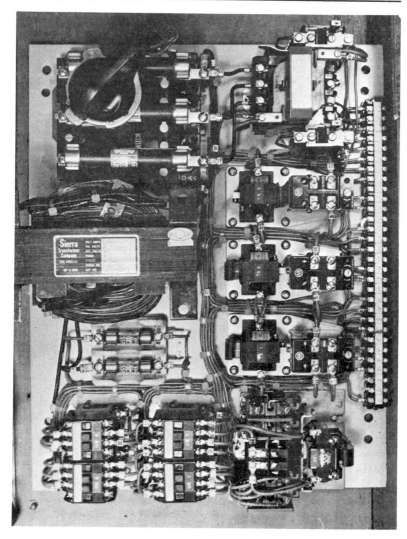

Courtesy of The Aro Corporation

Fig. 16. Electrical control panel.

Figs. 15 and 16.   The logic control circuit requires only about one-fourth of the space of that required by the electrical panel and is a replacement for it.

Figure 17 shows a logic control panel used on an air gaging machine. Note the clean appearance of the control box when the door is closed

Courtesy of The Aro Corporation

Fig. 17. Logic control panel.

as shown in Fig. 18. Various types of indicators and switches are included in the fluid logic hardware so as to give the control system designer a considerable latitude in his selection.

Figures 19 and 20 show a logic control panel which is being used on a mill which is completely automated. Functions include collet engagement-disengagement, table jog and automatic feed with return control. Also, an emergency stop capability is provided. Note that all tubing is placed in a common conduit leading out of the control box.

Pneumatic logic control systems are being employed to replace electrical control systems for many operations in fluid circuits. It must be remembered that the pneumatic logic control systems are controls and are not complete fluid power systems or circuits. These are instances where pneumatic logic control systems are used to actuate very small bore cylinders, but their big usage is to provide a signal to a pilot-operated control valve or valves. The valves would be air-pilot operated but the medium passing through these valves to function the circuit may be air, oil, water, vacuum, or some other fluid medium.

Courtesy of The Aro Corporation
Fig. 18. Logic control panel shown in Fig. 17, with door closed.

While some pneumatic logic control systems operate at very low air pressure and require amplifiers to actuate the control valves, other pneumatic logic control systems operate at pressures from 50 to 125 psi.

Courtesy of Clippard Instrument Laboratory, Inc.

Fig. 19. Pneumatic logic control panel.

Figures 21 and 22 illustrate a couple of pneumatic logic control systems. In Fig. 21 is a two hand anti-tie-down control system which provides a continuous output signal as long as both Inputs are ON. This Output could be connected to a pilot operated two, three, or four-way control valve that is spring offset; or it could be connected to a very small single acting cylinder-spring offset, gravity return or mechanical return. In Fig. 22 is a two hand anti-tie-down control system that has a four-way Output.

The ladder diagram approach is also used in pneumatic logic. In this, the conventional electric diagram is cut in half since the ground half of the diagram is not required because there is no return in air systems. Terms that are used are LV for limit valves, RV for relay valves, TS for time delay, TRV for timing relay valves,

Courtesy of Clippard Instrument Laboratory, Inc.

Fig. 20. Pneumatic logic circuit applied to a mill.

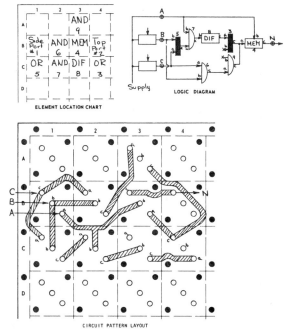

Courtesy of The Aro Corporation

Fig. 21. Element location chart, logic diagram and circuit pattern layout for a two-hand anti-tie-down control which provides a continuous output signal as long as both Inputs are ON.

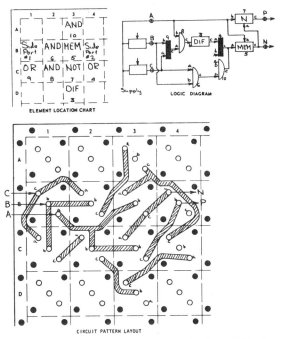

ELEMENT LOCATION CHART

LOGIC DIAGRAM

CIRCUIT PATTERN LAYOUT

Courtesy of The Aro Corporation

Fig. 22. Element location chart, logic diagram, and circuit pattern layout for a two-hand anti-tie-down control that has a four-way Output.

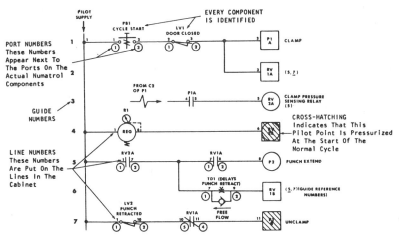

Courtesy of Numatrol Controls Div., Numatics, Inc.

Fig. 23. Typical Numatrol diagram.

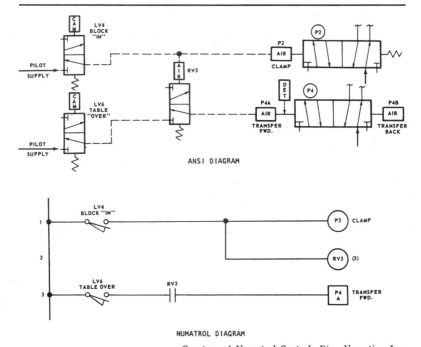

Courtesy of Numatrol Controls Div., Numatics, Inc.

Fig. 24. Comparison of ANSI and Numatrol diagrams.

etc. Figure 23 shows a ladder diagram. All control air comes from the left-hand supply line, and flows from left to right through the various control devices, until eventually it reaches the right end of the flow path, and enters either a pilot point on a relay valve, or a pilot point on a power valve. Circuit breaks are three-way breaks, and it is not necessary to show the exhaust ports. It is only necessary to show the working ports and the flow condition at the start of a cycle.

Figure 24 shows a comparison between an ANSI Diagram and a Numatrol Diagram. Note the simplicity in the Numatrol Diagram.

Figure 25 shows the ladder diagram design used in a complex pneumatic control system. Note the legibility of this layout.

Four-way air control valves that can be actuated by pneumatic logic control systems are shown in Figs. 2, 3 and 4 on pages 19-2, -3 and -4, respectively. Hydraulic valves of the air pilot type that can be used with pneumatic logic control systems are shown in Figs. 26 and 27. This eliminates electrics from the entire system.

Many other very small controls are available for use with logic systems. Figure 28 shows a miniature pilot sensor which can detect

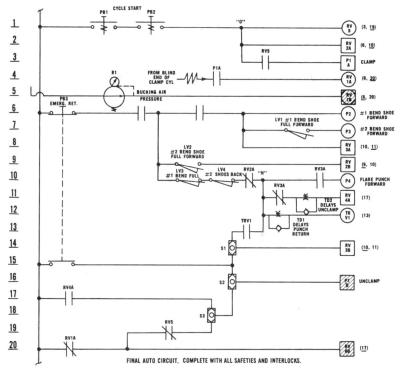

Courtesy of Numatrol Controls Div., Numatics, Inc.

Fig. 25. Complex pneumatic control circuit showing final auto circuit, complete with all safeties and interlocks.

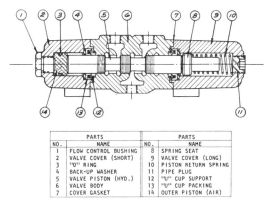

PARTS		PARTS	
NO.	NAME	NO.	NAME
1	FLOW CONTROL BUSHING	8	SPRING SEAT
2	VALVE COVER (SHORT)	9	VALVE COVER (LONG)
3	"O" RING	10	PISTON RETURN SPRING
4	BACK-UP WASHER	11	PIPE PLUG
5	VALVE PISTON (HYD.)	12	"U" CUP SUPPORT
6	VALVE BODY	13	"U" CUP PACKING
7	COVER GASKET	14	OUTER PISTON (AIR)

Courtesy of Logansport Machine Co., Inc.

Fig. 26. Cross section of a four-way spring-offset air-pilot operated hydraulic valve.

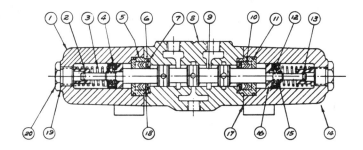

PARTS		PARTS	
NO.	NAME	NO.	NAME
1	VALVE COVER	11	"O" RING GASKET
2	SPRING RETAINER	12	CUP FOLLOWER
3	RETURN SPRING	13	SNAP RING
4	CUP PACKING	14	COVER GASKET
5	PACKING RETAINER	15	ELASTIC STOP NUT
6	PKG. COMPRESSION SPRING	16	PILOT PISTON
7	STOP WASHER	17	COVER SCREW
8	VALVE BODY	18	SPRING WASHER
9	VALVE PISTON	19	BUSHING GASKET
10	"V" RING PACKING	20	FLOW CONTROL BUSHING

Courtesy of Logansport Machine Co., Inc.

Fig. 27. Cross section of a four-way three-position air-pilot operated hydraulic valve.

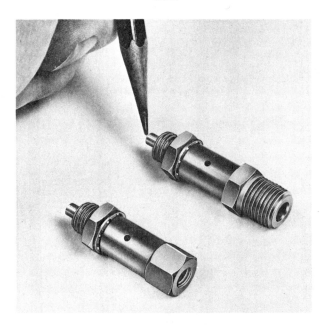

Courtesy of Clippard Instrument Laboratory, Inc.

Fig. 28. Miniature pilot sensor.

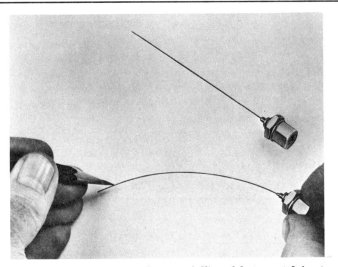

Courtesy of Clippard Instrument Laboratory, Inc.

Fig. 29. Miniature pneumatic "whisker" or two-way valve.

deviations of positions of as little as 0.005 of an inch with repeatability. Direct force on the stem to operate the sensor can be as low as 2 ounces.

Miniature type pneumatic "whisker" valves are used for detecting the presence of objects in out-of-the-way or remote places such as on conveyor lines, etc. Figure 29 shows a "whisker" or two-way valve.

CHAPTER 25

Fluidics

The word "Fluidics" is derived from the words "fluid" and "logic," and is used specifically to describe the technology of the control of fluid force components by means of devices utilizing fluid interaction phenomena to produce useful output signals. While the utilization of air and hydraulic circuitry has been traditionally controlled primarily by electronic and electro-mechanical means, the advent of fluidic devices makes it possible to control fluid power force components and fluid power master control components by the same media with which the force components are operated—either gases or liquids. To illustrate, control of the aerosol can filling machine, shown in Fig. 1, is performed by fluidic circuitry. The control box is shown at the lower center. A closeup view of the control box, with a person adjusting the pressure level for the fluidic circuitry, is illustrated in Fig. 2.

The basic characteristics of using fluids to control themselves offer some outstanding advantages over the traditional alternate methods. For example, fluidic devices offer exceptional thermal and physical stability and ruggedness, they are impervious to conditions of severe vibration and shock, and are completely insensitive to radiation, even of extremely high levels.

In addition to these ideal characteristics, the nature of the fluidic device and its construction is such that no maintenance problems are encountered with the deterioration of moving parts, inasmuch as fluidic devices incorporate no moving parts. This basic design characteristic in turn contributes considerable simplicity not only in the area of maintenance, but also in the area of functional reliability, and at a relatively low cost.

These devices perform their control function by a stream of gas or liquid seemingly directing itself through precision channels and networks quite similar to the apparent flow of electricity through wires and contact points. While their reaction time is somewhat slower than that of electronic systems, fluidic devices do respond in milliseconds, or speeds that are quite adequate for most requirements.

Although originally fluidic devices appeared to be quite expensive due to the nature of their manufacture, at this writing the cost has been reduced to where they are currently quite competitive with most of their electronic or electro-mechanical equivalents. Maintenance is simplified by simple replacement of modular components, some of which are shown mounted on a grid board in Fig. 3, and in-plant responsibilities are thereby limited to single rather than multiple job classifications. An assembled fluidic industrial control system in its metal housing is shown at the rear in Fig. 4. It presents a neat, compact appearance. Also pictured in Fig. 4 are the drilled aluminum grid board at the right, and an interface valve, pneumatic timer, fluidic-to-electrical switch, pneumatic pushbutton, selector switch, three industrial control modules, and two fluidic indicators, going clockwise from the left.

The technology of fluidics is creating an increasing number of practical applications in industry. Industry's high operating speeds and

Courtesy of Corning Glass Works

Fig. 1. Aerosol can filling machine that employs fluidic circuitry control. Control box is shown at the lower center.

Courtesy of Corning Glass Works
Fig. 2. Fluidic circuit control box, with open cover, facilitates adjustment of
pressure level.

ever-increasing complexity of machine functions, coupled with the
constantly rising costs of labor, demand higher machine reliability
and the elimination of down time due to component part failure where
ever possible.   In addition, industrial people are voicing increasing
demands for multi-functional equipment that offers flexibility in control circuits and provision for rapid change-over due to constantly
changing manufacturing methods.

The traditional electronic, electro-mechanical, pneumatic-mechani-
cal and straight mechanical methods can become quite cumbersome,
and in many cases prove to be totally inadequate for these requirements.   The technology of fluidics, however, promises to fulfill most

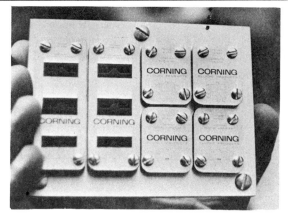

Courtesy of Corning Glass Works

Fig. 3. Grid board on which eight fluidic control modules have been mounted.

of these requirements, and is being looked upon with considerable interest by industry, not only from the standpoint of usage, but this interest is evidenced also by the growing number of manufacturers who are developing fluidic devices.

Although fluidic circuits operate at low pressures, highly reliable interface devices have been developed and are continually being developed to provide full airline pressure functions as well as expanding the use of normal airline pressures controlling high pressure hydraulic systems. Since air is normally used as the working fluid within these devices, electrical noises, vibration, fatigue and contact contamination problems are nonexistent. Since there is no arcing or sparking of switching elements, circuits employing fluidic devices can be operated quite safely in highly explosive or otherwise dangerous environments. In one application of fluidic integrated circuitry (see Fig. 5), the fluidic device used was substantially reduced in size and complexity as compared to the original control system for a water pumping station. The device senses the water level in a wet well, then starts and stops two pumps at pre-set levels, using air as both the sensing and control medium. The system contains no electrical or moving parts except two fluidic-to-electric switches that convert air signals to electrical signals to operate the pump motors.

Physically a fluidic device is a block of material with an internal network of passages. In manufacturing, it is usually produced by creating the channels in one layer of material, and then covering it with a flat piece. The two sections are then fused together to make one totally enclosed, tightly sealed unit. The internal network of passages

Courtesy of Corning Glass Works

Fig. 4. Assembled fluidic control system in its metal housing, at rear, with accessory hardware, in foreground. Drilled aluminum grid board is shown at right rear, and pictured clockwise from left are an interface valve, a pneumatic timer, a fluidic-to-electrical switch, a pneumatic pushbutton, a selector switch, three control modules, and two fluidic indicators.

varies, of course, depending on the nature of the specific device, and the functions required of it. One such fluidic device is illustrated in Fig. 6. In this illustration, showing a 36-gate uncerammed fluidic manifold, the power supply channels and permanently imbedded nuts are clearly visible. A fluidic module is attached to the manifold by four of the nuts. A single input flowing through the inner channels provides the power supply for all of the modules mounted on the manifold.

One of the basic underlying principles of the functioning of these flow passages within the fluidic element is what is known as the "Coanda effect,"* or "wall-attachment" effect. As can be seen in

* Named after Henri Marie Coanda, civil aeronautics engineer and inventor, born in 1885 in Bucharest, Romania.

Courtesy of Corning Glass Works

Fig. 5. Fluidic device that controls water level in a water pumping station.

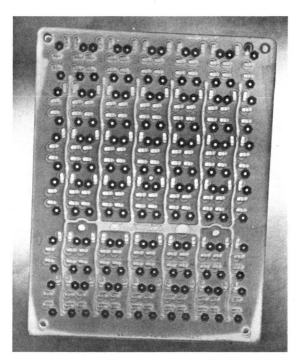

Courtesy of Corning Glass Works

Fig. 6. A 36-gate uncerammed fluidic manifold showing power supply channels and permanently imbedded nuts.

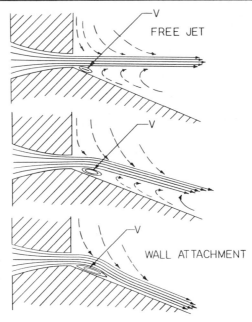

Fig. 7. A free jet of air passing through this orifice pulls with it more air from the underside of the jet, creating a small vortex area, V (low pressure area), which tends to pull the free jet towards the angled wall below. This is called the "Coanda or wall-attachment effect."

Fig. 7, a free jet of air passing through an orifice will continue in a given direction, pulling in with it any available fluid that surrounds the jet as it leaves the orifice. If there is a greater availability of this

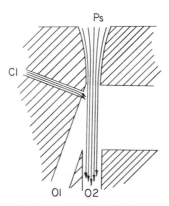

Fig. 8. Monostable flip-flop.

entraining air from one side than from the other, there tends to be created near the nozzle exit a small vortex area V, which is in reality a low pressure area. This low pressure area then tends to attract the free jet, distorting it and pulling it toward the angled wall. Once this free jet is pulled all the way or tangent to the exit wall, the continued existence of this vortex tends to hold the jet attached to that wall. This attachment continues until such time as a small air supply is fed to the general area of the low pressure vortex, thus relieving the attraction of the jet to the wall. When this signal is injected, the free jet then detaches itself from the wall and resumes its normal, uninterrupted flow path.

Monostable Flip-Flop

Figure 8 shows an additional downstream porting for this jet, and is shown in its straight path with the supply signal relieving the normal vortex area. The absence of a signal at point C1 would cause the jet to attach itself and deliver the free jet to the 01 port. With the presence of the signal, the flow of the free jet would be as shown in the diagram, or to the 02 port. This arrangement and function is known as a monostable flip-flop. In other words, it will pursue the path to the 01 port with the absence of a C1 signal, or it will flop to the 02 port path with the presence of a signal at C1.

Bistable Flip-Flop

The diagram shown in Fig. 9 illustrates a bistable flip-flop. In this illustration, the air flow is going to be from input Ps down either

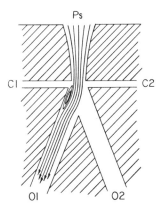

Fig. 9. Bistable flip-flop.

the 01 channel or the 02 channel, depending on the existence of a signal from either C1 or C2. This bistable flip-flop operates on the principle of the low pressure vortex being created on one side or the other of the element, creating the wall-attachment effect toward either the 01 or 02 channel. As the illustration shows, the most recent signal has been injected at the C2 signal source. This flow pattern will continue until such time as a signal is injected in the C1 signal port. At such time the vortex shown in the diagram will be relieved and the flow will flop to the 02 channel and remain there until a signal is received at the C2 signal source.

The bistable flip-flop derives its name from the fact that it has two stable conditions, 01 and 02, and it can be made to switch or flip from one to the other. This bistable flip-flop is normally used as a memory device, since the output will continue to be from the leg corresponding with the last control signal, although the signal is no longer in evidence.

Since the two signals described above primarily serve the purpose of supplying only that amount of air required to relieve the low pressure vortex previously described, the control pressure necessary to switch the free jet is substantially lower than the supply pressure. Thus the control signal is not only remembered by the bistable flip-flop, but is amplified, by as much as ten to one or even more, thus creating sufficient output pressures to operate interface devices, or control other fluidic elements in the circuit.

The symbol recommended by the National Fluid Power Association (NFPA) for the bistable flip-flop is shown in Fig. 10, with equivalent methods of control as they would appear in straight pneumatic or electrical disciplines. Table 1, based on Boolean Algebra and called a "truth table," may help to describe the functioning of the flip-flop. The symbol X represents the presence of pressure and the symbol 0 represents the absence of pressure.

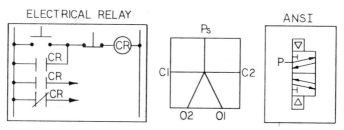

Fig. 10. NFPA symbol for bistable flip-flop (*Center*) with electrical (*Left*) and pneumatic (*Right*) equivalents.

TABLE I

TRUTH TABLE FOR BISTABLE FLIP-FLOP

C1	C2	01	02
X	0	X	0
0	0	X	0
0	X	0	X
0	0	0	X

OR/NOR Gate

Figure 11 depicts a second basic fluidic element, the OR/NOR gate. By comparing this symbol with the arrangement shown in Fig. 8, it becomes evident that the main variation lies in the adding of

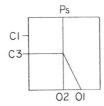

Fig. 11. NFPA symbol for OR/NOR gate.

one additional signal port. This element has outputs corresponding with two conditions: OR—pressure at one or any combination of the control ports; and NOR—pressure at none of the control ports. The OR/NOR gate is designed in such a manner that flow will always be in the NOR leg when no control signal is present, provided, of course, Ps is present. The 02 port represents the NOR output and the 01 port represents the OR ouput. Either a C1 or a C3 signal must be present to get an output at the 01 port; but neither C1 nor C3 can be present to get an output at the 02 port. The OR/NOR element is the basic logic building block. The OR/NOR block can be used, put together in various ways to achieve all of the other logic functions.

A variation of the OR/NOR gate is the turbulence amplifier depicted in Fig. 12. In this type element there is a laminar stream of air flowing from an input nozzle to an output nozzle. The laminar stream is capable of jumping the gap between nozzles and exerting a useful signal at the output nozzle unless this laminar air stream is disturbed. The injection of a small cross jet from one or more of the signal ports will cause the laminar air to become turbulent with the result that it will lose its ability to jump the gap and develop a use-

CONTROL NOZZLES

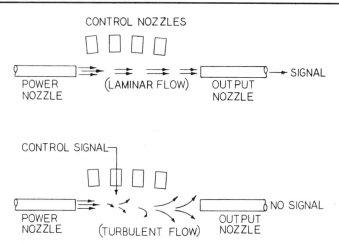

Fig. 12. Turbulence amplifier—a variation of the OR/NOR gate.

ful output. Although some modulation of the output signal intensity can be created by modulating the disturbing effect of the turbulence-creating signal, the device is basically a digital or on-off device.

When describing fluidic components or designing circuits, it is often desirable to use standard logic notation. A · (dot) means AND; + (plus sign) means OR, and a bar over a letter ($\overline{A}$) means that the condition represented by the "A" is absent. This notation is then read as "NOT A." Using this notation, then, the operation of the OR/NOR gate can be written as follows:

$$01 = C1 + C3 = OR$$
$$02 = \overline{C1} \cdot \overline{C3} = NOR$$

The OR/NOR gate is available with multiple input signals, any combination of which can be used to derive the OR output. Figure 13 shows the NFPA symbol for such a multiple signal element and Table 2 gives the truth table for the OR/NOR gate function.

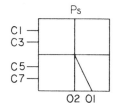

Fig. 13. NFPA symbol for 4-input OR/NOR gate.

TABLE 2

TRUTH TABLE FOR OR/NOR GATE FUNCTION

C1	C3	02 (NOR)	01 (OR)
0	0	X	0
0	X	0	X
0	0	X	0
X	0	0	X
0	0	X	0
X	X	0	X

AND Gate

Another fluidic element is the AND gate, which is similar to the NOR gate, except that the NOR gate is used to determine when none of the control signals is present, whereas the AND gate is used to

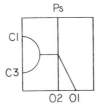

Fig. 14. NFPA symbol for AND gate.

determine when *all* the control signals are present. In Fig. 14, both the C1 and C3 signal sources must be present to get an output at the 01 port. The absence of either or both will result in the stable output

TABLE 3

TRUTH TABLE FOR AND GATE

C1	C3	01	02
		AND	NAND
0	0	0	X
X	0	0	X
0	0	0	X
0	X	0	X
0	0	0	X
X	X	X	0
0	0	0	X

at the 02 port. The relationships depicting the AND gate functions
would be:

$$01 = \underline{C1} \cdot \underline{C3} = AND$$
$$02 = \overline{C1} \cdot \overline{C3} = NAND \text{ (not AND)}$$

Figure 14 gives the NFPA symbol for the AND gate and Fig. 15 the
equivalent electrical and pneumatic symbols for the AND gate functions. Table 3 depicts the truth table for the AND gate.

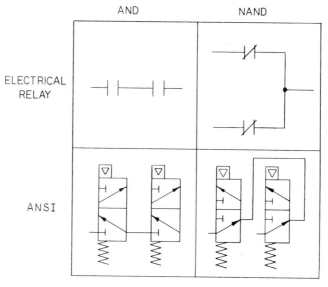

Fig. 15. Equivalent electrical (*Upper*) and pneumatic (*Lower*) symbols for
AND gate functions.

Schmitt Trigger

The Schmitt Trigger, the NFPA symbol for which is shown in Fig.
16, is a useful element in fluidic circuit design. With the Schmitt

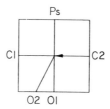

Fig. 16. NFPA symbol for Schmitt Trigger.

Trigger, it is possible to switch from one output to another, depending on which of the control signals, C1 or C2, is stronger or of a higher pressure value. For example, it may be desirable to have an 01 output signal when the pressure at the C1 signal source is greater than 3 psi, and conversely, an 02 output when the signal at C1 is less than 3 psi. In this case this operation would be performed by connecting the C2 input to a constant 3 psi source, and the C1 signal would come from whatever pressure value is to be measured. The desired switching function would then be performed as the 3 psi norm is passed.

The Schmitt Trigger is most valuable when used in conjunction with sensors, because the switching points can be adjusted over a very wide range by varying the bias pressure (the 3 psi) from 1% to 35% of the supply pressure Ps, and the "switching band" is very narrow. The switching band is the difference between the rising pressure to switch from 01 to 02 and the falling pressure that switches the element back from 02 to 01. It is possible, therefore, when using the Schmitt Trigger, to attach a sensor directly to C1 and very accurately detect when the input pressure (C1) is greater or less than the bias pressure at C2.

The truth table for the Schmitt Trigger is given in Table 4.

TABLE 4

TRUTH TABLE FOR SCHMITT TRIGGER

	01	02
C1 > C2	X	0
C1 < C2	0	X

Binary Counter

In many applications it is necessary to count the number of steps or operations which have occurred. It is possible to do this by using a circuit composed of one or several *Binary Counters*. In a binary

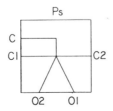

Fig. 17. NFPA symbol for binary counter.

TABLE 5

TRUTH TABLE FOR BINARY COUNTER

C Count	C1 Set	C2 Reset	01 One	02 Zero
0	X	0	X	0
0	0	0	X	0
0	0	X	0	X
0	0	0	0	X
X	0	0	X	0
0	0	0	X	0
X	0	0	0	X
0	0	0	0	X
X	0	0	X	0
0	0	0	X	0

counter, the output is switched from one output port to the other with each succeeding input signal or pulse. Figure 17 gives the NFPA symbol and Table 5 the truth table for a binary counter.

Where C is the "count" control input, C1 is the "set" control input, and C2 is the "reset" control input. The set control is used to set the counter with initial output at 01 and the reset is used to set the counter at 02. If 01 corresponded to the digit "1" and 02 corresponded to "0", it is possible to see how the reset causes "zero output" (02), and the set causes "one output" (01). If the counter initially has zero output (02), a single pressure pulse at C will cause the output to switch from 0 to 1, that is, from 02 to 01. The next pulse will cause the output to switch back from 1 to 0, and so on. Using only one binary counter, it is possible to count only to one. With a series of counters, "staging" them, higher numbers can be counted.

Although the foregoing is not intended to be an exhaustive presentation of all the functional uses and designs of fluidic elements, it is felt that the basic elements herein presented will afford the basic functional principles upon which pyramided fluidic logic circuits may be built. Specific design data as well as additional refined variations of these basic elements should be obtained from specific manufacturers of these devices.

CHAPTER 26

Hydraulic Servo Control Systems

Automating machines or processes to produce parts or products in the most efficient manner to meet industrial requirements often requires the application of automatic control systems of some type.

An Automatic Control System

For the efficient operation of many machines and devices it is important that they be equipped with automatic control devices. For example, a home water system, one not connected to the city water mains, consists of a deep well pump, an electrical control box, a water tank, a pressure switch and pressure gauge, all of the piping connected to the various outlets, and the electrical lines. The pressure switch may be set so that the contacts close when the pressure in the tank drops to 25 psig and the pump is called upon to deliver water to the tank. The tank has a head of air which the water compresses. When the water compresses the air to raise the water pressure to 50 psig the contacts in the pressure switch open and the pump stops pumping. When toilets are flushed or automatic dish washers are operated and showers turned on, etc., the water pressure drops to the point where the contacts in the pressure switch automatically close, directing electric current to the pump through the control box. This automatic operation assures that water under pressure will always be available. In this installation the pressure switch acts as a differential sensing device that gives a feedback signal to the pump controls telling the pump either to start pumping or to stop pumping.

Another automatic control device which is rather complicated, but whose operation almost everyone now takes for granted, is the automatic color hue and light intensity adjustment on the modern color television set. There are, of course, many other automatic control devices too numerous to mention.

General Features Found in a Servo Control System

A servo control system is one in which a comparatively large amount of power is controlled by small impulses or command signals and any errors are corrected by feedback signals. In the automatic control of many machine functions there are requirements such as the exceptionally high degree of accuracy in acceleration, velocity, and positioning. These requirements can be met by servo control systems. A servo control system can essentially be classed as a closed-loop or an open-loop system. A closed-loop system has a very rapid response in order to reduce the error signal to zero. An example which is very common is power steering on an automobile. An open-loop system in most instances has a slower signal response. An example would be an air conditioning system in which the thermostat temperature may rise several degrees before the signal would cause the air conditioner to function. Feedback is one of the principal elements of a closed loop servo system. Figure 1 shows a closed-loop system consisting of:

(*A*) A command signal which may be derived from a punched card, a tape, a tachometer, a potentiometer, or some other device.

(*B*) A servo amplifier which receives a low-power ac signal input from *A*, and amplifies the resultant output signal to a higher voltage and power level.

(*C*) A torque motor/servo valve which receives the output signal from the amplifier *B* to actuate the servo valve torque motor.

(*D*) An hydraulic actuator which could be linear (cylinder powered) or rotary (hydraulic motor powered).

(*E*) Load.

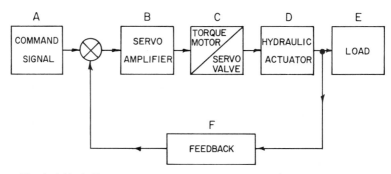

Fig. 1. A block diagram of an electrohydraulic basic closed-loop servo circuit.

(*F*) A feedback transducer which can be a linear variable differential transformer (LVDT), a tachometer, a potentiometer, or a synchro which is a rotary electromagnetic device generally used as an ac feedback signal generator which indicates position. This device measures the results at the load and sends a feedback signal to the amplifier input. The accuracy of the control system depends upon the accuracy of the feedback transducer. The feedback signal is compared with the command signal, and whenever a difference (or error) exists, a hydraulic correction is automatically made that will bring the output signal into correspondence with the input. The difference between the two electrical signals is called the "error signal." When the error signal is zero, the ideal condition exists. The purpose of the servo valve is to bring about this condition.

Hydraulic servo valves may be mechanically, pneumatically, or electrically operated. Mechanically operated servo valves are generally employed in the less complex systems. Application for mechanically operated servo valves are on: steering devices, test and training devices, copying devices (such as on machine tools), and heavy-duty mobile equipment. Most servo control valves are of the four-way type, although they may be connected for three-way or two-way control. The sliding spool design is employed as the main valving element in many four-way, servo control valves and is somewhat similar to the design used in industrial hand-operated, mechanically-operated and solenoid-operated directional control valves. One major difference is the closeness of fit required between the spool and the hardened sleeve or body (depending on construction) of the servo valve. Most industrial directional valves do not use a hardened sleeve due to the cost. In the servo valve, the diametral clearance between the spool and the sleeve could be as little as 80 millionths of an inch in order to keep leakage at a minimum. The shape of the porting in the sleeve or body may be square, rectangular, round, or full annulus.

As for the axial alignment between the lands on the spool and the porting in the sleeve or body, different "lap" conditions are employed. Lap is the length relationship between the metering lands of the spool and the port openings in the sleeve or body.

The *closed center* servo valve has an overlapped condition. Here the lands are slightly longer than the porting area and create a dead zone. A dead zone is not desirable in a servo valve as it makes the valve unresponsive to small signals.

The *open center* servo valve has an underlap condition in that the lands on the spool are somewhat narrower than the port holes in the sleeve or body. In the center position of the spool, there is a constant fluid flow from the pressure inlet, across the spool to the tank port. Pressure drop due to such restrictions causes an intermediate pressure at the cylinder ports.

The *zero lap* is the most desirable construction. This means that the edges of the spool lands are precisely made to line up with the metering ports in the valve sleeve or body. This axial alignment can be accomplished by a process known as "null grinding" which is performed to extremely close tolerances on each spool and sleeve assembly. The reason for the accurate axial fit is to eliminate a dead zone or dead band. Any movement of the spool in either direction from its centered position will cause a corresponding flow. The position of the spool in the sleeve controls both flow rate and direction of flow.

In this chapter the discussion is centered mainly around electrohydraulic servo controls which are usually employed in the more sophisticated control systems such as on tape-controlled machine tools, high-speed printing presses, press brakes, etc. Electrohydraulic servo valves may be of the single-stage, two-stage, and in some instances, three-stage configurations. The two-stage configuration uses a pilot stage and a main stage. The pilot stage makes use of such flow directors as spools, single flappers, double flappers, jet pipe, etc. (An open center hydraulic amplifier is created by a flapper, nozzles, and an orifice arrangement. A flapper could be considered a type of valve arrangement. The flow is modulated by the movement of the flapper.)

In a typical two-stage, electrohydraulic, servo control the first stage or pilot stage receives the electromechanical input, amplifies it and controls the second or main stage. In a two-stage, spool type servo control, with torque motor configuration, the torque motor actuates the spool of the pilot section which directs fluid to shift the spool of the second stage in response to the input signal.

Generally, control pressure for two-stage servo controls is taken from a source which is separate from the supply pressure but sometimes the supply pressure source is used for the control pressure by utilizing a pressure-reducing valve and an accumulator. The advantages of using a separate control pressure source are: (1) Less power is wasted. Usually, the maximum control pressure is 1000 psi, while the supply pressure may be as much as 3000 psi; (2) Load fluctuation will not have any effect upon the pilot spool response which becomes an important factor in certain applications; and (3) Separate filtration of the supply fluid is accomplished. Good filtration of the fluid is of the utmost importance in order to reduce or eliminate sticking of spools. In many applications of the spool type, electrohydraulic servo controls, *dither* is used to counteract static friction or "stiction" and keep the control more dirt tolerant. Dither is defined as a low-amplitude alternating signal superimposed on the dc signal to the valve to keep the pilot spool or main spool in motion within the dead zone of the valve. This minute oscillation of the

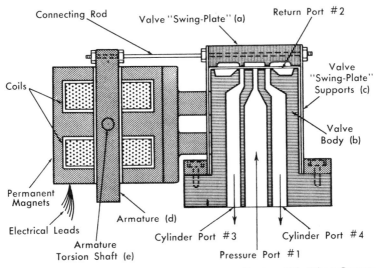

Courtesy of The Oilgear Company

Fig. 2. Single-stage, four-way servo control of "swing-plate" design.

spool does not affect the operation of the system, however, it does prevent the spools from sticking. Frequencies which are employed may range from 60 to 400 cps (cycles per second) but the lower frequency is used more often.

Figure 2 shows a cross section of the "swing plate" design of a four-way, single-stage, electrohydraulic servo-control valve. Use of the swing-plate principle allows the port length to be modified from valve to valve. The flow that is wanted from the servo valve can be specified which, in turn, permits specifying maximum pump-stroking velocity to control acceleration of the load.

Some of the fundamental features of this type of servo control which provides exceptional simplicity, high performance, and reliability on both closed and open-loop systems are: (1) Only two moving parts—torque motor armature and servo-valve swing plate are the only parts that move; (2) No metal-to-metal contact between swing plate and valve body, a condition which virtually eliminates sticking, scoring, or jamming and which assures a fast response to minute signals; (3) Flow rate, essentially a linear function of valve displacement giving high fluid-power gain with minute swing-plate movement; (4) Elimination of the need for pilot system pressure, centering devices, orifices or internal filters—valve automatically centers in case of power failure, as torque motor armature (*d*) actually twists torsion shaft (*e*) under signal power—centering is positive without external centering devices; and (5) Safety in hazardous atmospheres without special modifications.

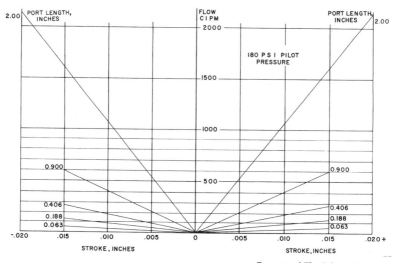

Courtesy of The Oilgear Company
Fig. 3. Servo valve flow curves.

Figure 3 shows the flow curves of the four-way servo-control valve of the swing-plate design which is shown in Fig. 2. Note the curves for the various port lengths.

Figure 4 shows a schematic view of pump and "V"-control components. Note that the four-way servo-control valve is mounted on the manifold plate that is connected to the pump. The servo valve positions the slide block of the pump by directing fluid to the stroking pistons. The pump output of fluid is determined by the position of the slide block. The pump cylinder contains the pumping pistons which are confined by a thrust ring in the slide block and is driven by a constant speed electric motor. The bores of the pumping pistons are connected to the outlet ports through a stationary pintle valve.

In Fig. 4, the slide block is positioned eccentric to the cylinder. Pistons move outward on the bottom and inward on the top. The pistons draw fluid from the bottom side of the pintle valve and force the fluid into the top side of the pintle valve.

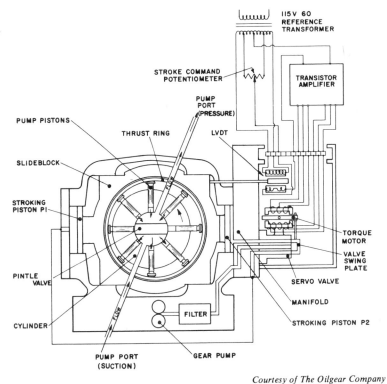

Courtesy of The Oilgear Company

Fig. 4. Schematic view of pump and "V" control components.

In Fig. 5, the slide block is positioned concentric with the cylinder and no fluid is pumped. In Fig. 6, the slide block is positioned with eccentricity opposite that of Fig. 4, and flow is reversed.

When the servo-valve swing plate is deflected downward (see Fig. 4), the left stroking piston (P_1) connects to gear pump pressure, the right stroking piston (P_2) connects to drain. The slide block moves to the right. When the swing plate is deflected upward, the slide block moves to the left. When the swing plate is centered, the slide stops and holds position.

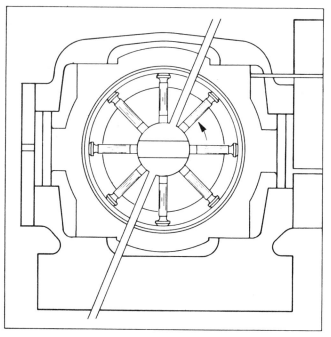

Courtesy of The Oilgear Company

Fig. 5. Schematic view of pump with slide block concentric to pump centerline.

The torque motor armature in the servo-valve assembly is spring centered and deflects the servo-valve swing plate proportional to the current from the amplifier. The amplifier provides the required torque motor power. The signal circuit uses low-power electrical transducers.

The linear variable differential transformer (LVDT) transmits an electrical voltage proportional to the slide-block position, thus the slide block can be electrically commanded remotely. A potentiometer gives a command voltage; it is wired to the amplifier in series with the pump LVDT. The slide block moves until the LVDT voltage equals the potentiometer voltage. The error voltage is then zero and the servo-valve swing plate centers.

A reference transformer for each size pump provides voltages for exactly the full flow of the pump in each direction. The reference transformer supplies excitation voltage for command potentiometers, calibrating them for zero to full potentiometer rotation. Figure 7 shows an electrohydraulic, servo-controlled relief valve used for pressure, force, and torque control in which the "V" type control is employed to adjust the spring compression of a precision relief valve. By the use of a servo-valve-operated pilot cylinder with LVDT feedback, the spring of the relief valve is accurately set at the command of a remote potentiometer, or other electrical transducer. Pilot pressure at approximately 150 psi must be supplied to actuate the servo. The lower part of Fig. 7 illustrates the schematic circuit.

Figure 8 shows a constant displacement motor equipped with an integral electrohydraulic servo-control valve. The motor is of the axial piston design for operation at 2500 psi. The hydraulic motor converts the pump flow to rotary motion due to the oil pressure acting upon the constant displacement pistons in the motor, which causes the motor shaft to rotate with high torque.

Closed-Loop Pressure Control

In closed-loop systems pressure, force, and torque control must be carefully analyzed to produce anticipated results. Position and speed controls absorb a considerable pump flow in positioning and holding speed. Pressure systems can snap to pressure quickly while

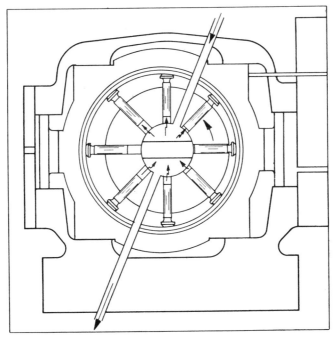

Courtesy of The Oilgear Company

Fig. 6. Schematic view of pump with slide block to left of cylinder centerline.

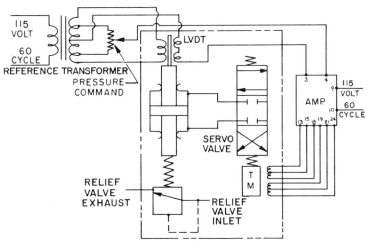

Courtesy of The Oilgear Company

Fig. 7. (Top) Electrohydraulic servo-controlled adjustable relief valve; (Bottom) Oil-electric circuit for servo-controlled relief valve.

absorbing only a small compressibility flow. Pressure control systems have an inherent difficult stability problem.

Figure 9 shows control circuitry for accurate pressure control of a large press where even at full pump stroke, several seconds are re-

Courtesy of The Oilgear Company

Fig. 8. Constant displacement hydraulic motor with integral electrohydraulic servo-valve.

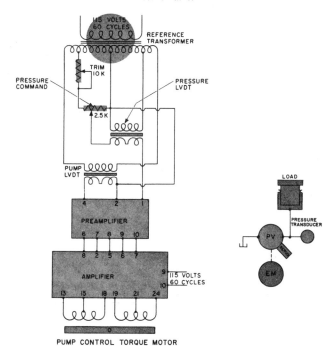

Courtesy of The Oilgear Company

Fig. 9. Signal circuit for typical pressure control on a large press.

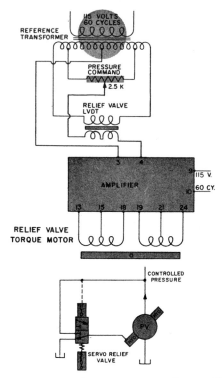

Courtesy of The Oilgear Company

Fig. 10. Signal circuit for typical pressure control on a translatory or rotary drive with low volume of compressible fluid.

quired to compress the system to full pressure. Pressure command is a potentiometer; feedback is a Bourdon tube LVDT pressure transducer. Pump feedback is used, making the system a regulator, but allowable gain is high to give accurate pressure control.

Figure 10 shows circuitry for a system with a small volume of compressible fluid. Pressure is set by a servo-controlled relief valve. A pressure-unloading type of pump is employed to limit flow to that demanded by the load. This circuitry adapts to cylinder systems and web-winding type of rotary drives.

CHAPTER 27

Hydrostatic Transmissions

Basically, a hydrostatic transmission is composed of a hydraulic pump, a control, a hydraulic motor, and the connecting piping. Figure 1 shows a typical hydrostatic transmission. It has a variable-volume hydraulic pump and a fixed-volume hydraulic motor in a closed-loop hydraulic circuit capable of overcenter operation. With a closed-loop circuit, the inlet and discharge ports of the hydraulic pump are connected to the discharge and inlet ports of the hydraulic motor.

The pump shown in Fig. 1A is a variable-volume unit containing nine pistons and has a manually operated volume control (servo) as shown. The pump also has a fixed volume Gerotor charge pump and two check valves. One type pump unit has a low-pressure relief valve. The hydraulic motor has nine pistons with fixed stroke. This unit also contains a shuttle valve, a low-pressure relief valve, and a high-pressure relief valve. When the motor is of the variable volume configuration it also contains a manually operated volume control. Figure 1B shows an ANSI schematic of the variable-volume pump and the fixed-volume motor. The pump control, which is a servo-mechanism, changes the volume and direction of flow by varying the length and direction of the piston stroke. In the neutral position there is no piston stroke and no flow from the pump. When the control is moved from neutral, oil flows from the pump to the motor through one of the closed-loop lines causing the shaft of the hydraulic motor to rotate. The greater the distance the control is moved off center, the faster the motor shaft will rotate.

When the control is moved back past neutral, in the reverse direction, the pump flow is reversed; this causes the motor shaft to rotate in the opposite direction. The overcenter capability of the pump, however, eliminates the need for a four-way valve to reverse the rota-

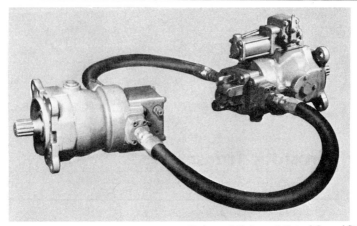

Courtesy of Hydreco, A Unit of General Signal

Fig. 1. A hydrostatic transmission.

tion of the motor shaft. The pump volume-control is actually a hy-draulically operated low-pressure servo. The charge pump furnishes the low-pressure oil to the control. An adjustable, pilot-operated, high-pressure relief valve in the motor protects the system against

Courtesy of Hydreco, A Unit of General Signal

Fig. 1A. Cross section of variable-volume pumping unit. 1. Control; 2. Piston; 3. Cam; 4. Charge pump (gerotor); 5. Charge pump relief.

pressure surges which may be encountered.   To prevent depletion of the closed system, the high-pressure relief valve discharges to the low-pressure side of the system.   A pressure-actuated shuttle valve is employed to direct high and low pressure to the respective relief valves.

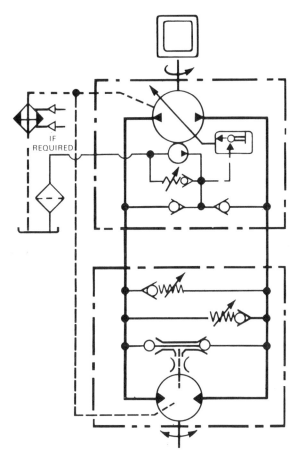

Fig. 1B. ANSI schematic of variable-volume pump and fixed-volume motor.

With the advent of extra-heavy-duty machinery in the construction industry, the agricultural industry, and the material handling field, the use of hydrostatic transmissions is growing at a very rapid rate. Hydrostatic transmissions are not a recent development; they were first designed in the early 1800's.   They were slow to receive acceptance due to certain design disadvantages some of which were the weight factor, the high cost of components, and their poor reliability, but there have been some vast improvements over the years. As components designs were improved and better manufacturing techniques established to speed up the manufacturing of the components, hydrostatic transmissions opened a whole new avenue for the manufacturers of heavy equipment as well as for those manufacturers of lighter mobile equipment such as garden tractors, mowers, snowmobiles, etc. The brute force control methods used in the past could be eliminated and replaced by torque control and speed control devices that could be manipulated by finger force.

A hydrostatic transmission functions on the principle that energy is imparted to the hydraulic fluid by a pump of the positive displacement design. Then the hydraulic fluid is directed to a motor of the positive displacement design. The transmission operates primarily on a change of pressure. Although the hydraulic fluid carries kinetic energy, this plays no part in the transmission of power since the fluid velocity is unchanged in passing through the hydraulic motor.

In the early hydrostatic transmissions use was made of low-pressure systems which necessitated large volume pumps and motors in order to meet the requirements of speed and torque. Heat generation and power loss became factors which were mainly the result of turbulence in the flow of the hydraulic fluid through the conduits. When a turbulent flow condition exists, the loss through heat generation varies as the cube of the velocity but the power transmitted in a hydrostatic transmission varies directly with the velocity of the flow.

With the advent of high-pressure pumps and motors which operate efficiently, low-volume components and smaller lines help maintain high efficiencies at velocities that often exceed 20 ft/sec.

Characteristics of a Hydrostatic Transmission

Before discussing the components of a hydrostatic transmission some of the advantages of such a system should be noted:

a. *Lack of complexity.* No clutch mechanism is required. There is no shifting of gears when load requirements increase or decrease. Speeds are infinitely variable in either the forward or reverse direction. Sudden and rapid reversals of speed, which are often necessary on earth-moving equipment and on other large mobile machinery, can be accomplished without damaging the drive. Dynamic braking is very simple.

b. *Maneuverability.* This is an important asset for any piece of moving equipment. When heavy loads are suddenly released there is no surge or speed change. Constant ground speed can be maintained without depressing or releasing an accelerator pedal.

c. *Effortless operation.* The operation is almost effortless in that response is immediate and quite positive to finger-tip operation. Unskilled workers can learn to operate hydrostatic transmissions in a very short time.

d. *Speeds can be varied.* Speeds can be changed "on the go" while the working mechanism is independently operated at its most efficient speed. The prime mover can also be run at optimum speeds regardless of the mechanism speed. A totally new and almost limitless control of all functions is achieved for maximum vehicle efficiency.

e. *Minimum maintenance.* There are no working parts to which sand, dirt, and grit are exposed. The components are well "packaged" to resist water and other contaminates.

Major Components of the Hydrostatic Transmission System

There are four basic components involved in hydrostatic transmissions:
   a. Variable displacement pumps
   b. Fixed displacement pumps
   c. Variable displacement motors
   d. Fixed displacement motors.

Figure 2 illustrates a cutaway of a variable displacement hydraulic pump. The variable displacement, reversing, swashplate pump is controlled and positioned by a servo system. The charge pump, which provides oil to the servo system, is mounted to the main pump housing. The purpose of the charge pump is to provide a flow of oil through the transmission for cooling purposes, to provide the required oil under pressure for control purposes, to provide internal leakage makeup, and to supply oil under pressure to maintain a positive pressure on the low-pressure side of the main pump/motor circuit.

Fixed displacement pumps are shown in Chapter 3.

Figure 3 shows a cutaway of a fixed displacement hydraulic motor of the axial piston design with high-pressure relief valve, shuttle valve, and charge pressure control valve.

Variable displacement hydraulic motors are similar in design to Fig. 3 except that the swashplate is not fixed.

All valves required for a closed-loop circuit are generally contained in either the hydraulic pump or hydraulic motor assemblies, or both.

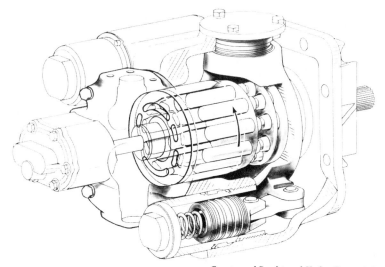

Courtesy of Sundstrand Hydro-Transmission
Fig. 2. Variable displacement hydraulic pump with servo control.

Minor Components of the Hydrostatic Transmission System

While the following components may be only minor in cost, they are important in the overall function of the system:

Filters. Depending upon the system, 3-micron to 10-micron filters are usually employed. The manufacturer of the hydrostatic transmission should be consulted for recommendations.

Heat Exchanger. In order to keep the oil at the proper recommended operating temperature it may be necessary to include a heat exchanger in the system.

Reservoir. The reservoir size will depend largely upon the fluid storage requirement in order to remove entrained air in the fluid and to take care of the fluid expansion.

Fluid Lines. The lines must be of sufficient strength to carry the high-pressure fluid and must be sized so that restrictions will be at a minimum when the system is producing its full volume. Both flexible and rigid conduit are employed in most systems.

Fluids. Fluids which have been employed successfully in various hydrostatic transmissions are: anti-wear hydraulic oil, automatic fluid Type "F," certain hydraulic transmission fluids, and certain fire-resistant fluids. Fluid suppliers can offer assistance.

Basic Circuits for Hydrostatic Transmissions

The pumps and motors listed previously under, "Major Components of the Hydrostatic Transmission System" can be combined into four basic circuits:

1. Variable displacement pump driving a variable displacement motor.

2. Variable displacement pump driving a fixed displacement motor.

3. Fixed displacement pump driving a variable displacement motor.

4. Fixed displacement pump driving a fixed displacement motor.

All of these circuits offer different performance characteristics. In the following descriptions, the input speed is constant. When the input speed is varied, and various control methods are introduced, the number of possible applications grows rapidly. The horsepower at any given working pressure is directly proportional to the pump flow. The torque of the motors depends upon the displacement and the working pressure established.

Circuit #1: Variable Displacement Pump—Variable Displacement Motor. (See Fig. 4.) Characteristics of this circuit are:

a. Maximum efficiency occurs near mid-speed.

b. The efficiency is highest over a broad part of the speed range.

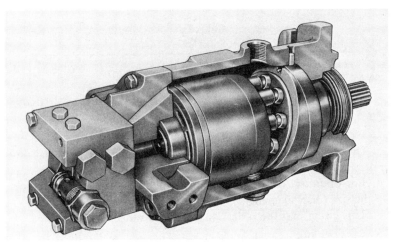

Courtesy of Sundstrand Hydro-Transmission
Fig. 3. Heavy-duty, fixed displacement hydraulic motor.

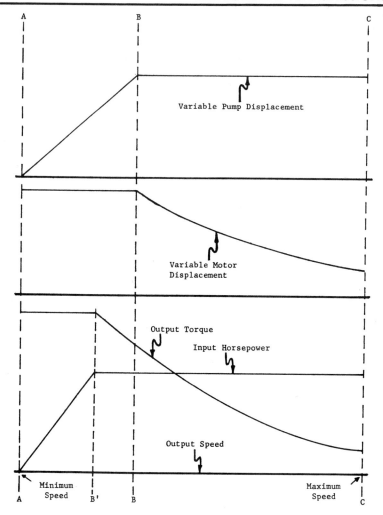

Courtesy of Sundstrand Hydro-Transmission

Fig. 4. Circuit #1 for variable displacement pump—variable displacement motor

c. Output speed is controlled by both pump displacement and motor displacement (and pump input speed).

d. For optimum hydraulic sizing, the pump is approximately one-half the size (displacement) of the motor.

e. This circuit has the widest output speed range.

At point *A* the variable displacement pump is at minimum (zero) displacement thus there is no flow and no transmission output. The variable displacement motor is at maximum displacement so that when fluid flow starts maximum starting torque is provided at the motor output shaft. This makes it possible to utilize the entire range of the transmission.

Point B' shows the point at which input horsepower reaches maximum. This shows that maximum horsepower can be reached before the pump reaches maximum displacement or flow. Pressure (not shown) remains at maximum up to B' then falls off between B^1 and B, and is constant from B to C. Between points B' and C the transmission delivers constant horsepower assuming constant working pressure conditions.

At point B the pump displacement has increased to maximum. The motor displacement has remained constant but the motor output speed has increased. For any given pressure, no further increase in output horsepower is possible since horsepower is a function of flow and pressure. From this point on, the motor displacement goes from maximum to minimum with a resulting decrease in output torque and increase in output speed.

At point C the transmission has reached its maximum outspeed and the horsepower output is the same as it was at point B'.

A good example of the use of a transmission with a variable displacement pump and a variable displacement motor is to propel a crawler tractor. Two transmissions of equal size could be used, one driving each track. This arrangement would provide:

1. Independent control of each track for steering.
2. Tracks can be operated in opposite directions for "spin" turns.
3. Infinite speed control by varying the pump and motor displacement.
4. High output torques for starting, bulldozing, and towing.
5. High speed for traveling.

The application of a variable displacement motor in a transmission, where the output speed will be varied, requires lower circuit flows than are possible with the fixed displacement motor, allowing smaller pumps to be used. The variable displacement pump-variable displacement motor combination reaches maximum efficiency at mid-speed range whereas the variable displacement pump-fixed displacement motor combination reaches maximum efficiency at maximum speed.

Circuit #2. Variable Displacement Pump—Fixed Displacement Motor. (See Fig. 5.) Characteristics of this circuit are:

a. Maximum efficiency occurs near top motor input speed.

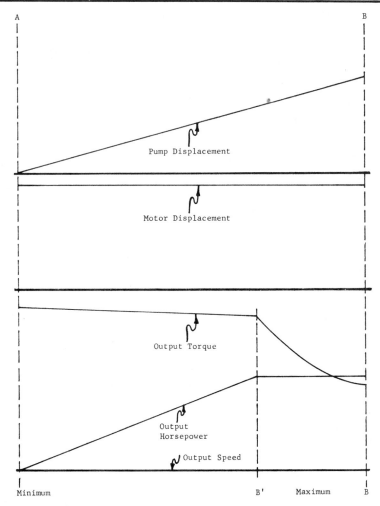

Courtesy of Sundstrand Hydro-Transmissions
Fig. 5. Circuit #2 for variable displacement pump—fixed displacement motor.

b. Output speed is controlled by pump displacement (and input speed).

c. For optimum hydraulic sizing, the pump size (displacement) is equal to the motor size (displacement).

·d. This is the simplest system providing infinite control. It is considered to be the most frequently used system.

At point *A* pump displacement is minimum (zero); therefore there is no output from the transmission. As the pump is put in stroke, maximum starting torque will be available, since the motor is in maximum displacement.

At Point B' the torque has remained basically constant, while the horsepower has increased from minimum to maximum, assuming constant working pressure.

At Point B the pump displacement is now at maximum. The motor displacement has remained constant, since its displacement is fixed. The motor is now at full speed since the pump displacement is at maximum. The torque is reduced because horsepower was constant while output speed increased from B' to B. Typical applications of this transmission are conveyor drives, combine drives, and road roller drives. Varying the pump displacement provides an infinite number of output speeds, while the fixed displacement motor provides a constant torque capability.

Circuit #3. *Fixed Displacement Pump—Fixed Displacement Motor.* The combination of these two components is the hydraulic equivalent of a mechanical shaft and gear drive. It can be employed to transmit power without altering the speed or horsepower between the engine and the load. This transmission is convenient when the power source is remotely located from the load.

Circuit #4. *Fixed Displacement Pump—Variable Displacement Motor.** With the pump driven at a constant speed, the circuit is a constant horsepower, variable speed, variable torque transmission. Output speed is controlled by changing the displacement of the motor. As the motor speed increases, output torque decreases. The system has no neutral because of the fixed displacement in the pump. The motor can be stopped by moving the squash plates to the position where the pistons will not move.

The foregoing illustrates basic arrangements of hydrostatic transmissions. Only a simple system of one pump and one motor was employed to illustrate the basic circuits. Many variations of these circuits are possible, such as using one pump with two motors, two pumps and two motors, etc., to get the required results. It is also common practice to use different sizes of pumps and motors in combination. Regardless of the pumps and motors that are used, the basic characteristics apply.

The Hydraulic Wheel

The heart of the hydraulic wheel is a high-torque, low-speed hydraulic motor that is ruggedly built to withstand the heavy service imposed on it in the agricultural, construction, marine, material han-

* Courtesy of Sundstrand Hydro-Transmission

Courtesy of Poclain Hydraulics
Fig. 6. Heavy-duty radial piston motor.

dling, and industrial fields. Such a hydraulic motor usually operates in pressure ranges above 3000 psi; some as much as 6000 psi. The radial piston design is most widely used although some designs employ the axial piston.

The radial piston hydraulic motor in Fig. 6 has small-diameter pistons for high-pressure operation, double roller bearings which

Courtesy of Poclain Hydraulics
Fig. 7. Three models of hydraulic wheels.

compensate for very high shaft side loads, and a rugged splined shaft designed to resist shock loads. These motors have a very low noise-level. They also have a level torque at start-up and low speeds, and provide smooth operating speeds.

Figure 7 shows three different hydraulic wheel units. These hydraulic wheels are compact drive units with lugs for mounting. The one shown at the left is a unit with a disc type brake, the unit in the center has a drum type brake, and the one at the right is without a brake. The diameters of these units, which have a capacity of from 37 to 146 cu in., range from 15 to 21 inches.

Review Questions

Chapter 1—Review Questions
1. What was Blaise Pascal's role in fluid power? Sketch an illustration showing his theorem.
2. Name three persons living around the 17th Century who carried on experiments that were related to fluid power.
3. What did W. G. Armstrong, of England, invent? What impact did it have on industry?
4. What use was made of the "Brotherhood" three-cylinder, constant stroke hydraulic engine? When was it patented?
5. What type of industries were represented in setting up the J.I.C. Standards for Industrial Equipment?
6. What did the J.I.C. Standards for Industrial Equipment cover?
7. What does the term "fluid" refer to in fluid power?
8. What name now applies to the graphic fluid power symbols?
9. What are some of the advantages of using fluid power devices? What are some of the disadvantages?
10. What effect does the use of fluid power equipment have on the lives of people throughout the world?

Chapter 2—Review Questions
1. What methods are used to determine which fluid should be used in a hydraulic system?
2. What types of fluid are available for hydraulic systems?
3. What are the advantages and disadvantages of water as the medium in a hydraulic system? What type of system usually employs water?
4. Give six important service properties of a hydraulic oil.
5. Explain what is meant by "viscosity" of a fluid. What is the importance of viscosity?
6. Name several ways in which foaming can be reduced in a hydraulic fluid.
7. Name six general standard specifications used in the selection of a hydraulic oil.
8. What can additive treatment do for a hydraulic oil?
9. What types of fire resistant hydraulic fluids are available? What are their advantages and disadvantages?
10. What steps should be followed in cleaning a hydraulic system?

Chapter 3—Review Questions
1. What is the function of a hydraulic power unit?

2. Basically, what is included in a hydraulic power unit?
3. What types of pumps are available for hydraulic power units? Give pressure ratings of each type.
4. What are some of the conditions which will cause a hydraulic pump to fail?
5. What conditions will cause cavitation of the pump?
6. How should the oil reservoir be sized in regard to the capacity of the pump?
7. What types of filters are most commonly used in conjunction with the hydraulic power unit?
8. What means are available to drive the hydraulic pump?
9. Give some of the simple rules to follow for installing a hydraulic power unit.
10. What are some of the disadvantages of using a single hydraulic power unit to operate the hydraulic circuits on several machines.

Chapter 4—Review Questions
1. What purpose does an accumulator serve?
2. Name four types of accumulators that might be found in hydraulic systems. Which type is generally used in a central system?
3. What are the advantages and disadvantages of a deadweight-type accumulator?
4. What will cause failure of a deadweight-type accumulator?
5. What will cause a spring-loaded-type accumulator to malfunction?
6. What will cause a bladder-type accumulator to malfunction?
7. Name five uses for accumulators.
8. Make a circuit sketch showing how an accumulator can be used as a leakage compensator in conjunction with a rotating chucking cylinder.
9. Make a circuit sketch showing the use of an accumulator as an ink dispenser.
10. Make a circuit sketch showing the use of an accumulator as an emergency source of fluid under pressure, other than as shown in Circuit 5.

Chapter 5—Review Questions
1. Name three of the basic requirements when making a piping layout for a hydraulic system.
2. What are some of the basic tools used when installing hydraulic tubing?
3. When hydraulic piping is installed, what are some of the things that should be avoided?
4. Name three classes of fluid power lines. Give an example where each can be applied.
5. Name four types of piping or lines used in air systems.
6. Name four types of piping or lines used in hydraulic systems.
7. Name two types of tube fittings. Where is each used? Explain the steps that should be followed in flaring copper or aluminum tubing.
8. Give the pressure ratings for three types of flexible lines. Give an example where each is used.
9. What are some of the reasons for hose, or hose assembly, failures?
10. Where are quick disconnect couplings employed in fluid power systems?

Chapter 6—Review Questions
1. What is the basic function of a hydraulic relief valve? How does it operate?
2. What types of hydraulic relief valves are available? How do they differ?
3. What are some of the causes of relief valve malfunction?
4. What is a three-way, three-position hydraulic control valve? What is a four-way, three-position open center hydraulic control valve?
5. What types of flow directors are found in four-way hydraulic valves? What are their methods of actuation?
6. What is meant by a "piggy back" hydraulic control valve?
7. What is a pressure-compensated, hydraulic flow-control valve? What is a cam operated flow-control valve? Where is each used?
8. Name two types of hydraulic sequence valves. Make a circuit sketch showing how a sequence valve can be used.
9. Make a sketch of a circuit showing use of a hydraulic reducing valve with a built-in check valve. Also make a sketch of a circuit with a hydraulic reducing valve without a built-in check valve.
10. Name two types of hydraulic check valves. Where are they most commonly used?

Chapter 7—Review Questions
1. What is the function of a hydraulic cylinder?
2. What is the difference between a ram-type hydraulic cylinder and a telescopic type hydraulic cylinder? Give examples where each is used.
3. What means may be employed to keep the piston rod from rotating in relation to the cylinder tube of a hydraulic cylinder?
4. Name several sealing methods that are employed to effect a seal between the cylinder piston and the cylinder tube.
5. What means may be employed to protect the precision finish of a piston rod of a hydraulic cylinder?
6. Name four types of seals employed to effect a seal between the piston rod and the cylinder cover. Give advantages of each.
7. If an 8-inch bore hydraulic cylinder is operated at 3000 psi what force will it develop on the forward stroke? If the cylinder has a 5¾-inch diameter piston rod, how much force will the cylinder develop on the return stroke?
8. In Chart 1, what would be the forward travel speed of the piston, in inches per minute, of a 4-inch bore cylinder being fed 1000 cu in. of oil per minute?
9. If a 6-inch bore by 18-inch stroke hydraulic cyclinder is to complete its forward stroke in five seconds, how much oil will be required, in gpm?
10. What is the function of a hydraulic intensifier? If the input end of an intensifier has an area of 16 sq in., and the output end has an area of 2 sq in., what will be the output pressure if the input pressure is 1500 psi? Make a list of applications where intensifiers are employed.

Chapter 8—Review Questions
1. What is the function of a heat exchanger in a hydraulic system?
2. What factors determine the capacity required of the heat exchanger?

3. What conditions in a hydraulic system would require a heat exchanger?
4. 'Make a circuit showing the location of a heat exchanger if the relief valve is the cause of heat.
5. What is a single-pass type heat exchanger? A two-pass type heat exchanger? A four-pass type heat exchanger?
6. What will cause a heat exchanger to malfunction?
7. What type of control is used in conjunction with the heat exchanger to keep the temperature constant within close limits?
8. What types of heat exchangers are used in conjunction with hydraulic systems?
9. What types of heaters are employed when hydraulic systems are used in cold locations?
10. What applications would require heaters to be placed in a hydraulic system?

Chapter 9—Review Questions
1. Name some applications where it would be desirable to have two cylinders synchronized in movement.
2. Name some of the difficulties encountered when trying to synchronize two cylinders.
3. What would be the disadvantages of mechanically linking two cylinders to effect synchronization? What are the advantages?
4. How can fluid motors be employed to effect synchronization?
5. What precautions should be used when employing fluid motors for synchronization?
6. How is a flow divider control valve used to effect synchronization? Make a sample sketch showing its use in conjunction with two cylinders.
7. What would be the disadvantages of using double-end cylinders to effect synchronization?
8. Make a sketch showing how four air-hydraulic cylinders might be connected to effectively provide synchronization.
9. What would be the disadvantage of using air-hydraulic cylinders where a large force is required in the synchronizing process?
10. When a two-pump system is employed to effect synchronization of two cylinders, name some of the problems that might occur.

Chapter 10—Review Questions
1. Name at least three advantages of using a dual-pressure system?
2. If dual-pressure systems were not available what problems might a designer encounter?
3. What is the function of a reducing valve? If the upstream pressure is 1500 psi what pressures can be obtained on the downstream side of the valve? Make a sketch of a simple circuit to show how the valve is used.
4. What is the principle of the intensifier or pressure booster?
5. If an intensifier has an input pressure of 1500 psi and the area of the input piston is 10 sq in., what would be the output pressure if the output ram has an area of 1.5 sq in.? How much intensified fluid would be available if the stroke of the intensifier is 20 inches?
6. Show the application of an intensifier in a sample circuit.

7. What are the advantages of using two pumps to effect dual pressures in a circuit?
8. Show how two relief valves can be employed in a single circuit to effect dual pressures.
9. What are the advantages of a high-low pump circuit?
10. Make a circuit sketch showing the use of a cam-operated relief valve to provide three different pressures in a single circuit.

Chapter 11—Review Questions
1. What are the important factors of safety controls in a hydraulic system?
2. What provisions can be made by the use of hydraulics to protect the operator where highly explosive or other adverse conditions are present in the work area?
3. What importance does the hydraulic relief valve play in protecting the workpiece and also the hydraulic system?
4. In a hydraulic press application, what mechanical means can be provided to protect the operator?
5. In a hydraulic press application, what hydraulic control means can be provided to protect the operator?
6. What type of application would require four-hand safety control? Explain.
7. How can a hydraulic circuit provide for safety to the tool that is used in a machining operation?
8. How can a hydraulic system using a hydraulic hand pump provide safe operation in case of an emergency?
9. Explain how an accumulator can be used as a safety device in a hydraulic system. Make a circuit sketch showing this.
10. What types of actuation for directional control valves readily lend themselves to safety controls?

Chapter 12—Review Questions
1. Name at least five ways in which a sequence of operations can be accomplished with hydraulics.
2. Explain how a hydraulic sequence valve functions. How can flow be accomplished through this valve in both directions?
3. What is the difference between a direct-acting and a direct-operated pilot type sequence valve?
4. What problems can be experienced when using a hydraulic sequence valve in a circuit to set up a sequence of operations.
5. What types of porting are available in sequence valves? What are the merits of each?
6. Make a circuit sketch showing the use of two air-pilot-operated hydraulic valves to effect the sequence operation of two hydraulic cylinders.
7. Make a circuit sketch showing the use of solenoid-operated hydraulic valves to effect the sequence operation of two hydraulic cylinders.
8. Design a circuit to show how two sequence valves can be used in conjunction with a four-way control valve to operate two cylinders in sequence.
9. Design a circuit to show how two sequence valves can be used in conjunction with a four-way control valve to operate a cylinder and a hydraulic motor in sequence.

10. Design a circuit using both air-pilot and solenoid-operated control valves to actuate cylinders as follows: Piston of cylinder *A* advances, piston of cylinder *B* advances and retracts, piston of cylinder *A* retracts to complete cycle.

Chapter 13—Review Questions

1. Why are packings and seals important in the function of a hydraulic system?
2. Where are packings and seals employed in hydraulic pumps, hydraulic cylinders, and hydraulic control valves?
3. What materials are employed in packings and seals for hydraulic components?
4. Name two general classifications of packings. Where is each generally applicable?
5. Make a rough sketch of the following: "O"-ring, Quad-ring, "V"-ring, block vee, and cup packing.
6. What is meant by a "static" seal? A dynamic seal? Illustrate where each is employed.
7. What are the factors that affect the selection of packings and seals?
8. What are the advantages of an automotive type piston ring used on the piston of a hydraulic cylinder? What are the disadvantages? How many piston rings should be used?
9. What are the advantages of a cartridge type hydraulic rod packing?
10. What steps should be used when installing rod scrapers on hydraulic cylinders?

Chapter 14—Review Questions

1. Why is it necessary to keep compressed air clean, lubricated and free of water?
2. What kind of oil should be supplied to the lubricator?
3. What conditions will cause condensation in the air lines? What are the results of condensation in the air lines?
4. What types of devices are employed to remove condensation from air lines?
5. What size particles should industrial air-line filters be able to trap? What types of elements are used in air-line filters?
6. What are the advantages of a unit which contains a regulator, filter, and a lubricator built into one housing?
7. What are the disadvantages of a unit which contains a regulator, filter, and a lubricator built into one housing?
8. What types of lubricators are available for use on industrial air lines? What are the advantages of each?
9. What types of air pressure regulators are employed in industrial applications? Discuss the advantages and disadvantages of each.
10. What is the importance of the air muffler? Where should they be applied?

Chapter 15—Review Questions

1. Name three types of pneumatic directional control valves and give an example of an application for each.

2. Name four types of actuators used to move the flow director of directional control valves.
3. What are some of the determining factors in selecting the proper type of actuator?
4. What are the port designations of a two-way pneumatic valve? Of a three-way pneumatic valve? Of a four-way pneumatic valve?
5. Name at least four types of seals used in pneumatic control valves. Give the merits of each.
6. What is a two-position, four-way pneumatic control valve? A three-position, closed-center, four-way pneumatic control valve? Give two applications of each.
7. Name two types of solenoid actuators used with four-way pneumatic control valves. Give the merits of each. Give two applications of each.
8. If the speed of the piston of an air cylinder is to be controlled on the retracting stroke, where should the speed control be placed? How can speed controls be used to produce the same speed in both directions of a cylinder piston?
9. Where two air cylinders are employed in a circuit, make a sketch to show how a pneumatic sequence valve can be used to eliminate a four-way valve when operation is as follows:
   a. Piston of cyl. 1 advances to the end of its forward stroke.
   b. Piston of cyl. 2 advances to the end of its forward stroke.
   c. Pistons of cyls. 1 & 2 return together.
10. Make a list of conditions which may cause pneumatic control valves to malfunction.

Chapter 16—Review Questions
1. Make a list of ten applications for air cylinders.
2. Make a list of the mounting types offered for air cylinders.
3. What is meant by the term "mill type air cylinders"? What is the difference between a mill type air cylinder and a standard industrial type air cylinder?
4. What types of seals are employed on the piston of air cylinders? Discuss their merits. What materials are used for low temperatures?
5. What types of seals are used to effect a seal between the piston rod and the rod end cover of an air cylinder? Give merits of each.
6. What methods are used to secure the cylinder covers to the cylinder? What sealing methods are used?
7. If a cylinder must move through an arc, what types of mountings should be used? Explain.
8. What methods are employed to protect a cylinder from adverse environmental conditions such as heat, high concentration of dirt, and highly corrosive atmospheres?
9. Should cushions be used if the full stroke of the cylinder is not employed? How long is the normal cushion on industrial cylinders? What problems are encountered if long cushions are needed?
10. If a cylinder has a bore of 12 inches, what force will be exerted at the end of the piston rod on the forward stroke if the operating pressure is 100 psi? If the cylinder has a 4-inch-diameter piston rod, what force will be exerted on the return stroke?

Chapter 17—Review Questions
1. What is a hydraulic accumulator?
2. What is a gravity type accumulator and how does it operate?
3. What are the advantages of a gravity type accumulator?
4. What are the advantages of a bladder type accumulator?
5. How does a piston type accumulator differ from a hydraulic cylinder?
6. In a hydraulic system, how does the use of an accumulator reduce heat?
7. How can an accumulator be employed in a hydraulic circuit so as to deliver fluid in case of a power failure?
8. What type of an accumulator is usually employed where a relatively large amount of hydraulic fluid is needed in a central hydraulic system such as in that used for a group of molding machines?

Chapter 18—Review Questions
1. What protective devices are employed to protect the pneumatic system? Why are these devices so important?
2. Why is it important to provide controls in a pneumatic system so that the operator will be protected?
3. What type of controls are available if there is a sudden drop in pressure in a pneumatic system?
4. Draw a pneumatic circuit which will provide a safety stop if an overload occurs.
5. What are the advantages of using pilot-operated pneumatic control valves in highly explosive atmospheres?
6. How can solenoid-operated pneumatic control valves be used in a safety circuit where the operator's hands must be protected?
7. What type of pneumatic controls can be employed to apply two different forces on a workpiece in a sequence of operations?
8. Where two operators are working on one machine, what type of controls can be used to afford each ample protection? Make a simple circuit sketch of such an application.
9. Draw a pneumatic circuit in which two cylinders have their movements interlocked.
10. Draw a pneumatic circuit which provides for an emergency return of the piston of an air cylinder.

Chapter 19—Review Questions
1. What type of applications make use of pilot-operated four-way, two-position, pneumatic air-control valves.
2. Name the two types of actuators employed in pilot-operated pneumatic control valves and explain how each functions.
3. What is a pressure-differential type pneumatic control valve? What is its principle of operation?
4. Name three applications where a pilot-operated, four-way, three-position, spring-centered, closed-center control valve might be employed.
5. Name three applications where a pilot-operated, four-way, three-position, spring-centered, cylinder ports open in neutral with inlet blocked pneumatic control valve might be employed.
6. Where would a three-way, air pilot-operated control valve be employed?
7. What are the advantages of a pilot-operated pneumatic system?

8. What type of controls can be applied to operate pilot-operated, four-way pneumatic control valves?
9. Make a circuit sketch showing the use of a pilot-operated, three-position, four-way control valve to actuate a pneumatic cylinder. Explain the operation.
10. Make a circuit showing the use of two pilot-operated, two-position, four-way control valves to actuate two pneumatic cylinders in sequence. Explain the operation.

Chapter 20—Review Questions
1. What are the advantages of using both air and oil in a single fluid-power circuit? What are the disadvantages?
2. What is an air-hydraulic cylinder unit? How does it function? What three designs are often used?
3. In an air-hydraulic cylinder unit, what would be the advantage in having the bore of the hydraulic cylinder larger than that of the air cylinder.
4. In an air-hydraulic cylinder unit, what would be the advantage in having the bore of the air cylinder larger than that of the hydraulic cylinder.
5. Make a circuit sketch showing the use of an air-hydraulic cylinder unit to actuate a tool slide. Explain the function of the circuit.
6. Make a circuit sketch showing the use of an air-hydraulic cylinder unit with a skip feed arrangement that might be applied to machine tables. Explain the operation of the circuit.
7. Show by a sketch how an air-hydraulic cylinder unit can be employed as a device for measuring fluid and filling vessels such as crankcases, transmissions, radiators, etc.
8. What degree of accuracy in feed can be accomplished by the use of an air-hydraulic cylinder unit? How does this compare with a hydraulic system?
9. What conditions might cause an air-hydraulic cylinder unit to malfunction?
10. Make a circuit sketch of a system in which an air pilot-operated, four-way hydraulic valve is used to actuate a double-acting hydraulic cylinder. Also, show the air system.

Chapter 21—Review Questions
1. What is meant by the term "high-pressure hydraulic system"?
2. What are some of the parameters that must be considered when designing a high-pressure hydraulic system?
3. Why are piston pumps usually employed in high-pressure hydraulic systems? Name two categories of high-pressure pumps.
4. How is the hydraulic hand pump protected from being overloaded by some husky operator?
5. Name some applications that make use of a hydraulic hand pump.
6. Explain how a two-stage, high-pressure hydraulic pump functions.
7. If the outlet pressure of a hydraulic pump is 10,000 psig, what will be the output force of a 2-inch-bore hydraulic cylinder operated by that pump?
8. What type of seals are employed in high-pressure hydraulic cylinders? What problems are often encountered if seals are not properly installed?

9. What designs are employed on high-pressure hydraulic directional control valves?
10. What type of piping should be used in high-pressure hydraulic systems? What precautions should be taken?

Chapter 22—Review Questions

1. What is a rotary actuator?
2. Name the basic types of rotary actuators. Which will provide the maximum degree of rotation?
3. What advantages do rotary actuators provide?
4. Name five applications for hydraulic actuators.
5. How does a tandem rotary actuator function?
6. What would be the advantage of a quad type rotary actuator?
7. What is the purpose of a two-fluid system for operating a tandem type actuator?
8. What would be the purpose of a hydraulic hand pump being employed in conjunction with a quad type hydraulic actuator?
9. Explain how the rotary actuator in Fig. 8 functions. Can this actuator be employed in either air or oil circuits?
10. Make a circuit drawing showing the use of a rotary hydraulic actuator which is controlled by a four-way, three-position, closed-center hydraulic control valve and two flow-control valves.

Chapter 23—Review Questions

1. What is an air motor?
2. What are some of the advantages of an air motor over an electric motor?
3. What are some of the advantages of an air motor over a hydraulic motor?
4. What internal designs are used in air motors?
5. At what speeds will air motors operate?
6. On what type of applications are air motors employed?
7. Make a circuit sketch showing the use of a non-reversing type air motor controlled by a two-way control valve.
8. Make a circuit sketch showing the use of a reversing type air motor controlled by a four-way, three-position control valve.
9. Make a circuit sketch showing the use of a non-reversing air motor used in conjunction with a double-acting air cylinder.
10. Make a circuit sketch showing the use of a non-reversing air motor to operate a hydraulic pump. Also show a two-way valve to operate the air motor, and a four-way, two-position valve to operate the double-acting hydraulic cylinder.

Chapter 24—Review Questions

1. What is pneumatic logic?
2. What is Boolean Algebra?
3. What are the two numbers used in Boolean Algebra?
4. What are some of the advantages of using pneumatic logic controls?
5. Explain the function of the logic elements OR and AND.
6. Draw a circuit showing the logic element OR controlling an air motor.
7. Explain the function of a memory element: a differentiator.

8. What operating pressures are used in pneumatic logic systems?
9. On what type of applications are pneumatic logic circuits employed?
10. Name some applications for miniature type pneumatic "whisker" valves. What advantages do they offer?

Chapter 25—Review Questions
1. What is fluidics?
2. What advantages does fluidics offer?
3. What is the difference between fluidic devices and moving parts logic devices?
4. How do fluidic devices compare in price to other control devices?
5. At what pressures do fluidic devices function?
6. What is the approximate size of a fluidic device? How does it compare in size with other control devices?
7. When a number of devices are mounted on a manifold how many inputs are usually required? What advantages do the manifold or grid board provide.
8. Name several important applications for fluidic devices.
9. What is a monostable flip-flop? A bistable flip-flop?
10. What is the OR/NOR gate? The AND gate? The Schmitt trigger?

Chapter 26—Review Questions
1. Name several control devices that would be found in the home; on an automobile; in a factory.
2. Define a command signal. What devices are employed to cause a command signal?
3. What is the purpose of a feedback transducer? What is the error signal?
4. Name six applications for mechanically operated servo controls.
5. Name six applications for electrohydraulic servo controls.
6. Name four major types of pilot or first stages employed in multistage servo valves.
7. Explain what is meant by dead zone or dead band in a servo control. Why is this objectionable?
8. What is meant by zero lap? How is this accomplished?
9. What are the advantages of using a separate control source instead of taking it from the supply pressure source?
10. What is dither? What is stiction? Why is a filter so important in a servo-control system?

Chapter 27—Review Questions
1. What is meant by "closed loop" hydraulic circuit?
2. What is the function of the volume control in Fig. 1A?
3. Explain how a basic hydrostatic transmission functions when a variable displacement pump and a fixed displacement motor are employed.
4. What is the purpose of the high-pressure relief valve in the motor circuit of a hydrostatic transmission?
5. What is the purpose of the shuttle valve in the motor circuit?

6. What are the advantages of a hydrostatic transmission?
7. What is the importance of the filter in a hydrostatic transmission? The reservoir? The fluid conduits?
8. Name the four major basic components involved in hydrostatic transmissions.
9. What kinds of fluids are employed in hydrostatic transmissions?
10. What is a hydraulic wheel? How does it function? Name five applications for the hydraulic wheel.

ANSI GRAPHIC SYMBOLS *For Fluid Power Diagrams*

CONDUCTORS		
Line, Working	————	
Line, Pilot	– – – –	
Drain Line	– – – – – –	
Line, Flexible		
Connector	●	
Direction of Flow (Air)		
Direction of Flow (Hydraulic)		
Line Crossing		
Line Joining (Tee, Cross, Etc.)		
Plugged Port	—×—	
Restriction, Fixed		
Line to Reservoir above Fluid		
Line to Reservoir below Fluid		
Vented Manifold		
Quick Disconnect	See top of next column, four items	

Without Check, Connected	
Without Check, Disconnected	
With Two Checks, Connected	
With Two Checks, Disconnected	

ENERGY STORAGE AND FLUID STORAGE	
Reservoir, Vented	
Reservoir, Pressurized	
Reservoir with Connecting Lines, Above Fluid Level	
Accumulator	
Accumulator, Spring Loaded	
Accumulator, Gas Charged	
Accumulator, Weighted	
Receiver, Air or Gases	

COMPRESSORS	
Single Fixed Displacement	
Single Variable Displacement	

29-1

ANSI GRAPHIC SYMBOLS *For Fluid Power Diagrams*

FLUID MOTORS		
Pneumatic Unidirectional		
Pneumatic Bidirectional		
Hydraulic Fixed Displacement		
Hydraulic Bidirectional		
Hydraulic Unidirectional Variable Displacement		
Hydraulic Bidirectional Variable Displacement		
Pump-Motor-Hydraulic—Operate in One Direction as Pump and in Other Direction as Hydraulic Motor		

LINEAR DEVICES

Cylinders, Hydraulic or Pneumatic

Single Acting	
Double Acting— Single End Rod	
Double Acting— Double End Rod	
Double Acting—Fixed Cushion Advance and Retract	

Double Acting— Adjustable Cushion Advance	
Double Acting— Oversize Rod	

PRESSURE INTENSIFIERS

Pressure Intensifier —Single Acting	
Intensifier — with Double Acting Piston	

SERVO POSITIONER (SIMPLIFIED)

Hyd.   Pneum.

ACTUATORS AND CONTROLS

Spring	
Manual	
Push Button	
Lever	
Pedal or Treadle	
Mechanical	
Detent	
Pressure Compensated	
Solenoid (Single Winding)	

ANSI GRAPHIC SYMBOLS For Fluid Power Diagrams

Reversing Motor

Pilot Pressure—
Remote Supply

Pilot Pressure—
Internal Supply

Actuation by
Released Pressure

Remote Exhaust

Internal Return

Pilot Controlled, Spring
Centered, Complete Symbol

Pilot Differential

Complete Symbol

Solenoid or Pilot—
External Pilot
Supply

Solenoid or Pilot—
Internal Pilot
Supply and
Exhaust

Solenoid
and Pilot

VALVES

Envel-
opes

Ports

Ports, Internally Blocked

Flow Paths, Internally Open

Two-Way On-Off
Valves (Simplified)

On-Off
Valve (Off)

On-Off
Valve (On)

Check Valve
(Simplified)

Check, Pilot
Operated to Open

Check, Pilot
Operated to Close

Two-Position, Two-
Way Valve—
Normally Closed

Two-Position, Two-
Way Valve—
Normally Opened

Infinite Position
Normally Closed

Infinite Position
Normally Open

Three-Way Valve—
Two-Position,
Normally Open

Three-Way Valve—
Two-Position,
Normally Closed

Four-Way Valve—
Two-Position

Four-Way Valve—
Three-Position,
All Ports Blocked
in Neutral

Four-Way Valve—
Three Position,
Ports Open in
Neutral

Four-Way Valve—
Three-Position,
Neutral Cylinder
Ports Blocked
Pump to Tank

Four-Way Valve—
Three-Position, In
Neutral Cylinder
Ports Connected,
Pump and Tank
Blocked

ANSI GRAPHIC SYMBOLS *For Fluid Power Diagrams*

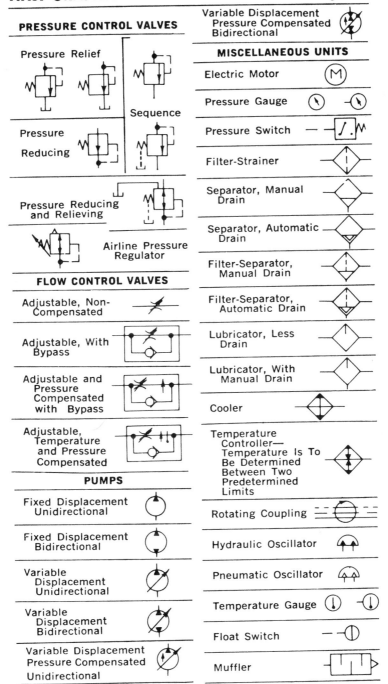

PRESSURE CONTROL VALVES

Pressure Relief

Sequence

Pressure Reducing

Pressure Reducing and Relieving

Airline Pressure Regulator

FLOW CONTROL VALVES

Adjustable, Non-Compensated

Adjustable, With Bypass

Adjustable and Pressure Compensated with Bypass

Adjustable, Temperature and Pressure Compensated

PUMPS

Fixed Displacement Unidirectional

Fixed Displacement Bidirectional

Variable Displacement Unidirectional

Variable Displacement Bidirectional

Variable Displacement Pressure Compensated Unidirectional

Variable Displacement Pressure Compensated Bidirectional

MISCELLANEOUS UNITS

Electric Motor

Pressure Gauge

Pressure Switch

Filter-Strainer

Separator, Manual Drain

Separator, Automatic Drain

Filter-Separator, Manual Drain

Filter-Separator, Automatic Drain

Lubricator, Less Drain

Lubricator, With Manual Drain

Cooler

Temperature Controller— Temperature Is To Be Determined Between Two Predetermined Limits

Rotating Coupling

Hydraulic Oscillator

Pneumatic Oscillator

Temperature Gauge

Float Switch

Muffler

Index